TOXIC HERITAGE

Toxic Heritage addresses the heritage value of contamination and toxic sites and provides the first in-depth examination of toxic heritage as a global issue.

Bringing together case studies, visual essays, and substantive chapters written by leading scholars from around the world, the volume provides a critical framing of the globally expanding field of toxic heritage. Authors from a variety of disciplinary perspectives and methodologies examine toxic heritage as both a material phenomenon and a concept. Organized into five thematic sections, the book explores the meaning and significance of toxic heritage, politics, narratives, affected communities, and activist approaches and interventions. It identifies critical issues and highlights areas of emerging research on the intersections of environmental harm with formal and informal memory practices, while also highlighting the resilience, advocacy, and creativity of communities, scholars, and heritage professionals in responding to the current environmental crises.

Toxic Heritage is useful and relevant to scholars and students working across a range of disciplines, including heritage studies, environmental science, archaeology, anthropology, and geography.

Elizabeth Kryder-Reid is Chancellor's Professor of Anthropology and Museum Studies and director of the Cultural Heritage Research Center, Indiana University, Indianapolis.

Sarah May is a Senior Consultant in Cultural Heritage at the sustainable development consultancy, Arup.

Key Issues in Cultural Heritage

Series Editors: William Logan and Laurajane Smith

Safeguarding Intangible Heritage
Practices and Politics
Edited by Natsuko Akagawa and Laurajane Smith

World Heritage and Sustainable Development
New Directions in World Heritage Management
Edited by Peter Bille Larsen and William Logan

Urban Heritage in Divided Cities
Contested Pasts
Edited by Mirjana Ristic and Sybille Frank

Migrant, Multicultural and Diasporic Heritage
Beyond and Between Borders
Edited by Alexandra Dellios and Eureka Henrich

Design and Heritage
The Construction of Identity and Belonging
Edited by Grace Lees-Maffei and Rebecca Houze

Toxic Heritage
Legacies, Futures, and Environmental Justice
Edited by Elizabeth Kryder-Reid and Sarah May

The following list includes only the most recent titles to publish within the series. A list of the full catalog of titles is available at: https://www.routledge.com/Key-Issues-in-Cultural-Heritage/book-series/KICH

TOXIC HERITAGE

Legacies, Futures, and Environmental Injustice

Edited by Elizabeth Kryder-Reid and Sarah May

Routledge
Taylor & Francis Group
LONDON AND NEW YORK

Cover image: Peter C. Little

First published 2024
by Routledge
4 Park Square, Milton Park, Abingdon, Oxon OX14 4RN

and by Routledge
605 Third Avenue, New York, NY 10158

Routledge is an imprint of the Taylor & Francis Group, an informa business

British Library Cataloguing-in-Publication Data
A catalogue record for this book is available from the British Library

ISBN: 978-1-032-42999-1 (hbk)
ISBN: 978-1-032-42997-7 (pbk)
ISBN: 978-1-003-36525-9 (ebk)

DOI: 10.4324/9781003365259

Typeset in Bembo
by MPS Limited, Dehradun

CONTENTS

CONTRIBUTORS

Grace Akese is a Geographer and Discards Studies scholar interested in waste geographies and political economies. She researches the geographies of e-waste. For Grace Akese, writing this chapter is possible through the Africa Multiple Cluster of Excellence at the University of Bayreuth. Deutsche Forschungsgemeinschaft (DFG) Excellence Strategy – EXC 2052/1 – 390713894.

Santiago Alzugaray is a PhD candidate in Social Anthropology, Universidad Nacional de San Martín, Argentina, and an Assistant Professor at the academic unit of the university research council at Udelar. He works in the field of science, technology, and social studies from an anthropological perspective and is particularly interested in the study of science, technology, and social and environmental inequalities.

Ana Baptista is an Associate Professor of Professional Practice in the Environmental Policy and Sustainability Management program at The New School. She also serves as the Co-Director of the Tishman Environment and Design Center at The New School. Her research and practice are focused on advancing environmental and climate justice.

Tobias Bausinger is an environmental scientist focused on the assessment of military-contaminated sites in Europe for more than twenty years. His special expertise is the analysis and preparation of chemical warfare agents and explosives from both World Wars and also environmental metabolites of these compounds.

Matteo (Teo) Benussi is a social anthropologist specialising in religion, ethics, politics, and heritage in Eurasia and Central/Southern Europe. He received his training in Venice (IT), Cambridge (UK), and Berkeley (CA), with research projects on memory and ritual in Chernobyl (Ukraine) and the Islamic revival in Tatarstan (Russia).

Julia Brenan is a PhD candidate in Archaeology at Memorial University of Newfoundland and Labrador. Her research focuses on the intersection of heritage and contamination from military activity in Labrador, Canada through theory, mapping, and dendrochemistry.

Elizabeth Grennan Browning is an Assistant Professor of History at the University of Oklahoma.

Matthew Carter is the Research Director at The Major Projects Foundation, Honorary Associate, Department of Archaeology and History La Trobe University, Melbourne, Australia.

Holly Cusack-McVeigh is an Associate Professor of Anthropology and Museum Studies, Public Scholar of Collections and Community Curation, and adjunct in Native American and Indigenous Studies. Her research focuses on repatriation, cultural heritage, social justice, community collaboration, and toxic heritage. In *Stories Find You, Places Know: Yup'ik Narratives of a Sentient World* (2017), she explores the concept of place in Indigenous communities.

Owen Dwyer is a cultural geographer and professor of Geography, IU School of Liberal Arts at Indiana University, Indianapolis. His research focuses on intersections of public space and collective memory with race, racism, and the Civil Rights Movement in the American South.

Victoria Evia holds a PhD in Anthropology from the Centro de Investigaciones y Estudios Superiores en Antropología Social (Mexico), where she specialised in Anthropology of Health and Illness. She is an Assistant Professor at the Institute of Anthropological Sciences of the Faculty of Humanities and Educational Sciences at the University of the Republic of Uruguay, where she teaches undergraduate and postgraduate courses and conducts research. Her lines of work include environmental health, social studies of science and technology, and political ecology.

Daniela Figueiredo is a researcher in Biology at the University of Aveiro (Portugal), with a post-graduation in Applied Ecology, a Master's degree in Molecular Microbiology, and a PhD in Biology. Over the past years, she has focused her research on freshwater quality issues, but has also been increasingly involved in science communication, environmental education, and citizen science approaches. Her scientific path can be found at https://orcid.org/0000-0003-3760-4142.

Valentina Figueroa is an archaeologist and Associate Professor at the Instituto de Investigaciones Arqueológicas y Museo-San Pedro de Atacama at Universidad Católica del Norte (Chile) and researcher at the Núcleo de Investigación TraGeMa -Trabajo, Género y Minería en el Desierto de Atacama-. Her lines of study are related to the social construction of mining landscapes in the Atacama Desert, to the contemporary archaeology of mining extractivism and to artisanal fishing in sacrificial zones in the hyper-arid coastal desert.

Gabriel Filippelli is a Chancellor's Professor of Earth Sciences at Indiana University-Purdue University Indianapolis and the Executive Director of the Indiana University Environmental Resilience Institute. He studies and writes on climate change, environmental contamination, and community science.

David M. Finch is a PhD candidate in Archaeology at Memorial University of Newfoundland and Labrador. His doctoral research is titled Practicing Heritage: Community-based Archaeology and Learning in Labrador, in which he is examining models of cultural resource management in Labrador. This work is community-driven and involves working with Innu interns and engaging with regulators and researchers.

Jonas Fischer is a graphic designer, illustrator, and comic artist from Kiel, Germany. He collaborates with friends, classmates, peers and designers, and scientists from a variety of disciplines. His work has led him to adventures in Moldova and Ecuador as well as in Otterndorf, Brunsbüttel, and Schleswig, Germany.

Amelia Fiske is a Senior Research Associate at the Institute for History and Ethics of Medicine at the Technical University of Munich. Her work is situated between cultural anthropology, feminist science and technology studies, and environmental humanities. She has a particular interest in integrating graphic art and ethnography.

Jonathan Gardner is an archaeologist and heritage researcher. He is currently a Leverhulme Early Career Fellow at Edinburgh College of Art where he researches waste landscapes. He completed his doctorate at the UCL Institute of Archaeology; this research was published as, *A Contemporary Archaeology of London's Mega Events: from the Great Exhibition to London 2012* (2022).

Carmem Regina Giongo, Psychologist, PhD in Social and Institutional Psychology, professor with a Master's degree in Psychology at Feevale University (Brazil), conducts research in the field of the implementation of large hydroelectric enterprises, investigating mental health and environmental justice. Her academic scientific research can be found at http://lattes.cnpq.br/3074416863232933.

Mike Hannis is Senior Lecturer in Ethics, Politics, and Environment at Bath Spa University, where he leads a Masters programme in Environmental Humanities. With a disciplinary background in environmental ethics, his empirical research interests include land use planning and energy policy.

Paul Heersink is based in Toronto, Canada and is a professional cartographer, amateur historian, author, and Esri Canada Community Maps Program Manager (BA (Alberta), PMP).

Cornelius Holtorf is Professor of Archaeology and holds a UNESCO Chair on Heritage Futures at Linnaeus University in Kalmar, Sweden. He also directs the Graduate School in Contract Archaeology (GRASCA) at Linnaeus University. Holtorf recently co-edited (with Anders Högberg) the volume *Cultural Heritage and the Future* (Routledge, 2021).

Gareth Hoskins is a senior lecturer in Geography at Aberystwyth University. His research interests include the politics of memory, heritage, and preservation. He has written widely on issues of identity, race, cultural value, narrative, and environmental histories of industrial development in the US and the UK.

Daniel Hubé is a senior geologist at the French geological survey with expertise in historic soil and groundwater industrial contamination characterization and risk assessment. His research focuses on military and environmental history, particularly the environmental consequences of WWI and post-war ammunition disposal activities. His publications include two books and many articles.

Bill Jeffrey is a maritime archaeologist with over 40 years of experience. He worked with the FSM National Historic Preservation Office from 2001 to 2008, during which time he implemented projects with Chuuk State HPO on the WWII shipwrecks. This led to an intensive study of the shipwrecks from a more local perspective.

Rosemary A. Joyce, a Distinguished Professor of Anthropology at the University of California, Berkeley, received a PhD in Anthropology in 1985 from the University of Illinois, Urbana-Champaign. She was previously Assistant Curator and Assistant Director of the Peabody Museum, Harvard University, and Director of the Hearst Museum of Anthropology at Berkeley.

Augustine C. Kohler is Assistant Director and FSM National Historic Preservation Officer at the Federated States of Micronesia National Archives, Culture and Historic Preservation (NACH).

Elizabeth Kryder-Reid is Chancellor's Professor of Anthropology and Museum Studies and director of the Cultural Heritage Research Center, Indiana University, Indianapolis. Her interdisciplinary research investigates intersections of landscape and power, including landscape gardens in the Chesapeake, California missions (*California Mission Landscapes: Race, Memory, and the Politics of Heritage*), and toxic heritage sites.

Manuelle Lago, Geographer, PhD in Sociology, is a researcher at Federal of Paraná University, Brazil. She works on social organisation, local leadership, risks perception, and protection of environment. Her work describes and analyses the relations between land insecurity, symbolic changes, and forms of resilience of traditional communities. Her scientific path can be found at http://lattes.cnpq.br/0824004305185761.

Peter Little, Associate Professor and Chair of the Department of Anthropology at Rhode Island College, is an anthropologist and political ecologist with general interests in electronics, environmental justice, and environmental health. He is the author of *Toxic Town: IBM, Pollution, and Industrial Risks* (2014), *Burning Matters: Life, Labor, and E-Waste Pyropolitics in Ghana* (2022), and his forthcoming book is *Platforms, Pathologies, and Plunder: Critical Zones of Technopower and Global Political Ecology*.

Loretta I.T. Lou is Assistant Professor in Social Anthropology at Durham University. She has a DPhil in Anthropology from Oxford University and her research specialises in the study of environment, health, healing, and social movements in East Asia. She is the co-Editor-in-Chief of *Worldwide Waste: Journal of Interdisciplinary Studies*.

Alice Mah is Professor of Sociology at the University of Warwick, with research interests in urban and environmental sociology. She is the author of *Petrochemical Planet* (2023), *Plastic Unlimited* (2022), *Toxic Truths* (with Thom Davies, 2020), *Port Cities and Global Legacies* (2014), and *Industrial Ruination, Community, and Place* (2012), winner of the British Sociological Association Philip Abrams Memorial Prize.

Sarah May is Senior Consultant in Cultural Heritage at the sustainable development consultancy, Arup. She has over 27 years of experience advising on, developing, and managing archaeological and cultural heritage strategies. She has substantial experience working with communities and government agencies to use cultural heritage as an asset in transformation. She is a member of the UNESCO Chair on Heritage Futures based at Linneaus University in Sweden. Her research here draws on her time as a Senior Lecturer in Public History and Heritage at Swansea University.

Arthur McIvor is a Professor of Social History and Co-Director of the Scottish Oral History Centre at the University of Strathclyde, Scotland. McIvor is a specialist in oral history, work, deindustrialization, and occupational and environmental health, and has published widely in these areas. He is currently working on an environmental history of Glasgow: Toxic City.

Bridget McKenzie is a researcher and creative curator. After roles such as Education Officer for Tate and Head of Learning at the British Library, she founded Flow Associates in 2006. In 2019, Bridget founded Climate Museum UK and co-founded Culture Declares Emergency.

Ashley Meredith is a National Cultural Anthropologist and Deputy National Historic Preservation Officer at the Federal States of Micronesia Office of National Archives, Culture and Historic Preservation (NACH).

Tiago Silva Alves Muniz is a post-doc researcher in Anthropology at Exorigins project, based at Centre Alexandre-Koyré, École des hautes études en sciences sociales, Centre national de la recherche scientifique (CAK, EHESS, CNRS) - Paris, France. His major research interests are botanical colonialism, Amazon rainforest, and archaeology of rubber.

Scott Neilsen is an Associate Professor in the School of Arctic and Subarctic Studies at the Labrador Campus of Memorial University. His current research is coordinated through the Laboratory for Applied Archaeological Research and Community Heritage, and is undertaken in response to requests from communities, organisations, and individuals in Labrador.

Leila Papoli-Yazdi is an archaeologist based at Malmö University, School of Arts and Communication. Her main interest is studying subordinated communities, poverty, and discrimination by investigating waste disposal and garbage-making behaviours.

Thomas W. Pearson is Professor of Anthropology and chair of the Social Science Department at the University of Wisconsin-Stout, and author of *When the Hills Are Gone: Frac Sand Mining and the Struggle for Community*.

Celmara Pocock is the Director of the Centre for Heritage + Culture, and Professor of Anthropology and Heritage Studies at the University of Southern Queensland, Australia. Her research foregrounds community attachment, aesthetic and sensuous experience of heritage places and landscapes. Her work on the heritage values of the Great Barrier Reef is published widely and includes her monograph *Visitor Encounters with the Great Barrier Reef: Aesthetics, Heritage, and the Senses.*

Lisa Rankin is a Professor and Research Chair in the Department of Archaeology at Memorial University of Newfoundland and Labrador. For over twenty years she has worked in collaboration with Indigenous communities to understand the past. Her publications have focused on culture contact, household archaeology, community archaeology, and ethical archaeological practice.

Johnny Reis is an Environmental Engineer, Postgraduate in Civil Protection, PhD Student in Risks Science, and Researcher Fellow at the Centre for Environmental and Marine Studies, Department of Environment and Planning, University of Aveiro, Portugal. He is a hazardous materials expert and Volunteer Firefighter. His academic scientific research can be found at https://www.researchgate.net/profile/Johnny-Reis.

Daniel Renfrew is Professor of Anthropology and acting chair of the Department of Sociology and Anthropology at West Virginia University, and author of *Life without Lead: Contamination, Crisis, and Hope in Uruguay.*

John Schofield is Professor of Archaeology at the University of York where he specialises in cultural heritage and archaeologies of the contemporary world. His research is currently focussed on the various ways archaeology and heritage practice can contribute to solving some of the world's so-called wicked problems, including plastic pollution. This is the subject of his forthcoming book.

Liz Ševčenko is Founding Director of the Humanities Action Lab, a coalition of universities, issue organisations, and public spaces in 40 cities that collaborate to produce community-curated public humanities projects on urgent social issues including Climates of Inequality: Stories of Environmental Justice. She is the author of *Public History for a Post-Truth Era: Fighting Denial through Memory Movements* (Routledge, 2022).

Paul A. Shackel is Professor of Anthropology, University of Maryland. His research projects have focused on the role of archaeology in civic engagement activities related to race and labour. His recent research investigates labor and migration in northeastern Pennsylvania (*Remembering Lattimer: Migration, Labor, and Race in Pennsylvania Anthracite Country,* 2018) and the history of unchecked capitalism in the region and its effects on the anthracite region's communities (*The Ruined Anthracite: Historical Trauma in Coal Mining Communities*, 2023).

Sophia Stamatopoulou-Robbins is Associate Professor of Anthropology at Bard College. She is the author of *Waste Siege: The Life of Infrastructure in Palestine* (2019). Her second book, *Controlled Alienation: Airbnb and the Future of Home*, examines changing attachments under austerity in Greece.

Sian Sullivan is Professor of Environment and Culture at Bath Spa University and UK lead for an AHRC-DFG funded project called Etosha-Kunene Histories (www.etosha-kunene-histories.net). She researches conservation and cultural landscapes of north-west Namibia, and the financialisation of nature (www.the-natural-capital-myth.net).

Javier Taks holds a PhD in Social Anthropology from the University of Manchester (United Kingdom). He is an associate professor at the Institute of Anthropological Sciences of the Faculty of Humanities and Educational Sciences of Udelar and an adjunct professor in the Research Program on Sustainable Development of the Faculty of Social Sciences of Udelar. His research revolves around ecological anthropology, energy and climate change studies, agrarian studies, and water anthropology.

Ana Valderrama is founder of Matéricos Periféricos, a collective for spatial and social justice. She is Full Professor at the School of Architecture, Planning, and Design and Director of the Master of Landscape Architecture of Universidad Nacional de Rosario in which she served as Associate Dean from 2015 to 2019. She is currently a PhD student in Landscape Architecture, University of Illinois at Urbana Champaign.

Anatolijs Venovcevs is a PhD candidate at the Institute of Archaeology, History, Religious Studies, and Theology at UiT: The Arctic University of Norway with a focus on twentieth-century single industrial mining communities in northern Canada, Norway, and Russia. He works within contemporary archaeology, industrial and historical archaeology, historical geography, and GIS.

Ranger Walter currently works as State Historic Preservation Officer for the Chuuk Historic Preservation Office, Federated States of Micronesia, USA.

Fabienne Wateau is a Social Anthropologist and Full Research Professor CNRS (France). She works on resource management, alternatives, and commons, mainly in the Iberian Peninsula and also directs documentaries. She teaches in the Department of Social Anthropology at the University of Paris Nanterre and at the Quai Branly Museum. Her academic scientific research can be found at https://cnrs.academia.edu/fabiennewateau.

Marina Weinberg is an anthropologist and Assistant Professor at the Instituto de Investigaciones Arqueológicas y Museo-San Pedro de Atacama at Universidad Católica del Norte (Chile) and researcher at the Núcleo de Investigación TraGeMa -Trabajo, Género y Minería en el Desierto de Atacama-. Her current work examines how material and social dynamics of mineral extractivism (lithium and copper) convey in the Atacama Desert, connecting and transforming landscapes and societies, and affecting and shaping life and death.

FOREWORD

Reckoning with toxic heritage is an urgent collective task. It is also unsettling work. It requires confronting painful truths about the roots of toxic injustice with courage, honesty, and humility. This collection takes up this task, delving into questions of legacy and memory in the context of pervasive toxic harm to multiple species and environments. The concept of toxic heritage is framed in this collection as heritage relating to the "materiality of toxic substances" (Kryder-Reid and May), exploring tensions between preservation and destruction in inherited material realities. Bringing together perspectives from critical heritage studies and interdisciplinary studies of environmental toxicity, the contributions demonstrate how the roots of toxic injustice stem from the extractive and dispossessing logics of capitalism and colonialism, yet how they also provoke forms of resistance.

When I first heard the term "toxic heritage," I must admit that it made me uneasy. In my research on deindustrialized and polluted communities, I have always thought about toxicity in a negative light, as pernicious harm which must be stopped. While attentive to the complexities of lived experience, my research takes a decisively anti-toxic stance. Thus, it was difficult to accept toxicity as part of heritage. However, as I have come to appreciate, engaging with toxic heritage in its myriad meanings opens possibilities for critical intervention and healing.

Few people want to own toxicity as a part of their heritage, especially if ownership implies accepting responsibility for toxic production and harms. But for people who experience illness or loss from toxic harms, the suppression of toxic histories represents a form of violence. Several years ago, when I was researching the declining chemical industry in Niagara Falls, a university friend from England moved to the nearby city of Hamilton in Canada for work. She associated Canada with "nature" and was dismayed to find that Hamilton was an old steel city. If she had known that it was such a polluted industrial city, she confessed, she would have thought differently about moving there. She sent a photograph of herself next to Lake Erie to her friends, cropping out the steelworks in the background. At the time, I thought this was a strange thing to do. However, as several authors in this volume detail, the impulse to erase toxic histories is common, and it can have long-standing consequences for public memory and environmental health.

The undeniable material reality of toxic harm across the planet underscores the importance of recognising toxic heritage. According to scientists, toxic chemical pollution has recently crossed a "planetary boundary," posing a risk to the stability of Earth systems and intersecting with the climate crisis, biodiversity loss, and a range of overlapping ecological problems (Persson et al. 2022). Toxic substances are found in bodies, places, and ecosystems, enmeshed in the fabric of life. Yet around the world, toxic hazards are disproportionately concentrated in racialized and marginalized communities (Pellow 2017). Kryder-Reid and May (in this volume) situate the unequal planetary impacts of toxic harm within the context of the "patchy Anthropocene" (Tsing, Mathews, and Bubandt 2019). This idea adds nuance to the Anthropocene narrative, which many scholars have criticized for being too universalistic and human-centered (see Haraway 2015), and for suggesting that "all humans are implicated in and affected by colonialism, capitalism and industrialization in the same ways" (Whyte 2017: 259).

From extractive mining debris to military waste; contaminated industrial cities to toxic agricultural land; arsenic-laced Indigenous sacred objects to climate justice museum practices, this collection speaks to the planetary scale and scope of toxic heritage while acknowledging its deeply unequal effects. One of the most difficult challenges of acknowledging toxic heritage in many places is stigmatisation, which exacerbates social and environmental inequalities. Moreover, when histories of contamination come to the surface, new risks and responsibilities emerge. Places with newly revealed toxic legacies – sacred lands, public parks, rolling countryside – face the challenge of confronting the implications. In some cases, local responses to toxicity are to find ways of living with pollution and re-imagining their relationships with place. In other cases, residents, activists, and community organizations work together on collaborative interventions to reframe toxicity and waste in their struggles for environmental justice.

If recognition implies a kind of acceptance of material reality, then what are the political and ecological implications? What aspect of toxic heritage, if any, should be preserved, especially if the harmfulness endures? What should be done with toxic heritage? The beauty of this collection is that it addresses such profound questions, offering insights grounded in specific contexts, while resisting easy answers. The methodological thread that unites the array of contributions is a shared commitment to grapple with difficult toxic legacies, across different perspectives, cultures, and scales. From an ethical standpoint, the collection explores toxic heritage in a spirit of humility, with the aim of recovering from toxic harms and moving toward just futures. It does not resolve the tensions between heritage and toxicity, but it does offer new ways of thinking about the entangled relationships between toxic pasts, presents, and futures.

Alice Mah

References

Haraway, Donna. "Anthropocene, Capitalocene, Plantationocene, Chthulucene: Making Kin." *Environmental Humanities* 6, no. 1 (2015): 159-165.

Pellow, David N. *What is Critical Environmental Justice?* Cambridge, UK: Polity, 2017.

Persson, Linn, Bethanie M. Carney Almroth, Christopher D. Collins, Sarah Cornell, Cynthia A. de Wit, Miriam L. Diamond, Peter Fantke et al. "Outside the Safe Operating Space of the Planetary Boundary for Novel Entities." *Environmental Science & Technology* 56, no. 3 (2022): 1510-1521.

Tsing, Anna Lowenhaupt, Andrew S. Mathews, and Nils Bubandt. "Patchy Anthropocene: Landscape Structure, Multispecies History, and the Retooling of Anthropology: An Introduction to Supplement 20." *Current Anthropology* 60, S20 (2019): S186-S197.

Whyte, Kyle Powys. "Indigenous Climate Change Studies: Indigenizing Futures, Decolonizing the Anthropocene." *English Language Notes* 55, no. 1 (2017): 153-162.

SERIES GENERAL CO-EDITORS' FOREWORD

This series of edited books was conceived as a platform for disseminating innovative research in heritage studies around specific research themes. The series aims to help identify and set agendas in new research areas in heritage studies. It showcases areas and issues often overlooked in more mainstream research that has tended to narrowly focus on technical heritage management and preservation issues. While challenging the traditional research focus, the series, nonetheless, is concerned with critical and reflective engagements with the practical and policy applications of research. With the development of the critical heritage studies movement, the ability to question and interrogate the nature of heritage and the social and political consequences it has on policy and practice and on people's lived experiences has become increasingly important. Interest in heritage preservation is never apolitical and simply technical, the developing field continues to explore "heritage" as a social and political construct encompassing all those places, artefacts, and cultural expressions inherited from the past which, because they are seen to reflect and validate our identity as nations, communities, families and even individuals, are worthy of some form of respect and protection.

What is named "heritage" results from a selection process; a process that intersects with the practices of remembering and forgetting, commemoration, leisure and tourism, and the development and expression of historical consciousness. This selection can often be 'top down' and regulated by experts, policy, and legislation, but it is equally a process 'from below' that can challenge or ignore established forms of heritage making and the selection and thus foster new ways of thinking about heritage and its relationship to the past and present.

Heritage has a complex relationship with both the past and present, and can be used in both positive and negative ways. For example, it may be used to foster respect for cultural and social diversity, provide community and individual self-esteem, and challenge prejudice and misrecognition. Equally, it may be used to foster exclusion, and the maintenance of misrecognition and prejudice, or it can be used to promote undemocratic political agendas and rally people against their neighbours in civil and international wars, ethnic cleansing, and genocide. Heritage is highly political and entwined with contemporary struggles over social justice and equity and is, as a consequence, integral to the setting and reimagining of aspirations for the future.

Gönül Bozoğlu and Laurajane Smith

TOXIC HERITAGE: AN INTRODUCTION

Elizabeth Kryder-Reid and Sarah May

Premise and Genesis

This collection represents a cross-disciplinary conversation about the intersection of heritage as a field of discourse and practice, and toxicity as a material and social reality. It brings together authors from varied disciplinary perspectives and methodologies to explore toxic heritage as both a material phenomenon and a concept. In case studies, visual essays, and substantive chapters, scholars draw on research around the world to provide an in-depth examination of toxic heritage as a global issue. Our definition of toxic heritage has two components. First, it includes the history of the processes and substances (toxins and toxicants) that create or threaten physical harm to environments and the life supported within them (Liboiron 2017). Second, it includes the intersections of that history of harm with both formal heritage institutions and informal memory practices. This focus on the materiality of toxic substances stands in contrast to other definitions that also include social and political toxicity and other forms of "dark heritage" (Wollentz et al. 2020: 299), although the physical and metaphorical are often entangled. The attention to memory practices, processes of heritagisation, lived experiences, and the ways in which interpretations of past environmental harm are entangled with fields of power center this work within critical heritage studies.

The genesis of this volume is the confluence of two research projects. For May, the work was part of the Heritage Futures project (Harrison et al. 2020), particularly her contributions to the essay "Toxic Heritage: Uncertain and Unsafe" (Wollentz, May, Holtorf, and Högberg 2020), which was, in turn, an extension of her longstanding investigation of industrial heritage. For Kryder-Reid, the catalyst was participating in the Climates of Inequality project (see Ševčenko, this volume) and her community-based, collaborative research on the pollution of Indianapolis waterways and its impact

DOI: 10.4324/9781003365259-1

on marginalized communities. Her previous research on stakeholder-defined values of heritage (Kryder-Reid et al. 2018), led to questions about how sites of environmental harm are treated as heritage and how heritage sites attend to their histories of environmental harm. May and Kryder-Reid (hereafter "we") connected to develop a session on toxic heritage at the 2020 London Association of Critical Heritage Studies (ACHS) meetings. In spite of the need to switch to a virtual session due to the pandemic, the conversation was substantive, and we were encouraged to reach out to other contributors for a publication. It quickly became clear that scholars from a variety of disciplinary, geographic, cultural, and institutional perspectives were working on the topic, but not necessarily in conversation with each other.

As the responses to our call for papers came in, we found them both affirming and devastating. The intersections and frictions among the contributions raised important issues and illuminated emerging lines of inquiry into toxic heritage. Clearly, this was an important, even urgent, topic. At the same time, reckoning with the pervasiveness and insidiousness of the manifestations of harm explored by the authors was at times overwhelming. The illnesses, exposure to noxious substances, and living conditions amidst waste were heartrending. In this global snapshot, the lines of privilege are starkly drawn by contrasting exposomes, and yet no one is spared the consequences of human impact on our planet. The shadow of toxic heritage now extends in a quite material sense across every continent and community.

The other context for the genesis of this volume is the mounting evidence for the scope and scale of our intersecting environmental crises – climate change, biodiversity loss, and pollution. As many of the contributions attest, the increasing urgency of these crises is often met with persistent complacency, deliberate indifference, denial, and even paralysis at individual, community, and governmental levels. The volume seeks to respond to both the planetary emergency and the inertia by examining the place and role of heritage in meeting the existential crises of our era. It asks, what is the role of critical heritage studies in probing the politics of toxic heritage and illuminating the fields of power in which these sites operate? In this sense, the volume is activist scholarship that seeks not just to understand, but also to spark conversation and spur change.

Themes

The project is international in scope and interdisciplinary in its approach to the cumulative environmental burdens of modern human history. The combined weight of the studies offers a chilling view of the scope and scale of the toxicity and the unevenness of its impacts across lines of social inequality in what has been described as the "patchy Anthropocene" (Tsing, Mathews, and Bubandt 2019). The consideration of toxic heritage as a planetary phenomenon highlights the scale and complexity of issues that transcend the typical national boundaries and chronologies of heritage. Instead, toxic heritage must consider eco-centric narratives and account for "the great acceleration" of the post-WWII years (McNeill and Engelke 2016).

This range of perspectives also offers insight into the particular manifestations of toxic heritage in diverse materialities and cultural contexts. It therefore raises interesting questions about the similarities and differences of transnational trends such as deindustrialization, waste management, and the introduction of synthetic chemicals, as they are negotiated in particular fields of power and experienced by people in unique cultural contexts.

Another theme running through the volume is the exploration of what it means to bring the ideas of toxicity and heritage together and how it advances thinking about the relationships of the present to the past and to the future of a damaged world. Authors draw on various intellectual traditions such as Michel Foucault's idea of biopolitics, Ron Nixon's (2011) work on slow violence, and Haraway and Tsing's work on other-than-humans and multi-species alliances, to locate the intersections of toxicity and heritage. The result is an array of ways to bridge the gap that typically exists between thinking about places as toxic and thinking about them as heritage. Bringing the concepts together helps make connections across scales of impact (cells, bodies, families, communities, ecosystems, nations, oceans, planet) and the parallel scales of entangled social and economic relations. It explores the specific ways that legacies of toxicity, contamination, and pollution intersect with the formal and informal heritage practices in different cultural contexts. Conversely, it tackles the resistance to, and even rejection of, the idea that our human toxic legacy can be understood as heritage (Holtorf; Wateau et al., this volume).

Another contribution of the collection is that it situates toxic heritage squarely in political and social arenas. In Palestine waste is wielded as an instrument of political oppression (Stamatopoulou-Robbins). Papoli-Yazdi shows that the status of both waste and heritage are conditioned by the power that changes the every day to the toxic in Tehran. Evia et al. explain how the bioaccumulation of pesticides and nutrients in soils not only affects animal, human, and ecosystem health but has had an impact on the intangible heritage of traditional dairy farming practices and the struggle for a sense of place and value of their lifeways. As McIvor notes, "The industrial past continues to linger on in the present in polluted soil, rivers, dirty buildings and in the bodies and memories of Glasgow's people." In Cusack-McVeigh's examination of contaminated museum collections, we see the effects on living communities and their social relations with their ancestors as the harms of colonialism are perpetuated even amidst efforts to repatriate toxic museum collections. Schofield and Pocock show how the materiality of plastic offers a socially useful metaphor that connects pasts with futures. Both McKenzie and Valderrama offer their experience of using the structures of heritage to provide a focus for action.

Central to this grappling with the impacts and harms of toxic heritage is a recognition of the disparate effects across lines of social inequalities. The downstream effects across geopolitical boundaries, as e-waste recycling contaminates communities and bodies in Ghana (Little and Akese) and across generations, as with WWI chemical weapons contamination (Hubé and Bausinger) or gold mining in California (Hoskins). They highlight the impacts on the meaning and significance of place and land, particularly impacts on Indigenous communities (Rankin et al., Joyce). These framings of toxic heritage also center communities not just as populations with adverse health effects but as communities with agency who weild heritiagisation as a tool for advocacy and environmental justice, as in the toxic tours described by Baptista and Fiske and Fischer, Filippelli's connection between heritage and activism surrounding lead contamination citizen-science, and Benussi's account of people caring for their families' graves in the shadow of Chernobyl.

A theme, or perhaps tension, running through the volume is the relationship between history and heritage. This was also something we considered when putting the volume together. Many of our contributors come from disciplines focussed on researching the past – establishing what happened. This volume builds on environmental humanities exploration of toxicity (e.g. Müller and Nielsen 2023; Sarathy, Hamilton, and Brodie 2018) with a particular focus on how we engage with those pasts today – hence the sections structured around framing, politics, activism, narrative, and interventions. All of these are contemporary practices that engage with toxic pasts. How do histories of environmental damage resonate with the proud narratives of industrial heritage (e.g. Shackle, Hannis and Sullivan, Weinberg and Figueroa) or the economically powerful roles of local industry (Lou, Pearson and Renfrew, Gardner, Muniz)? Where in the valorized narratives of war memory, national pride, and sacrifice is there space for accounting for enduring ecological impacts (e.g. Carter et al., Hubé and Bausinger)? Conversely, how, in contexts that struggle against negative stereotypes and marginalization, such as the public housing projects investigated by Elizabeth Browning and the copper industry heritage in de-industrialized Swansea explored by May, do histories of heavy metal contamination connect with public memory? These conflicts are particularly challenging when the perpetrators are not simply a profit-driven, extractivist corporation, but include the role of workers who applied the chemicals as part of their standard practice, as with agrotoxins (Evia et al.). In the case of retail dry cleaning (Kryder-Reid et al.), the complicity extends throughout the supply chain from manufacturer to consumer.

Organization and Format

The volume is organized into sections exploring five themes:

1 "Framing toxicity" explores fundamental issues of conceptualizing environmental harm as heritage including considerations of the nature of toxicity and its implications for understanding human heritage in the Anthropocene.
2 "The politics of toxic heritage" considers the role of policy, stakeholders, politics, and fields of power in which the sites and stakeholders operate.
3 "Affected communities, activism, and agency" focuses on the impact of environmental harm on communities and the ways in which people, as biological citizens, workers, community activists, and environmental justice advocates, have responded.

4 "Narratives of toxic heritage" examines the narratives and discourse practices around toxic legacies.
5 "Approaches and Interventions" highlights the ways in which individuals, organizations, and industries operate in the creation and enduring consequences of toxicity, including reflections on the role of heritage studies and organizations in the construction of toxic heritage.

Within this thematic organization, the book includes three formats of contributions – chapters, case studies, and visual essays – that each have distinct purposes. Chapters present substantive research on a variety of toxic heritage materials and contexts from a range of disciplinary perspectives. Case studies focus on a specific set of empirical data that exemplify issues or point to interventions. These analytical case studies help to connect the concepts of toxic heritage to practical applications, advocacy, and activism. Visual essays of images with extended captions offer an alternative format of scholarly communication that gives authors an opportunity to address the visual logic of sites, explore the aesthetics of toxic heritage, and reflect on the meaning of visualization (and invisibility) of toxic heritage. The visual essays also address important emerging strategies to democratize data, such as a graphic essay and a citizen science-generated digital platform. In addition to the chapters, case studies, and visual essays, and to this introduction and Alice Mah's foreword (xix-xxi), we offer section introductions as a "connective tissue" to frame questions and integrate the diverse contributions in each section.

We thank the contributors for sharing their important and timely research and for advancing the exploration of the work of heritage in the damaged world.

References

Harrison, Rodney et al. 2020. *Heritage Futures: Comparative Approaches to Natural and Cultural Heritage Practices*. London: University College of London.

Kryder-Reid, Elizabeth, Jeremy W. Foutz, Elizabeth Wood, and Larry J. Zimmerman. 2018. "'I just don't ever use that word': Investigating Stakeholders' Understanding of Heritage." *International Journal of Heritage Studies* 24 (7): 743–763. 10.1080/13527258.2017.1339110

Liboiron, Max. 2017. "Toxins or Toxicants? Why the difference matters." *Discard Studies*. September 11, 2017. https://discardstudies.com/2017/09/11/toxins-or-toxicants-why-the-difference-matters/

May, Sarah. n.d. "Dark heritage and toxic waste" Heritage Futures Lexicon - Dispatches." https://heritage-futures.org/dark-heritage-toxic-waste/

McNeill, J. R. and Peter Engelke. 2016. *The Great Acceleration: An Environmental History of the Anthropocene Since 1945*. Cambridge: Harvard University Press.

Müller, Simone M. and May-Brith Ohman Nielsen, eds. 2023. *Toxic Timescapes: Examining Toxicity Across Time and Space*. Athens, OH: Ohio University Press.

Nixon, Ron. 2011. *Slow Violence and the Environmentalism of the Poor*. Cambridge, MA: Harvard University Press.

Sarathy, Brinda, Vivien Hamilton, and Janet Farrell Brodie, eds. 2018. *Inevitably Toxic Historical Perspectives on Contamination, Exposure, and Expertise*. Pittsburgh: University of Pittsburgh Press.

Tsign, Anna Lowenhaupt, Andrew S. Mathews, and Nils Bubandt. 2019. "Patchy Anthropocene: Landscape Structure, Multispecies History, and the Retooling of Anthropology: An Introduction to Supplement 20." *Current Anthropology* 60:S20, S186–S197.

Wollentz, Gustav, Sarah May, Cornelius Holtorf and Anders Högberg. 2020. "Toxic heritage: Uncertain and unsafe." In Harrison, Rodney, et al. *Heritage Futures: Comparative Approaches to Natural and Cultural Heritage Practices*, 294–312. London: University College London.

SECTION 1

INTRODUCTION: FRAMING TOXICITY

Elizabeth Kryder-Reid and Sarah May

This section explores fundamental issues around thinking of environmental harm as heritage. When we chose the title of this volume, we were aware that both 'toxic' and 'heritage' are widely used and variously defined. The term 'toxic heritage' is therefore immediately recognisable, but what it means (and doesn't mean) is harder to pin down. For this volume we are using the term in a more narrowly defined manner than Wollentz et al. (2020) who include socially harmful legacies in their discussion. That broad definition was framed in order to create conceptual links between the management of toxic materials and the management of cultural heritage, but it was only the beginning of a deeper understanding of the challenges that toxic materials bring to cultural heritage. Our working definition of toxic heritage has two components. First, it includes the history of the processes and substances – including toxins (produced by plants, animals, and bacteria), toxicants (synthetic, human-made, toxic chemicals), and anthropogenic pollution from natural materials such as lead, arsenic, and mercury – that create or threaten physical harm to environments and the life supported within them (Liboiron 2017). Second, it includes the intersections of that history of harm with both formal heritage institutions and informal memory practices. While we focus on material which is harmful to health, we understand that notions of toxicity vary over time and among cultures. Investigating toxic heritage allows us to focus not just on the history of environmental harm, but on heritage as a set of practices that work with the past in the present.

The papers in this section focus on the different ways that heritage can deploy the past, such as valorisation, remembrance, forgetting, as a spur to action, and as a mechanism for forgiveness. These authors explore how we experience toxicity - as waste, bioaccumulation, post-mining landscapes, nuclear imaginary, and poisoned places. Hoskins, Joyce, Pearson, and Refrew examine how risk is perceived and what senses are engaged, especially when toxic materials can be invisible, silent,

DOI: 10.4324/9781003365259-2

odorless, and yet deadly. The authors illuminate issues of accountability and responsibility and examine how heritage can sequester environmental harm in the past or neutralize criticism of ongoing harm. Hannis, Sullivan, and Gardner examine how the material remains, particularly landscapes and landforms, represent and reproduce these harms. The papers in this section also explore the complex temporality of toxic heritage. There are materials like PFAS which will persist in their harms forever, and others, like nuclear waste which will decay slowly. Other forms of toxicity are rendered relatively benign from mitigation or natural dispersal, but their harm persists in inequality and generational health impacts. Schofield and Pocock offer a model for thinking about the temporality of environmental damage, particularly relevant in this time of the Great Acceleration (McNeill & Engelke 2014). Together, the papers help to frame ways of thinking about toxic heritage and its import.

References

Liboiron, Max. 2017. "Toxins or Toxicants? Why the difference matters." *Discard Studies* September 11. https://discardstudies.com/2017/09/11/toxins-or-toxicants-why-the-difference-matters/

McNeill, J. R., and Peter Engelke 2014. *The Great Acceleration: An Environmental History of the Anthropocene Since 1945*. Cambridge, MA: Belknap Press (an imprint of Harvard University Press).

Wollentz, Gustav, Sarah May, Cornelius Holtorf, and Anders Högberg. 2020. "Toxic heritage: Uncertain and unsafe." In Harrison, Rodney, et al. *Heritage Futures: Comparative Approaches to Natural and Cultural Heritage Practices*, 294–312. London: University College London.

1

TOXIC LEGACIES OF SLICKENS IN CALIFORNIA: A MOBILE HERITAGE OF HYDRAULIC MINING DEBRIS

Gareth Hoskins

Introduction: critical pedagogies of the toxic

Toxic is often discursively positioned in opposition to purity. It signals infiltration, boundary crossing and a disturbing of the natural order of the world. Toxic presents itself as a problem to be solved, a deviance to be put right. Toxic's origins from ancient Greek and Byzantine medical literature refer to a poison causing death or serious debilitation. This negative valance is reinforced in its use in the couplet 'toxic waste' – an abjection managed through containment and isolation, and through its use as a prefix in 'toxic masculinity'. Toxic is clearly something to be avoided.

This chapter argues, that notwithstanding its often-tragic consequences, the toxic can be generative and productive. Toxic can disrupt prevailing patterns of thought embedded in heritage and conservation that secure 'nature' and 'culture' in binary opposition. This mission of purification (Latour 1993) finds expression in all sorts of conservation-related endeavours from early notions of wilderness (Cronon 1996), wildlife photography (Franklin 2006) to early approaches underpinning the designation of World Heritage Sites as 'natural', 'cultural' or 'mixed' (ICOMOS 2005). Although many operations within UNESCO have subsequently sought to undermine this binary (Cultural landscapes, Geoparks) examples of untouched nature are much more likely to be championed compared to sites with an overt human presence deemed more conducive to excavation, development, or the dumping of waste (Kuletz 2016). Toxic troubles those boundaries, mixing categories, mutating and reorganising bodies in harmful ways at the edge of our scientific understanding. Toxic points to the disruptive agency of matter (Bennett 2010). It confronts us with a post-human 'geontology' (Povinelli 2016) and disturbs the linear temporality of heritage as a generational 'passing on' where inheritance aligns with essentialism and entitlement (Landzelius 2003).

DOI: 10.4324/9781003365259-3

The toxic is useful in critical pedagogy (Freire 2020; Giroux 2010) prompting learning about social justice and uneven exposures to harm. Work within industrial archaeology and environmental interpretation (Hardesty 2001; ICOMOS with Heyes 2004) has made use of this potential, as has the eco-anarchist petro-photography of Richard Misratch and Kate Orff (Scott 2019). In the context of post-Chernobyl Ukraine, Petryna (2009, 2013) tracks the emergence of new populations of sufferers of radiation exposure who participate in a 'biological citizenship' to secure welfare payments. Pezzullo in *Toxic Tourism,* examines 'toxically assaulted' communities 'turning to tours as a tactic of resistance' (2009, 11). Inspired by these two projects, this chapter focuses on Malakoff Diggins State Historic Park near Nevada City, California and site of what once was the largest hydraulic gold mine in the world, to explore how mercury contamination, as a form of mobile toxic heritage, intersects with other forms of heritage in California to prompt questions about environmental and social justice (Figure 1.1).

There are important benefits to understanding local consequences of global industrial processes as toxic heritage because western capitalism has been so effective in screening those consequences from our imagination. In his review for the journal *Public Historian*, industrial archaeologist Donald Hardesty speculates on the ways hazardous landscapes could instead be managed as a warning: 'The power of toxic waste as a real artefact in its original setting should not be overlooked as a way of conveying to visitors the impact of industrial technologies upon workplaces, communities, and landscapes' (2001, 20).

Remediation, in this sense, is erasure hiding more profound dysfunctions, especially when done superficially. Nowhere is this more expertly outlined than in

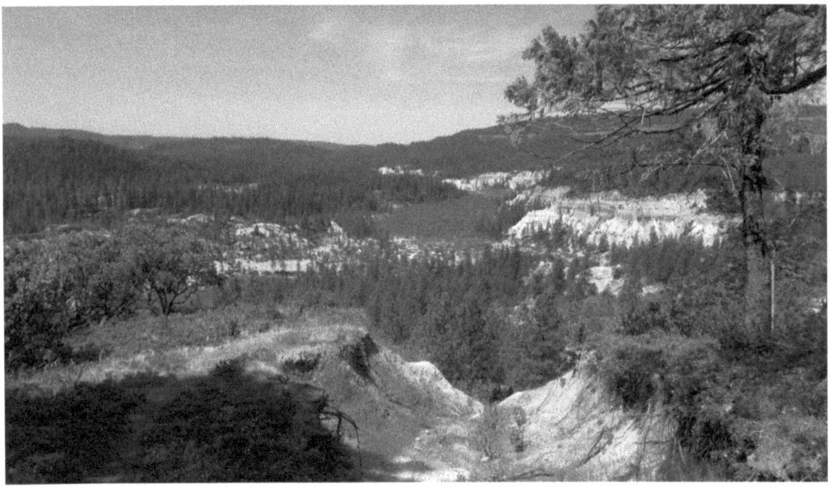

FIGURE 1.1 Picture of the view of Malakoff Diggins taken from the overlook of chute hill campground at Malakoff Diggins State Historic Park, Photograph by the author.

Shiloh Krupar's research (2011, 2012) on Rocky Flats plutonium factory near Denver Colorado or 'Rocky Flats National Wildlife Refuge' to give its more recent title. Krupar deploys the toxic as a metaphor to critique the mixing of nature and culture within contemporary industrial capitalism invoking a novel vocabulary of transnatural ethics, mutant ecologies (Masco 2004), queer ecologies (Mortimer-Sandilands 2005) and alien still-life.

In his examination of a Belgium creosote yard made safe for visitors, Stefaan Hayes, coordinator of the Flemish Association of Industrial Archaeology, questions:

> Is there a better way to illustrate that the soil grows leaner under the influence of unrestrained industrialization? Is there a better way to visualise the evolution in our environmental awareness, or to demonstrate the regenerating capacity of nature, than by means of authentic evidence? After all, the sterilised ground that is left behind after decontamination hardly lends itself to an enjoyable nature experience (2005, 49).

Notions of memorials to environmental damage or 'toxic ruin as classroom' have antecedents (Misrach and Misrach1990; Hoskins and Whitehead 2013; Dixon et al. 2016). There is interest in establishing what might be called 'heritage sites of environmental conscience' that could include Three Mile Island in Pennsylvania, the Love Canal in New York, or Pripyat, Chernobyl. Done poorly, this presents the risk of aestheticizing industrial ruin as a visual spectacle for middle-class consumption where communities bearing health burdens of contamination are reduced to a component in the scene (Strangleman 2013; Wells 2018; Pohl 2021).

Photographer Richard Misratch avoids this by deploying toxic aesthetics to confront us with the effects of industry on our environments. His proposal for a new National Park designation at Bravo 20 – a 64 square-mile naval bombing range on public land in the Nevada Dessert – chimes with a notion of toxic heritage as critical pedagogy.

His book *Bravo 20* makes the case:

> Bravo 20 National Park would be a unique and powerful addition to our current park system. In these times of extraordinary environmental concern, it would serve as a permanent reminder of how military, government, corporate, and individual practices can harm the earth. In the spirit of Bull Run and the Vietnam Memorial, it would be a national acknowledgment of a complex and disturbing period in our history. Although we pay homage to, and protect, geologically spectacular landscapes such as Yosemite and the Grand Canyon, since the 1970s we have established other criteria for the creation of state and national parks, such as the inclusion of wetlands and urban recreation areas. A contemporary version of a Civil War battleground, Bravo 20 National Park would not only provide a graphic record of our treatment of less celebrated landscapes but also help deter their destruction in the future (1990, 95).

Plans include a visitor centre devoted to the history of military abuse in peacetime with displays and exhibits about radioactive experiments on residents and a gift shop selling maps of radioactive landfills and postcards, caps and bumper stickers embossed with images of mushroom clouds and bomb sites.

There is a productive tension between the toxic and conventional landscape aesthetics (Storm 2014). Hardesty notes: 'Certainly toxic waste or other hazardous sites are not pretty, but no one said that history has to be pretty' (2001, 24). Edward Burtynsky's photography, however, depicts beauty in scenes of environmental disaster in a troubling juxtaposition of content and form. A recent review of Burtynsky work captures this well:

> The interesting paradox, though, is the epic allure he finds in such scenes. Looking at vertiginous quarries in Portugal ("inverted skyscrapers", he's called them) or green farmland somehow conjured out of the arid Texas plains, you don't always know whether to feel awe at the ingenuity by which such resources are extracted, or despair at whether the damage can ever be healed.
>
> (Dixon 2022)

By the end of the 1860s, the landscapes emerging as a result of hydraulic mining in California were provoking similar emotional unease. Many scenes were captured by Carleton Watkins, a renowned photographer of the American West, whose mammoth prints of hydraulic mining operations were displayed in galleries alongside images of Yosemite Valley and the Columbia River. In 1868 one of the nation's leading journalists Samuel Bowles visited the Sierra mining districts on a trip across the continent describing the effects of hydraulic mining:

> Tornado, flood, earthquake and volcano combined could hardly make greater havoc, spread wider ruin and wreck, than are to be seen everywhere in the track of the larger gold-washing operations. None of the interior streams of California, though naturally pure as crystal, escape the change to a thick yellow mud from this cause, early in their progress from the hills. The Sacramento River is worse than the Missouri. Many of the streams are turned out of their original channels, either directly for mining purposes, or in consequence of the great masses of soil and gravel that come down from the gold-washing above. Thousands of acres of fine land along their banks are ruined forever by the deposits of this character. A farmer may have his whole estate turned into a barren waste by a flood of sand and gravel from some hydraulic mining up stream; more, if a fine orchard or garden stands in the way of the working of a rich gulch or bank, orchard or garden must go. Then the tornout, dug-out, washed to pieces and then washed over side-hills, masses that have been or are being subjected to the hydraulics of the miners, are the very devil's chaos indeed. The country is full of them among the mining districts of the Sierra Nevada, and they are truly a terrible blot upon the face of Nature.
>
> (Bowles 1869, 422)

Slickens

Miners and farmers in the foothills of the Sierra Nevada in the late nineteenth century termed the hydraulic mining debris slickens. We might now understand this slickens as mobile toxic heritage; a material expression of an industrial past that shaped the social history of California precisely because of its capacity to migrate. Slickens is the extraction, processing, transport, use, discard and dispersal of thousands of tonnes of material that will remain hazardous for tens of thousands of years. Hydraulic mining began after the Gold Rush years as means to recover gold dust from fine-grained auriferous gravel and continued (with some regulation) through to the 1950s. Hydraulic mining required high-pressure jets of water gravity-fed by an extensive network of dams, ditches, and pipes as well as large quantities of liquid mercury, or quicksilver used to concentrate gold particles into an amalgam. The amalgam would be boiled leaving behind the gold. Much of the mercury escaped into the environment.

This letter from the state archives in Sacramento shows a request made in December 1889 from a hydraulic mining operator at North Bloomfield California (Figure 1.2).

> Dear Sir, could you either lend or sell to us three tanks of quicksilver. We need it for immediate use. Should you be able to spare it, please send down by first staffing tomorrow. Yours truly R. McMurray. P.S if you can't spare three, send one or two.
>
> (California Department of Parks and Recreation Archives, Sacramento, California).

The amount of mercury in a standard thermometer is around one gram which is enough to contaminate a 20-acre lake. The total amount of mercury released in California in years spanning 1850 to 1981 was over 220 million pounds (Eagles-Smith C. A.). A 1905 study discovered minute particles of quicksilver floating on surface water as far as 20 miles downstream of mining operations (Bowie 1905, 313). Much of what escaped into the ecosystem converts to monomethyl mercury (MMHg), which is, according to a recent study on hydraulic mining sediment along the lower Yuba River, 'subsequently being taken up by aquatic migratory biota including algae, aquatic insects, bivalves, forage fish, salmonids, sportfish, and waterfowl throughout the geographical region downstream of hydraulic mining sites' (Nakamura et al. 2018, 2). Monomethyl mercury passes easily through the blood-brain barrier compromising child development, creating deformities, and other acute health problems (Mergler et al., 2007; Cristol et al., 2008).

Complaints about slickens during the hydraulic mining era, however, focussed on their damage to agriculture and transportation. In his ecological history of mining in California, Andrew Isenberg (2010) describes the late nineteenth-century conception of slickens as a problem to be overcome: 'Imperceptibly moving

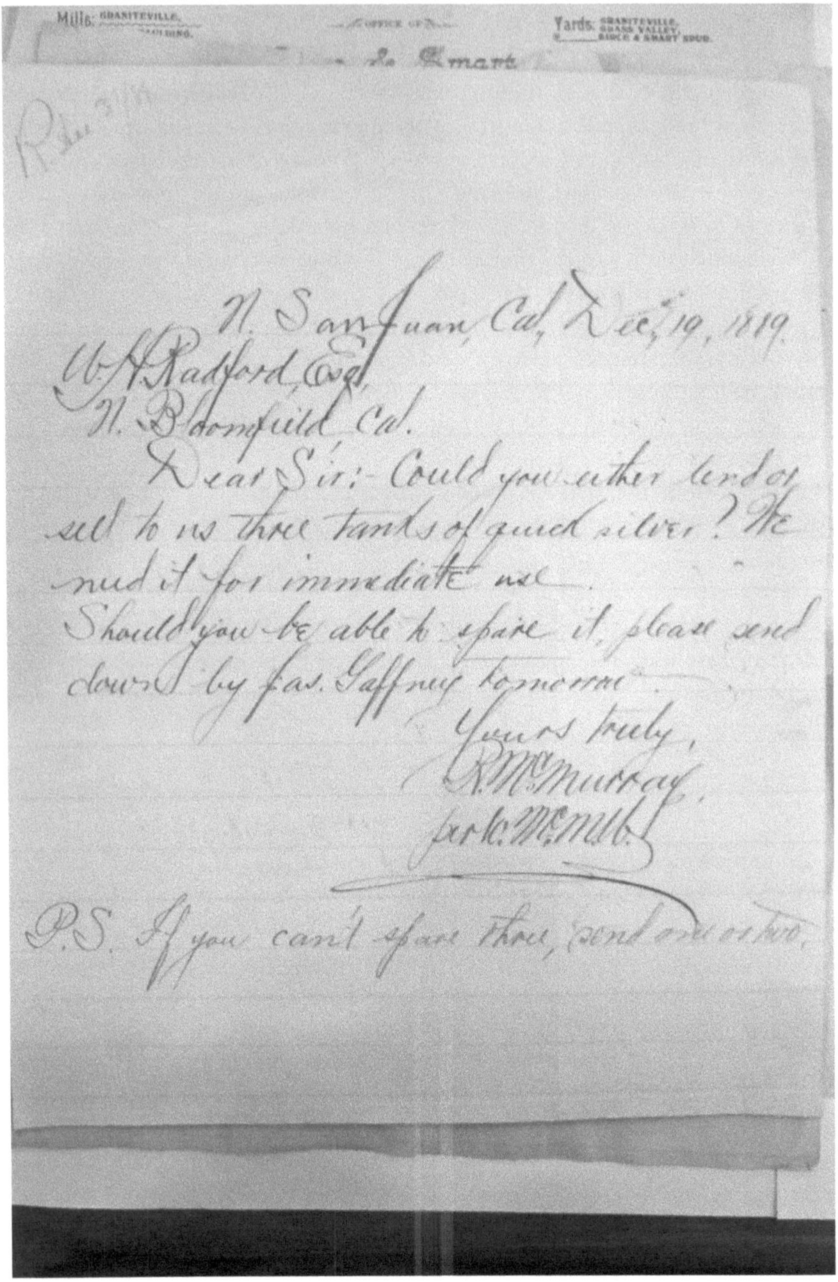
N. San Juan, Cal, Dec. 19, 1889.
W. H. Radford, Esq,
N. Bloomfield Cal.
Dear Sir:- Could you either lend or sell to us three tanks of quicksilver? We need it for immediate use.
Should you be able to spare it, please send down by Jas. Gaffney tomorrow.
Yours truly,
R. McMurray.
per [illegible]
P.S. If you can't spare three, send one or two.

FIGURE 1.2 Picture of correspondence sent from North San Juan California on 19th December 1889 requesting quicksilver (metal mercury) for use in the hydraulic mining operations at Malakoff Diggins. Photograph by the author of item held within North Bloomfield Gravel Mining Company Records 1870–1903, California State Archives Sacramento.

downward, glacier-like in its inexorable spread, fanning out over increasingly broad areas of valley land and sliding down river-beds to obliterate their channels, this "moving avalanche" as George Ohleyer called it, created an insistent demand. The river system had to be reclaimed. The Sacramento had to be placed under government control (2010, 243–4)'.

In the now famous 225-page ruling against the North Bloomfield Gravel and Mining Company operating out of Malakoff Diggins Judge Lorenzo Sawyer described how: 'the waters of the Yuba are so charged with debris that they are wholly unfit for watering stock, or for any of the uses, domestic or otherwise, to which water is usually applied without first being taken out of the stream and allowed to stand in some undisturbed place and settle' (1884, 10). Sawyer imposed a permanent injunction on the dumping of mining waste providing a legal foundation for subsequent regulatory control by the newly formed California Debris Commission (Hagwood 1981; Mitchell 1994).

The 1893 Caminetti Act required vast amounts of waste to be impounded behind dams but these structures were often ineffective and ultimately made the enterprise of hydraulic mining in California unprofitable. A great deal of the mercury that is now problematic escaped and moved downstream along with thousands of tonnes of sand and gravel blocking streams, raising riverbeds, flooding farms and settlements all the way down the Sacramento valley (Figure 1.3).

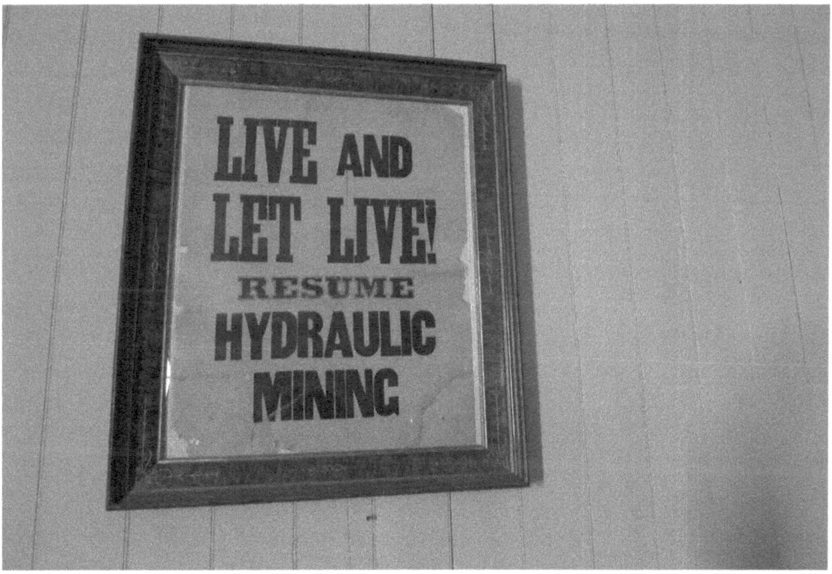

FIGURE 1.3 Picture of political campaign poster supporting hydraulic mining against plans to regulate the industry in the 1880s. Photograph by the author taken in restored historic old town of North Bloomfield.

As Robert Kelly outlines in his book about the dispute, *Gold vs. Grain*, slickens prompted the 'first successful attempt in modern American history to use the concept of general welfare to limit free capitalism' (1959, 2) and marked a transfer of power and influence away from miners towards farmers and their rapidly expanding agricultural economy. Slickens is baked into the social, economic and environmental history of modern California and yet its presence in the collective memory remains faintly eclipsed by more dramatic sites and events within the history of hydraulic mining (the carved-out cliffs, the water cannons, abandoned towns, and rusted pipes and gullies across the San Juan ridge) that are more easily apprehended as 'significant' (Hoskins 2015, 2016, Scott 2017).

While slickens has been examined scientifically as a pollutant (Prokopovitch 1984; Alpers et al. 2005), its more ambiguous social-political relations are less well accounted for. In contrast to previous environmental histories that depict slickens as passive, animated always by something else (miners, rainfall, or gravity for example), we might think of this mobile toxic heritage following Hird's conception of Canadian waste as a socio-material force that "occasions particular material and political mobilizations" (2017, 1). As an agent with the dynamic capacity to respond to its surroundings, to self-organise, form relationships and "participate" in environmental effects, slickens can guide us toward new ways of working with toxic heritage.

Making visible

A relatively unseen aspect of hydraulic mining heritage is its racial dimension. Environmentalist perspectives embraced by California State Parks and activist groups (e.g. the SierraFund) have only recently started working through critiques of white settler colonialism. Many natural, cultural and historical resources preserved to tell positive stories about California are in some way bound up with the state-sanctioned, indeed state-sponsored, extermination of indigenous peoples (Castillo 1998). The European appropriation of Native American land, taken for example, from thirteen thousand members of the Nisenan Tribe of Nevada County since the onset of the Gold Rush, is still marginal to better-funded initiatives conserving fish and wildlife (Hurtado 1988). It is only recently that indigenous activism has aligned with environmental activism in cultural and ecological reclamation (see for instance the Nisenan Cultural Reclamation Corridor in Nevada City and the California Heritage Indigenous Research Project).

The mercury used so liberally in hydraulic mining is often still visible in globules amongst the gravel in streams and creeks. Indeed, gold panning in the streams once popular on heritage tours and environmental living programmes at Malakoff Diggins State Historic Park is now prohibited along much of the local watercourse due to dangerous health risks of contamination and panning's potential to mobilise yet more contamination downstream. Gold panning instead takes place in a separate experience staged as a pre-industrial romantic pioneer Gold Rush activity where contemporary participants seem oblivious to the hydraulic mining that came in its wake. Some of this contradiction comes through in a diary entry I made while conducting fieldwork during one of the parks' annual 'Humbug Days'.

Saturday 11 June 2011, North Bloomfield, Malakoff Diggins State Historic Park, Lunchtime.

> At the edge of a paddock next to Skidmore House a line of children and their parents walk in a circle with lengths of string dipping them periodically in a cast iron pot of molten wax. Candle making is a popular event at Humbug Day alongside the wagon rides, craft making and family games. None seem as popular though as the 8-foot steel trough full of muddy water and gravel that comprises the gold panning activity. A visiting docent is providing help to those that can fit their pan in the trough explaining that gold is much heavier than the minerals that make up the sand and gravel so when the water is shaken and swirled the grains and flakes of gold collect on the bottom of the pan – the lighter materials are winnowed off with the water left to escape over the rim. The docent tells me the panning has been very popular as usual and that it would be about time to "salt" the mix with the other half of her dust. She shows me a small ziploc plastic bag with gold flakes. Each year the Malakoff Diggins Park Association provide this experience to visitors because it is part of an essential elements of being a prospector and only by trying it yourself can you glimpse the kind of labour involved and the anticipatory sense of striking it rich. There have been many complaints sent to the parks administration over the years about restrictions on gold panning along Humbug Creek. The trough now provides visitors with that trip back in time. And this year, with gold prices being their highest ever the parks association paid $2000 for the salting purchasing from a dredging operation lower downstream.
>
> (Research Diary, entry by author)

As a tourist attraction that recruits the body to acquire a skill and bring a sense of excitement, panning for gold in troughs like that described above (although often without the gold) can be found all over the world: from Old Sacramento Waterfront, through to Nutty Jakes Gold Mine experience in Wales, UK, from the reassembled 'old town' at The Big Hole Experience in Kimberly South Africa to 'ShantyTown' Heritage Park in Greymouth New Zealand and GoldRushTown in Jindu Shandong Province China. In these tourist attractions, a very distinctive imagined geography of small-scale ramshackle towns with benign techniques for striking it rich obscures the longer and more harmful legacy of hydraulic mining and its future damaging effects.

There is now widespread awareness that the mercury-laced heritage of hydraulic mining is part of ongoing environmental injustice. Studies have shown that the exposure to mercury contamination in California is socially uneven. The burden is most heavily felt by indigenous communities with Asians, Pacific Islanders and Native Americans having the highest prevalence of elevated blood mercury (Hightower et al. 2006, Nriagu et al. 2012). This toxic heritage works against not just the bodies of California Indian Peoples but also the traditions and cultural practices that define them. Fish, an important vector for mercury to the human population, is an important dietary staple for indigenous communities and crucial in forging cultural and spiritual connections to the land. The indigenous people's advocacy group Cultural Survival has explained how California Indians are forced to regularly balance toxic exposure with traditions. Sherri Norris, a member of the Osage Nation who works with the International Indian Treaty Council in San Francisco describes the dilemma in regards to fishing traditions and solutions such as eating smaller fish lowers down the food chain that is safer: 'Some people may have to abstain,' 'I don't like saying this. It's a form of cultural genocide, when children are born with learning disabilities. But those of us that are alive today are the strongest of our people' (Cultural Survival 2022).

The California Indian Environmental Alliance (2009, 2013) and the Mercury Tribal Health Program have produced guidelines, factsheets and pamphlets that include the 'Mercury Health Toolkit'. This might be thought of as part of an emerging biological citizenship that secures new relations with the state on the basis of adverse health effects. In reference to radiation-exposed citizens of Ukraine in the 1990s Petryna terms this 'illness as a counterpolitics'. The parallels are chilling:

> This dimension of illness as a counter politics suggests that sufferers are aware of the way politics know and do not know about their illnesses and that they are put in a role of having to use these politics to curb further deteriorations of their health, which they see as resulting, in part, from a collapsing state health system and loss of adequate legal protections.
>
> (Petryna 2004, 262)

If it passes through Congress, The Comprehensive National Mercury Monitoring Act (S. 1345) introduced on 22 April 2021 by Sen. Susan Collins (R-ME) will go some way to establishing a new more comprehensive picture of hydraulic mining's toxic heritage. The Bill proposes to authorise ninety-five million dollars over three years to task a number of federal agencies to track and report on long-term changes to mercury concentrations in air, water, soil, fish and wildlife as well as establish an online database for mercury data. It makes no direct mention of indigenous peoples or Native American tribal concerns but it does focus research on 'human communities with highly exposed and vulnerable populations'. In an Arctic context, Houde et al. (2022) provide a model example of how indigenous communities might be included as researchers in monitoring programs that build relationships and draw on traditional environmental knowledge in a collaborative process. It may be that this form of biological citizenship can be established when engaging with a toxic heritage of all kinds. Contamination is an important component of ongoing racial injustice that ultimately cannot be hidden but might instead provide an instructive moment and a platform to galvanise support for reform (Figure 1.4).

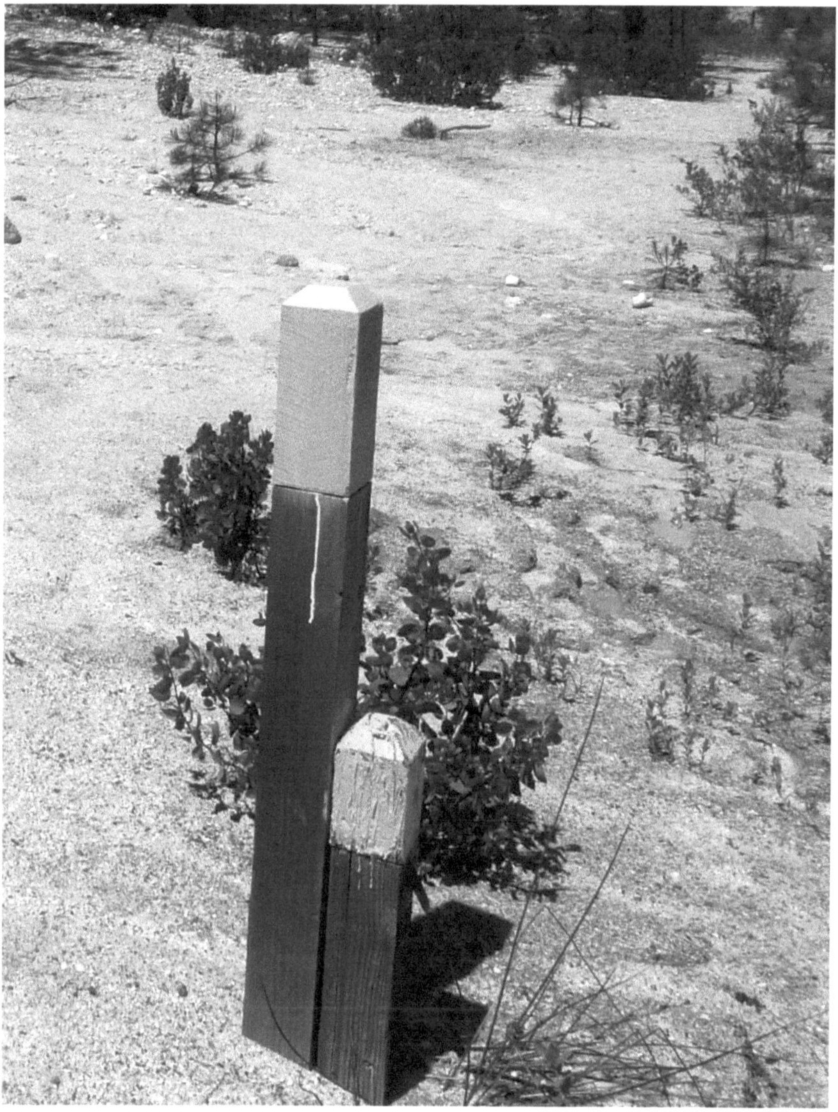

FIGURE 1.4 Picture showing two wooden path-marker posts with painted yellow tops positioned amongst gravel and manzanita bushes on the Diggins Loop Trail at the Malakoff pit. The posts appear at different heights because of movement and burial by the sediment. The taller post is more recent. Photograph by the author.

Looking forward

Hydraulic mining's toxic heritage is distinctive compared with other industrial heritage due to the temporal extent of its toxic legacy. Scholars have engaged with nuclear waste and the peculiar dilemmas associated with interpreting its hazard for generations far in the future Davis 1993; Holtorf and Hogberg 2021). The shelf life of hydraulic mining's toxic heritage is ten thousand years as deep reservoirs of mercury are disturbed by floods and tectonic movement and re-animated on their journey across California to the sea (Alpers et al 2005, Singer et al 2013). This lifespan is greater by orders of magnitude than the time frames typically dealt with in the heritage sector and presents a land management and ecosystem issue of bewildering complexity (Mergler et al., 2007; Cristol et al., 2008). In ten thousand years how will this toxic heritage of hydraulic mining be encountered and how will the people who created it, and those who fought to prevent it, be ultimately understood? We have no template for ethics that sustains for that long. In the meantime, we have to engage with the obscurities of the toxic and its capacity to disrupt and cause harm. Toxic heritage shines a light on social and environmental injustice and should become more central to our concerns.

References

Alpers, C., Hunerlach, M., May, J., & Hothem, R. 2005. Mercury Contamination from Historical Gold Mining in California. *United States Geological Survey Fact Sheet Version 1.1.* Available at http://pubs.usgs.gov/fs/2005/3014/. accessed 27/3/2023

Bennett, J. 2010. *Vibrant Matter: A Political Ecology of Things.* Durham, North Carolina: Duke University Press.

Bowie, A. J. 1905. *A practical treatise on hydraulic mining in California.* New York: Van Nostrand, p. 313

Bowles, S. 1869. *Our New West: Records of Travel between the Mississippi River and the Pacific Ocean.* Harford Conn: Hartford Publishing Co.

California Indian Environmental Alliance. 2013. *Mercury Health Toolkit: Information to Identify, Reduce, and Prevent Mercury Toxicityin the Human Body.* Berkeley California: California Indian Environmental Alliance. Available at https://www.cieaweb.org/brochures/

California Indian Environmental Alliance. 2009. '"Eating Fish Safely," brochure. Information on safe fish consumptions, resources, advisory locations'. Available at www.cieaweb.org

Castillo, E. 1998. 'Short overview of California Indian history'. Available at http://www.nahc.ca.gov/califindian.html

Cristol, D. A., Brasso, R. L., Condon, A. M., Fovargue, R. E., Friedman, S. L., Hallinger, K. K., & White, A. E. 2008. 'The movement of aquatic mercury through terrestrial food webs'. *Science*, 320(5874), 335–335.

Cronon, W. (ed.). 1996. *Uncommon Ground: Rethinking the Human Place in Nature.* New York: W.W. Norton and Company.

Cultural Survival. 2022. 'California Indians Balance Toxins and Traditions'. Available at https://www.culturalsurvival.org/news/california-indians-balance-toxics-and-traditions accessed 27/3/2023

Dasmann, R. F. 1998. 'Environmental changes before and after the gold rush'. *California History*, 77(4), 105–122.

Davis, M. 1993. 'The Dead West: Ecocide in Marlboro Country'. *New Left Review*, 200, 49–73.
Dixon, A. 2022. "We all participate" – Edward Burtynsky on photographing the epic ravaging of Earth'. *Guardian Newspaper*, 22 April, 2022. https://www.theguardian.com/artanddesign/2022/apr/13/all-participate-edward-burtynsky-photographing-epic-ravaging-of-earth-ukraine
Dixon, D., Pendleton, P. & Fearnley, C. 2016. 'Engaging Hashima: Memory Work, Site-Based Affects, and the Possibilities of Interruption', *GeoHumanities*, 2(1), 167–187.
Franklin, A. 2006. 'The humanity of wilderness photography'. *Australian Humanities Review*, 38, 1–16.
Freire, P. 2020. 'Pedagogy of the oppressed'. In Beck, J., Jenks, C., Keddie, N. & Young, M. F. D. (Eds.), *Toward a Sociology of Education* (pp. 374–386). London: Routledge.
Giroux, H. A. 2010. 'Rethinking education as the practice of freedom: Paulo Freire and the promise of critical pedagogy'. *Policy Futures in Education*, 8(6), 715–721.
Hagwood, J. J. 1981. *The California Debris Commission: A History of the Hydraulic Mining Industry in the Western Sierra Nevada of California, and of the Governmental Agency Charged with its Regulation*. Sacramento: US Army Corps of Engineers.
Hardesty, D. 2001. 'Issues of Preserving Toxic Wastes as Heritage Sites'. *The Public Historian*, 23(2), 19–28.
Hightower, J. M., O'Hare, A. & Hernandez, G. T. 2006. 'Blood mercury reporting in NHANES: identifying Asian, Pacific Islander, Native American, and multiracial groups'. *Environmental health perspectives*, 114(2), 173–175.
Hird, M. J. 2017. 'Waste, environmental politics and dis/engaged publics'. *Theory, Culture & Society*, 34(2-3), 187–209.
Holtorf, C. & Högberg, A. 2020. 'What lies ahead?: Nuclear waste as cultural heritage of the future'. In Holtorf, C. & Högberg, A. (Eds.), *Cultural Heritage and the Future* (pp. 144–158). London: Routledge.
Holtorf, C., & Hogberg, A. 2021. *Cultural Heritage and the Future*. London: Routledge.
Hoskins, G. & Whitehead, M. 2013. *Mining Memories: Placing the Anthropocene*. Aberystwyth: Aberystwyth University
Hoskins, G. 2015. 'People like us: Historical geographies of industrial-environmental crisis at Malakoff Diggins State Historic Park'. *Journal of Historical Geography*, 50, 14–24.
Hoskins, G. 2016. 'Vagaries of value at California State Parks: towards a geographical axiology'. *Cultural Geographies*, 23(2), 301–319.
Houde, M., Krümmel, E. M., Mustonen, T., Brammer, J., Brown, T. M., Chételat, J. & Whiting, A. 2022. 'Contributions and perspectives of Indigenous Peoples to the study of mercury in the Arctic'. *Science of the Total Environment*, 841.
Hurtado, A. L. 1988. 'Indians in Town and Country: The Nisenan Indians' Changing Economy and Society as Shown in John A. Sutter's 1856 Correspondence'. *American Indian culture and research journal*, 12(2), 31–51.
ICOMOS. 2005. "'Belgium: Heritage and the Environment" ICOMOS World Report 2004-2005 on monuments and sites in danger'. Available at http://www.international.icomos.org/risk/2004/belgica2004.pdf accessed 13/5/11
ICOMOS with Heyse, S. 2004. 'Belgium, Environment and Heritage'. In *Heritage at Risk 2004/2005*. pp. 47–49. Available at https://journals.ub.uni-heidelberg.de/index.php/heritage/article/view/21008/14781 accessed 27/3/2023.
Isenberg, A. C. 2010 *Mining California: An Ecological History*. New York: Hill and Wang.
Kelly, R. R. 1959. *Gold Vs. Grain: The Hydraulic Mining Controversy in California's Sacramento Valley*. Norman Oklahoma: The Arthur H. Clark Company.

Krupar S. R. 2011. 'Alien Still Life: Distilling the Toxic Logics of the Rocky Flats National Wildlife Refuge'. *Environment and Planning D: Society and Space*. 2011, 29(2), 268–290.

Kuletz, V. L. 2016. *The tainted desert: Environmental and social ruin in the American West*. London: Routledge.

Landzelius, M. 2003. 'Commemorative dis (re) membering: erasing heritage, spatializing disinheritance'. *Environment and planning D: society and space*, 21(2), 195–221.

Latour, B. 1993. *We Have Never Benn Modern*. Cambridge: Harvard University Press.

Masco, J. 2004. 'Mutant Ecologies: Radioactive Life in Post–Cold War New Mexico'. *Cultural Anthropology*, 19, 517–550.

Mergler, D., Anderson, H. A., Chan, L. H. M., Mahaffey, K. R., Murray, M., Sakamoto, M., & Stern, A. H. 2007. 'Methylmercury exposure and health effects in humans: a worldwide concern'. *AMBIO: A Journal of the Human Environment*, 36(1), 3–11.

Misrach, R. & Misrach, M. W. 1990. *Bravo 20: The bombing of the American west*. Baltimore: Johns Hopkins University Press.

Mitchell, M. D. 1994. Land and water policies in the Sacramento-San Joaquin Delta. *Geographical Review*, 84(4), 411–423.

Mortimer-Sandilands, C. 2005. Unnatural passions? Notes toward a queer ecology. Invisible Culture: An Electronic Journal for Visual Cultures. http://ivc.lib.rochester.edu/unnatural-passions-notes-toward-a-queer-ecology/ accessed 27/3/2023.

Nakamura, T. K., Singer, M. B. & Gabet, E. J. 2018. 'Remains of the 19th Century: Deep storage of contaminated hydraulic mining sediment along the Lower Yuba River, California'. *Elementa: Science of the Anthropocene*, 6. 10.1525/elementa.333

Nriagu, J., Basu, N. & Charles, S. 2012. 'Environmental Justice: The Mercury Connection'. In Bank, M. (Ed.), *Mercury in the Environment: Pattern and Process*. Oakland, CA: California Scholarship Online.

Petryna, A. 2013. *Life exposed: Biological citizens after Chernobyl*. Princeton: Princeton University Press.

Petryna, A. 2004. 'Biological citizenship: The science and politics of Chernobyl-exposed populations'. *Osiris*, 19, 250–265.

Pohl, L. 2021. 'The sublime object of Detroit'. *Social & Cultural Geography*, 22(8), 1063–1079.

Povinelli, E. A. 2016. *Geontologies: A requiem to late liberalism*. Durham N.C: Duke University Press.

Prokopovich, N. P. 1984. 'Occurrence of mercury in dredge tailings near Folsom South Canal, California'. *Bulletin of the Association of Engineering Geologists*, 21(4), 531–543.

Scott, C. 2017. 'Photographing Mining Pollution in Gold Rush: California'. *Photographies*, 10(2), 189–209.

Scott, C. 2019. 'The eco-anarchist potential of environmental photography.' Misrach, Richard, and Kate Orff's Petrochemical America'. In Tormey, J. & Durden, M. (Eds.), *The Routledge Companion to Photography Theory*. London: Routledge.

Singer, M. B., Aalto, R., James, L. A., Kilham, N. E., Higson, J. L., & Ghoshal, S. 2013. 'Enduring legacy of a toxic fan via episodic redistribution of California gold mining debris'. *Proceedings of the National Academy of Sciences*, 110(46), 18436–18441.

Strangleman, T. 2013. 'Smokestack Nostalgia,"Ruin Porn" or Working-Class Obituary: The Role and Meaning of Deindustrial Representation'. *International Labor and Working-Class History*, 84, 23–37.

Storm, A. 2014. *Post-industrial landscape scars*. London: Springer.

Wells, K. 2018. 'Detroit was always made of wheels: Confronting ruin porn in its hometown'. In Lyons, S. (Ed.), *Ruin porn and the obsession with decay* (pp. 13–29). Cham: Palgrave Macmillan.

VISUAL ESSAY 1

EXTRACTION OLD AND NEW: TOXIC LEGACIES OF MINING THE DESERT IN SOUTHWESTERN AFRICA

Mike Hannis and Sian Sullivan

Roots and routes of extraction

This visual essay draws on a 2017 journey from the South African Cape to the Khan valley of Namibia, tracing toxic (deter)mining legacies of roots and routes of mineral extraction. Copper, ilmenite, diamond, zinc and uranium exploitations encounter indigenous presence and resistance as colonial and corporate mineral 'rushes' intersect with local realities and cultural landscapes. Environmental mitigation efforts and 'offsetting' schemes may also leave toxic heritage in their wake as they greenwash profit-driven extractive agendas.

In 1685, after seeing green copper-bearing rocks brought to the Cape by Nama people, the Dutch East India Company (VOC) dispatched Cape Governor Simon van der Stel on an expedition 400 miles north, to investigate the viability of mining copper near the !Gariep or Orange River, now the border between South Africa and Namibia. Demonstrating the historical interplay of intangible cultural heritage with extractivist colonial agendas, such journeys relied on local Nama(qua) guides who shared their knowledge of springs with colonisers who would displace them from their ancestral lands. On a 2017 field research journey tracking early colonial expeditions northwards from the Cape to the Khan River, we found these routes peppered with past and present mining sites (Figure 1).[1]

DOI: 10.4324/9781003365259-4

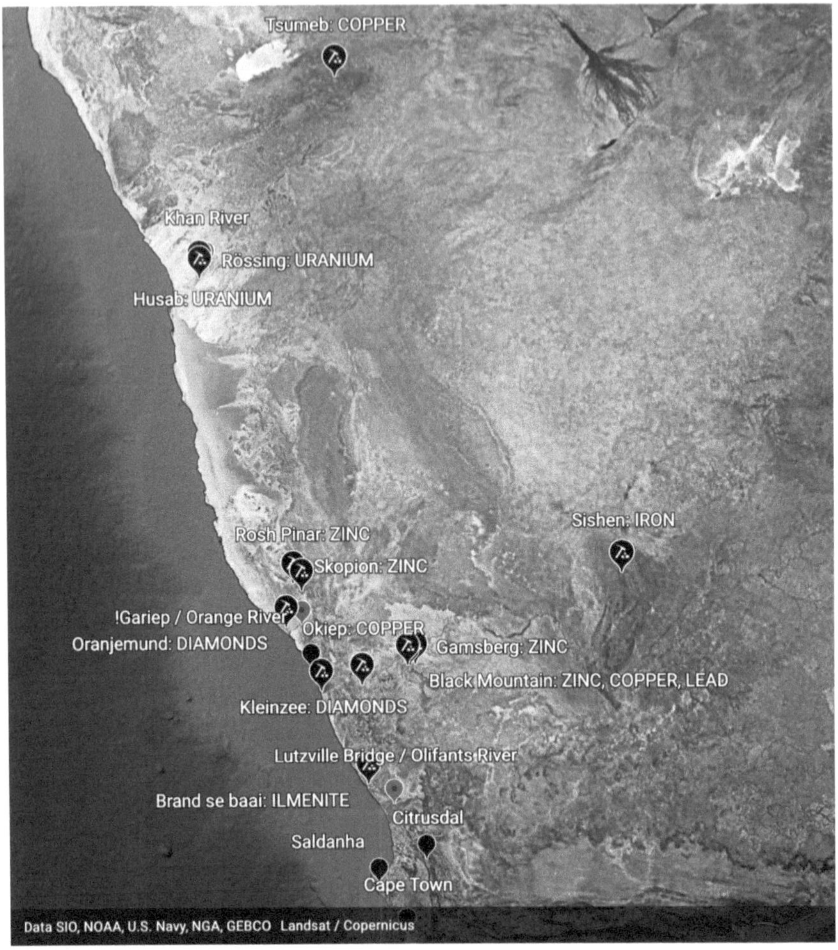

FIGURE 1 Mining sites mentioned in the text.

Ilmenite

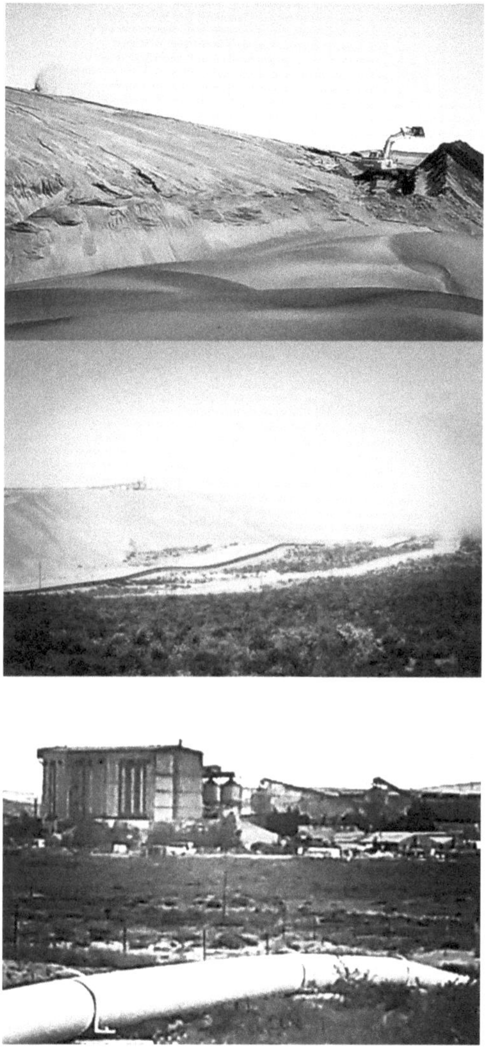

FIGURE 2 Ilmenite mined from fog-shrouded coastal sands around Brand se Baai near the Olifants river mouth produces titanium dioxide, a brilliant white pigment used in products from paint and paper to toothpaste and sunscreen. Operated by US company Tronox and Australian firm Mineral Resource Commodities, a legacy of environmental harm is created as large areas of sandy coastal habitat are excavated to access underlying heavy mineral sands, which are then processed by acid leaching.

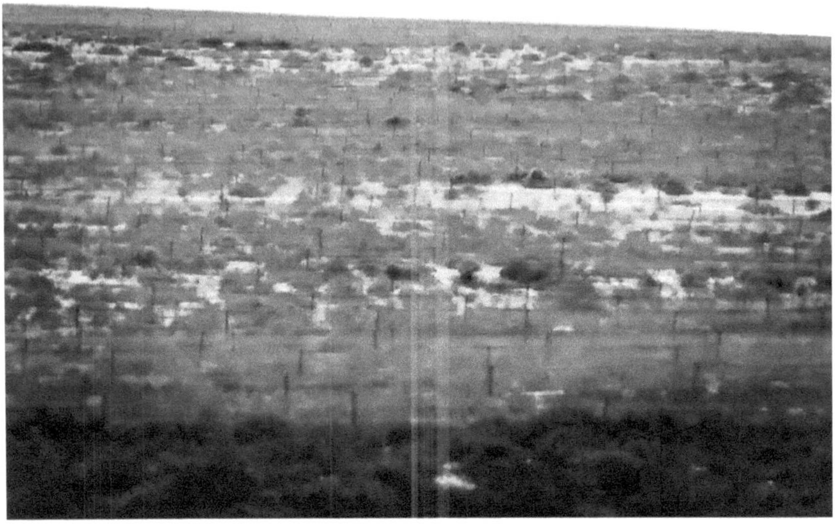

FIGURE 3 Aware of environmental impacts, the companies attempt to 'rehabilitate' mined areas. Miles of green plastic netting, intended as windbreaks to encourage re-establishment of vegetation and invertebrate biodiversity, instead become a new layer of toxic heritage and environmental harm. An environmental officer for one of the Brand se Baai mines told us, however, that few previously existing species have yet returned, nonetheless sharing their belief that 'Mother Earth' (their words) is very forgiving of destruction caused by mining.

Diamonds

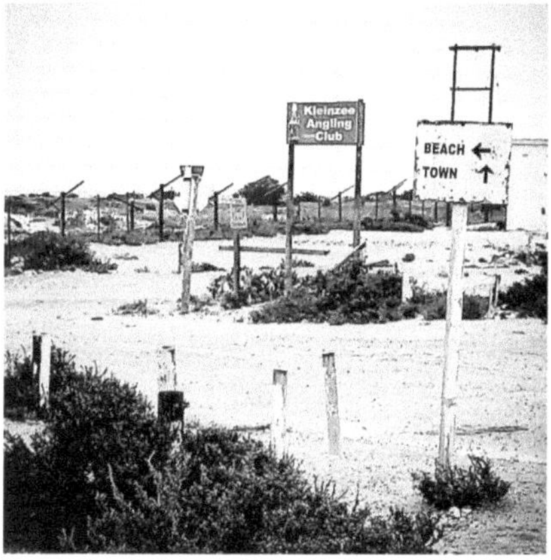

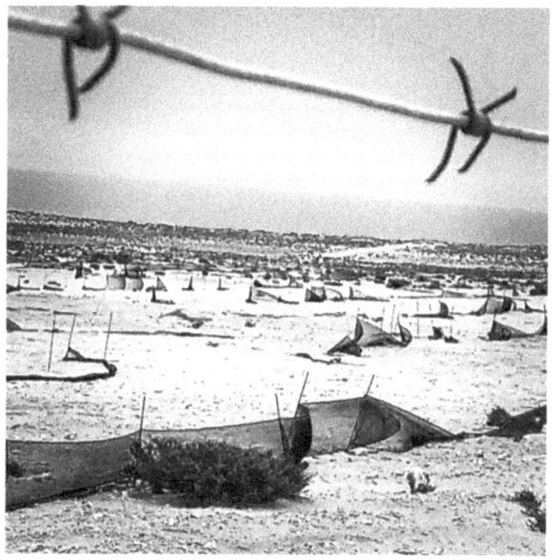

FIGURE 4 Not far beyond Brand se Baai, excavations and spoil heaps appear whose scale and extent dwarf the ilmenite workings. The next 500 miles of coastline have been massively reshaped by over a century of diamond mining. At Oranjemund, just north of the Orange River, decades of digging have pushed the Atlantic Ocean back hundreds of metres behind giant walls of sand and concrete, allowing the shoreline to be excavated down to the bedrock where diamondiferous gravels are found. These land operations are now winding down as the focus shifts to marine mining, leaving semi-abandoned towns such as Kleinzee (left) set in vistas of green plastic fencing purporting, as at Brand se Baai, to shelter re-establishing desert flora (right).

Copper

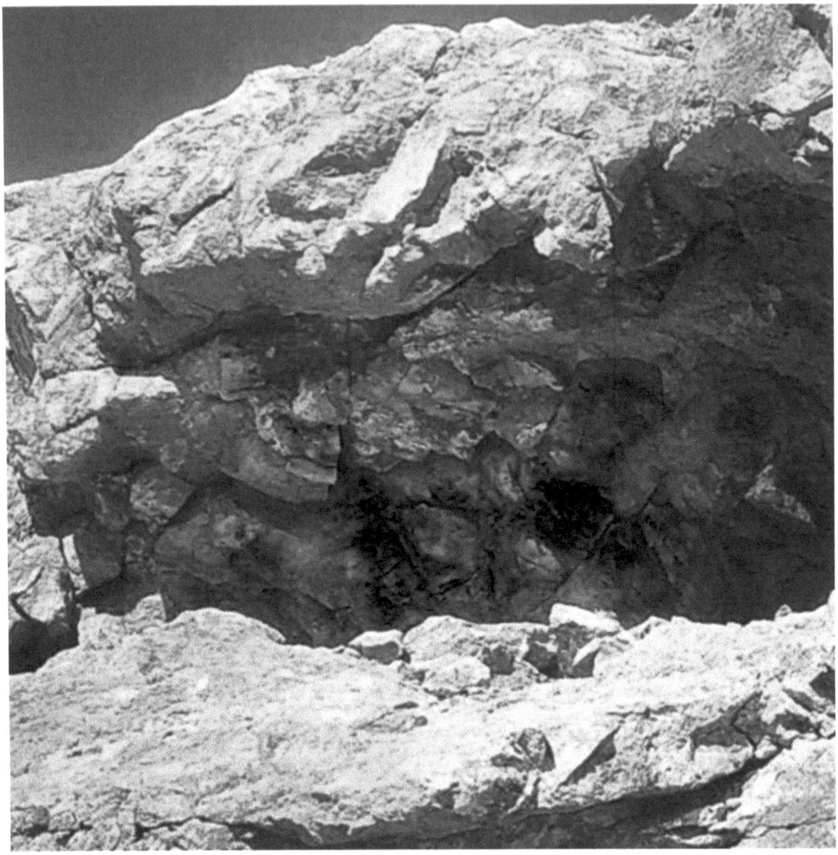

FIGURE 5 Van der Stel did find copper in the Northern Cape. This picture shows his exploratory diggings, burrowing into a rocky outcrop that is indeed remarkably green. In the 1850s, with the Cape Colony now under British control, high-grade ores were found nearby, and Okiep, near Springbok, eventually became 'the richest copper mine in the world', causing immense environmental excavation and landscape reshaping.

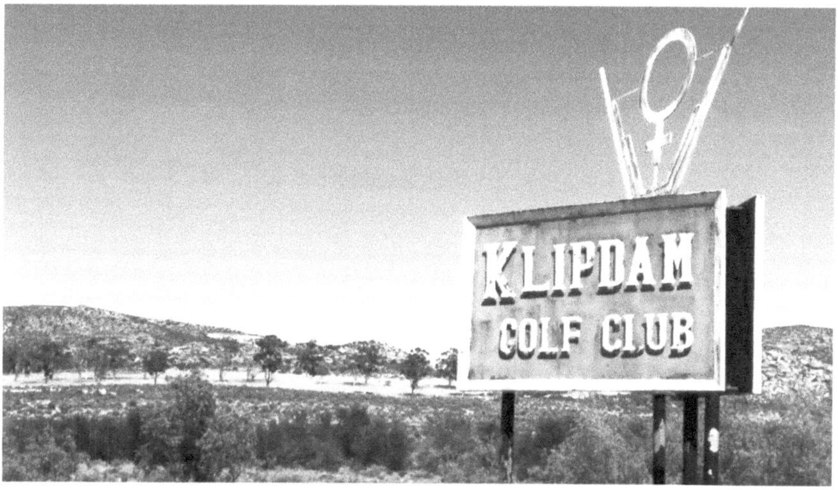

FIGURE 6 In areas like this, mining companies come and go over long periods of time, driven by fluctuations in market prices, technologies, labour costs, tax incentives, environmental regulations and the general compliance (or otherwise) of governments. The last pits in the Okiep area have recently closed, leaving a devastated landscape heritage of spoil heaps, closed mines, emptied towns and unemployed communities. This figure shows an abandoned desert golf club for former local elites, illustrating how the landscape of ruination has become a heritage of both the inequity and limited sustainability of mining ventures.

Zinc

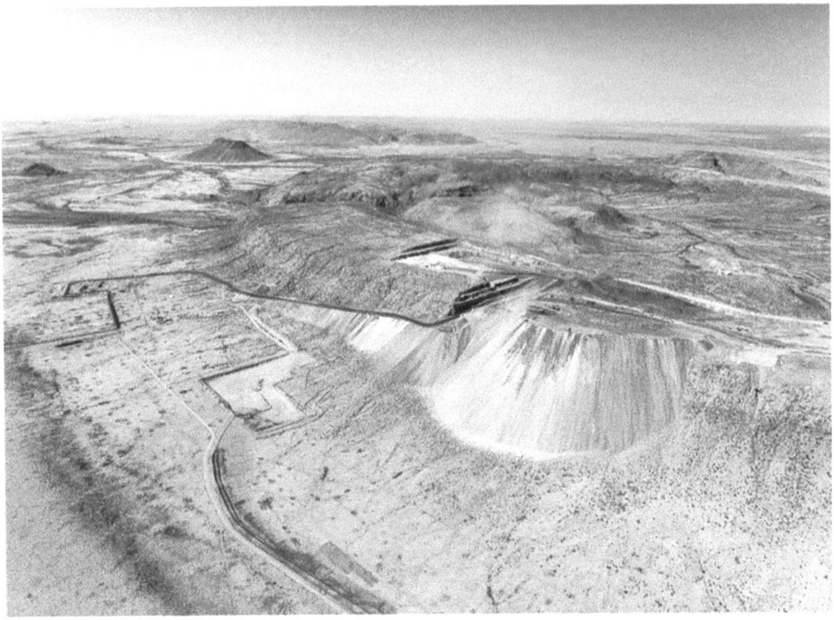

FIGURE 7 The nearest current mining boom to Okiep is happening a hundred miles east near Aggeneys at the recently opened Gamsberg zinc mine, whose Indian owners Vedanta hope to extract 214 million tonnes of ore over a 30-year period. This figure shows an aerial view of Gamsberg in mid-2017.[2] The Gamsberg mountain sits within the Succulent Karoo Biome, one of the world's 36 'biodiversity hotspots'. Vedanta's mine is hollowing out this spectacular inselberg, the acknowledged core of the Critical Biodiversity Area identified in the Namakwa District Bioregional Plan and thus an area central for sustaining biodiversity heritage. Toxic for biodiversity, such impacts will allegedly be 'offset' by enhanced conservation of similar habitats nearby. Under these offsetting plans Vedanta will take control of further large areas of land in the area (Hughes et al. 2015), but how effectively these areas will themselves be protected from future mining remains unclear.

Acid and Arsenic

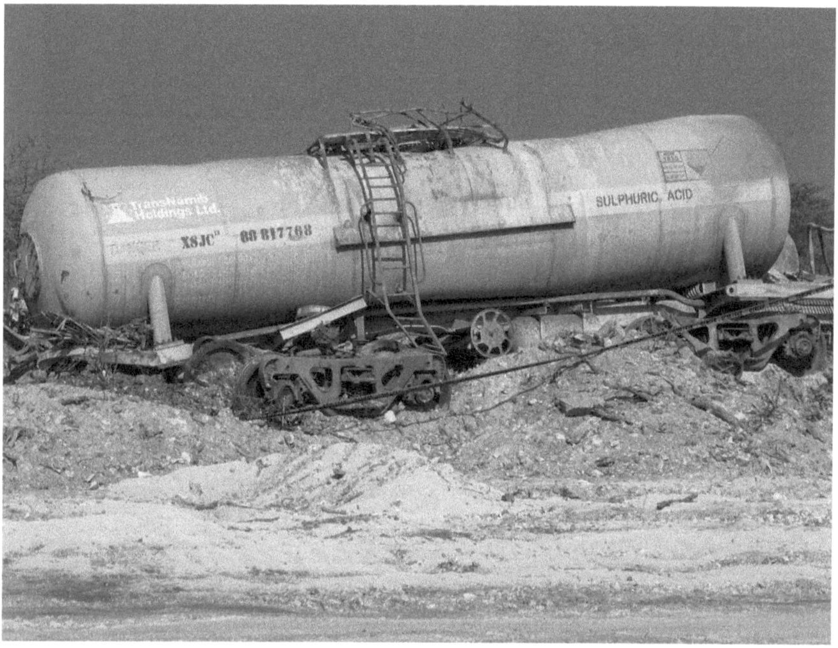

FIGURE 8 The Tsumeb copper smelter in northeast Namibia is owned by Canadian company Dundee Precious Metals (DPM). Originally serving a now-closed mine, it specialises in producing copper from ore containing high levels of arsenic and sulphur, a process banned in many countries on environmental grounds and thus indicative of the global displacement of environmental harms. It processes around 250,000 tonnes of copper ore concentrate a year, half of which is imported from a DPM mine in Bulgaria, the rest coming largely from Chile and Peru. Arsenic trioxide is stockpiled in the open air and sulphur is processed into sulphuric acid, shipped by train to Rössing, for use in leaching uranium ore. Derailments of these acid trains, as shown in the figure, can lead to major toxic spillages (see eg Simiyasa 2022: image by M. Hannis, 2022), leaving legacies of post-extractive waste rather than wisdom that 'sits in places' (Basso 1996; Baird 2022).

Uranium

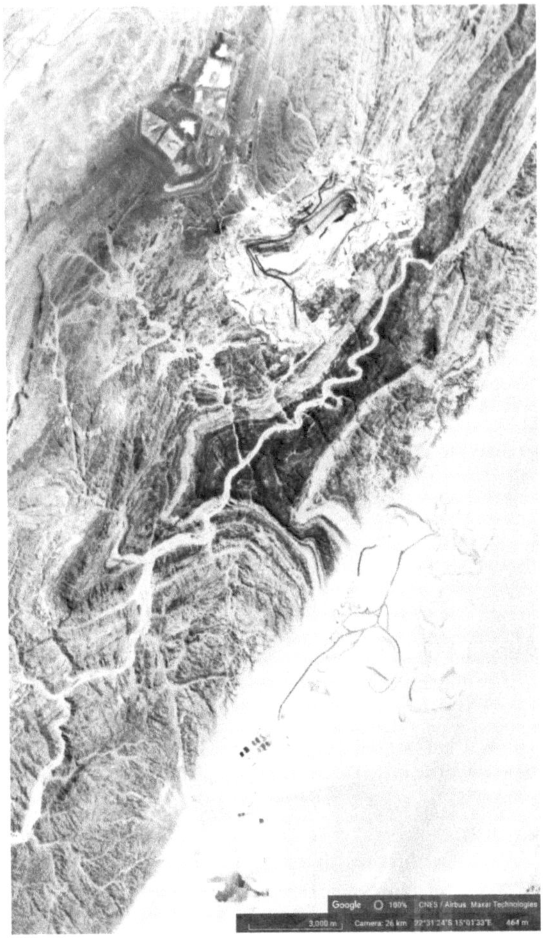

FIGURE 9 Namibia is one of the world's top five uranium producers. Owned since 2019 by China National Uranium Corporation, the Rössing uranium mine in the Namib desert (above) has produced more uranium than any other mine in the world. It was opened in 1976, while Namibia was controlled by apartheid South Africa. Just across the ephemeral Khan River is the Husab uranium mine (below), also majority-owned by China. When opened in 2017, Husab represented the single largest Chinese investment in Africa. As well as sulphuric acid and other chemicals, processing uranium ore into 'yellowcake' (the basis for nuclear reactor fuel) requires electrical power (produced mainly by diesel generation) and enormous volumes of water, piped into these desert mines from a desalination plant on the coast. Biodiversity offsetting is again proposed to mitigate the legacy of environmental harms caused by these processes (Sullivan 2013), although it is unclear how these practices can offset the radioactive future heritage entangled with uranium mining.

Future mining

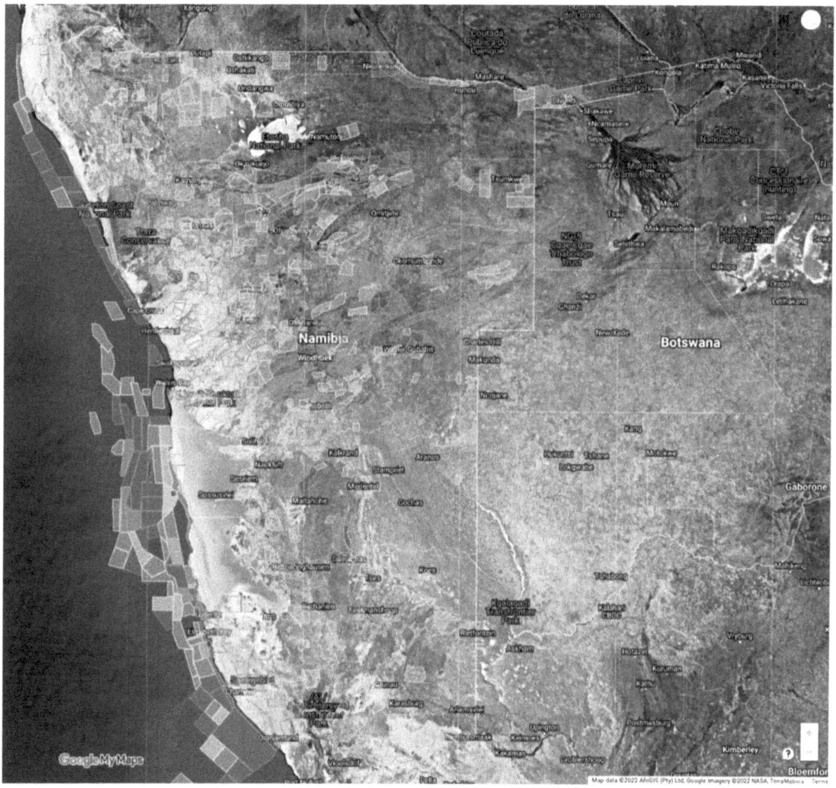

FIGURE 10 The next mineral rush in Namibia is likely to involve minerals such as tantalum and lithium, required for 'green' technologies. Gold mining is also expanding rapidly, and offshore oil exploitation seems imminent. Prospecting and extraction licenses cover Namibia's land- and sea-scapes, as shown in this 2017 snapshot of the online cadastral mining map for Namibia, showing Active Mining Licenses (blue) and License Applications (pink).[3] Clearly, the future heritage of this mining impact is likely to intensify in its toxic burdens and disruptions to human and other-than-human health.

Post-apartheid governments in South Africa and Namibia welcome international mining companies as agents of growth in their mineral-rich but cash-poor countries. But tax revenues, while substantial, are reduced by the companies' ability to use their influence to negotiate favourable terms. Often the lion's share of the profits goes abroad, just as it did in the days of the VOC or the British Empire. Historical processes of extraction, and their inherited legacies, help make sense of present mining trajectories and associated future toxic heritages in southern Africa, and beyond.

Acknowledgements

This research was supported by a research grant from the UK's Arts and Humanities Research Council (AH/K005871/2) for the project *Future Pasts* (www.futurepasts.net).

Notes

1 This essay develops from a longer referenced blog post at https://www.futurepasts.net/post/2018/03/04/extraction-old-and-new. Except where otherwise stated all photographs are by Sian Sullivan, September 2017, originally shared on https://www.futurepasts.net/instagram
2 Source: https://im-mining.com/2017/11/28/vedanta-zinc-international-ge-south-africa-collaborate-ground-breaking-greenfields-digitalisation-initiative-gamsberg/
3 Adapted from https://maps.landfolio.com/Namibia/

Bibliography (all URLs last accessed 17 July 2022)

Baird, M. 2022. Waste sits in places: post-extractive landscapes as heritage, pp. in Pettenati, G. (ed.) *Landscape As Heritage: International Critical Perspectives*. London: Routledge.

Basso, K. 1996. *Wisdom Sits in Places: Landscape and Language Among the Western Apache*. Albuquerque: University of New Mexico Press.

Hughes, J. *et al.* 2015. *Using biodiversity plans to guide mitigation and offsets for a zinc mine in Northern Cape, South Africa*. Paper presented to 35th Annual Conference of International Association for Impact Assessment. https://conferences.iaia.org/2015/final_papers.php

Simiyasa, M., 2022. 'Train derails with acid near Otavi.' *Namibia Press Agency*, https://www.nampa.org/index.php?model=featured&function=display&id=175743

Sullivan, S. 2013. After the green rush? Biodiversity offsets, uranium power and the 'calculus of casualties' in greening growth. *Human Geography* 6(1): 80–101.

2

OF BLAES AND BINGS: THE (NON)TOXIC HERITAGE OF THE WEST LOTHIAN OIL SHALE INDUSTRY

Jonathan Gardner

In this chapter, I discuss how an industrially produced material – oil shale waste (blaes) – has shifted over time from being understood as a valued raw material, a useless by-product and, latterly, a contested form of natural and cultural heritage. Today, 150 million cubic metres of blaes remain in enormous heaps (bings) near to Edinburgh (Scotland) in the district of West Lothian.[1] These are the most obvious remnant of the shale oil industry that operated here from 1851 until 1962. While the blaes ultimately derive from the area's Carboniferous strata (ca. 355–295 mya), they are also a by-product of Scotland's industrial past. Though the blaes are not chemically toxic (Figure 2.1) – as in poisonous to life – they are nonetheless evidence of past toxic human behaviour, not least of all the mining, processing, and consumption of hydrocarbons. Though less exuberantly burnt than coal, conventional oil or natural gas, shale oil nonetheless played an instrumental role in developing the petrochemical foundations of contemporary society. Its technologies, extraction, and usage thus also remain part of the 'Victorian problem' (Morgan 2016, 610) of the climate crisis we now face. Tracing the 'itineraries' of blaes (c.f. Joyce and Gillespie 2015), I explore how valuations of this industrial material can radically shift between notions of value and waste, and toxic and non-toxic.

Toxic language

Like the concept of 'negative' or 'dark' heritage (e.g. Meskell 2002, Rico 2016), calling something 'toxic heritage' involves a value judgement. While this is not to deny the literal toxicity of materials or sites in a biological sense, as with heritage more broadly, determining definitively what is or is not toxic heritage is not always straightforward.

The word toxic has its roots in the Ancient Greek *toxikon* and, though originally related to archery, it was later associated with *pharmakon*, a word that can mean both

DOI: 10.4324/9781003365259-5

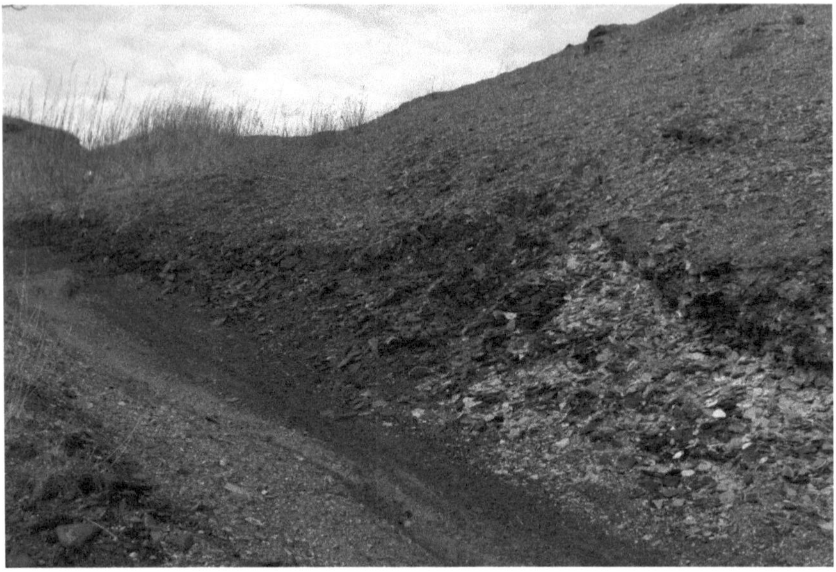

FIGURE 2.1 Oil shale blaes that make up the southern slopes of Greendykes Bing. The different colours of the blaes relate to their mineralogical makeup, the degree of heating during the retorting process, and the extent of weathering the rock has undergone. Photograph by Jonathan Gardner, CC BY-NC 4.0.

'drug' and 'poison'. This connection is significant in light of (Beverley Butler's argument that heritage itself can be understood as a pharmakon 2011); something that remains in a state of uncertainty between being 'poison' or 'cure', blessing or curse. In other words, we must recognise that relationships and responses to the past can be both positive and negative depending on one's perspective. Waste materials are also increasingly recognised as being of heritage value (e.g. Harrison 2021), and, in discard studies, similar arguments assert the *relative* valuation of waste, rather than assuming a fixity of ideas of what is 'wasted' definitively, and recognising that the concept is almost infinitely malleable (Reno 2018, 3, Moore 2012) and temporally contingent (Viney 2015).

Geosocialities

Relating to such shifting valuations is the useful concept of the *geosocial* – how 'geological strata might be seen to condition and enable specific social formations' (Clark and Yusoff 2017, 6). For example, understandings of Scotland's hydrocarbon deposits of peat, coal, oil shale, gas and North Sea oil are situated in a particular discourse where the Earth is passively there for the taking, a 'storehouse' for an extractivist existence. The raw material removed in this extraction is both product and producer of social and material effects (i.e. producing 'geopower' in Elizabeth

Grosz' terms: 2011): mining villages, refineries, networks of petro-capital, and patronage. This is not to assert a 'geological determinism' where our actions are shaped only by our environment. Rather, the opposite: recognition of such a new 'geopolitics' enables us to envisage opportunities outside of established biopolitical formations that have been dominated by ordering principles grounded purely in terms of human agency and dominance (Yussoff 2020). Hence, I seek to explore the blaes as a geosocial actant, something not merely reflective of our human desires and values – whether 'raw material' or 'waste' – but which, in our relations with it, continues to co-produce the worlds we inhabit.

Blaes and bings

Nine kilometres west of Edinburgh is a cluster of four of unexpectedly steep hills (Figure 2.2). Up to 95 metres high and punctuating the otherwise gently rolling West Lothian agricultural landscape, they loom over former mining villages and commuter towns. From some angles, these rust-tinged mounds resemble Near Eastern Tells,

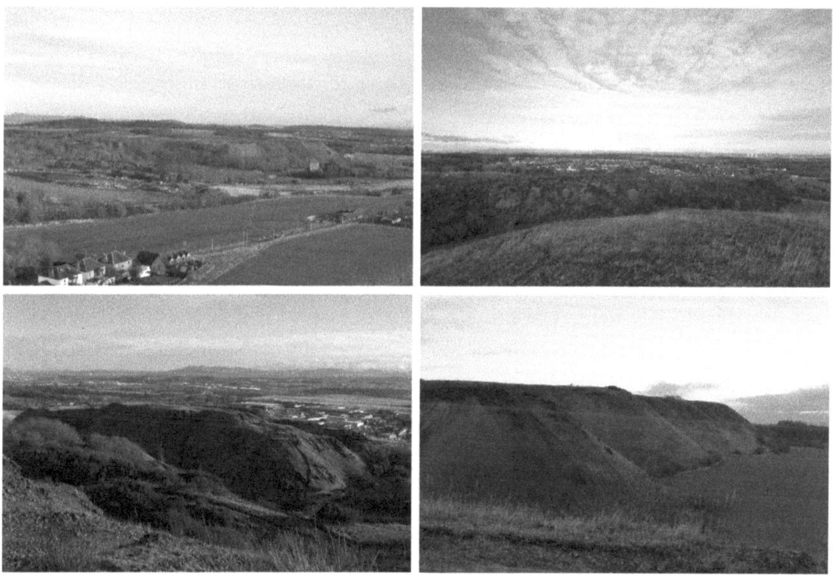

FIGURE 2.2 The four oil shale bings in West Lothian that are discussed in this chapter. From top left, clockwise: Niddry (under extraction for blaes as aggregate); Faucheldean (now a wooded nature reserve, with the Forth Bridges in the distance); Albyn (with Edinburgh's Arthur's Seat at extreme left in the background); Greendykes (the largest of the four bings and, along with Faucheldean, a Scheduled Monument). Photographs by Jonathan Gardner, CC BY-NC 4.0.

while from others they take on the mysterious character of Neolithic monuments like Silbury Hill. Like these predecessors, the bings are a work of human artifice, though of entirely 'unconscious design' (Richardson 2012), the product of tip-lines that disposed of spent shale. This waste – blaes – comprises billions of laminated, flaking fragments that vary in colour from a dusky blue to a pale gold (Figure 2.1). Each is a by-product of the shale oil industry that flourished in West Lothian between 1851 and 1962, an endeavour that has been called the world's first commercially successful oil industry.

The bings I discuss here, Albyn, Faucheldean, Greendykes and Niddry, are amongst the largest out of a total of nineteen remaining in West Lothian (Harvie 2005, 14). In some cases, these are now recognised as heritage sites but this status remains tenuous. To understand why this is the case, we must begin with the origins of the material itself.

Emergence

The oil shale has its origins in Lake Cadell, a shallow body of water that existed in the Carboniferous around 340 million years ago. At this time, the area of the crust that now forms Scotland lay at the equator and was home to a rich variety of plants, proto-reptiles, amphibians, insects, fish and microorganisms. It was the mass deaths of such organisms and the agglomeration of their remains in strata over a kilometre thick that produced the West Lothian Oil Shale Formation. These compressed organisms – mainly algae and some land plants – formed a hydrocarbon-rich kerogen within a matrix of shale, 'a fine-grained indurated clay rock … free from grit' (Knox 2013, 18).

Even here, hundreds of millions of years ago, humans have extended their presence. The Lake is named after H. M. Cadell, a man whose extensive research into the West Lothian oil shale helped to kick-start the industry (Cadell 1894). Cadell was a geological surveyor and aristocratic owner of numerous coal mines, iron works and part of a long line of Cadells who had profited from central Scotland's rich hydrocarbon and iron resources (Mendum 2010). Here, not for the first time, the science of geology, ancient life forms and industrial and social capital are inextricably mixed (see also Brown et al. 2021, Yusoff 2015).

Transformation

The transmutation of kerogen-containing shale to oil, paraffin and other 'products', including the spent blaes – a movement from the mineralogical, to geological to commodity – is attributed to James Young, who had successfully distilled paraffin from heating (retorting) cannel coal (a type of coal-like shale) in the 1840s. Having taken out a patent in 1851, he then established an oil works at a cannel coal seam near Bathgate in West Lothian (Meighan 2012, 165–6). This plant was exceptionally successful and is said to have produced around 25% of London's lamp oil in the early 1860s (Harvie 2005, 3).

After his patent expired in 1864, Young (by now nicknamed 'Paraffin Young') formed his own company and bought up mining rights in a large area of West Lothian to produce paraffin from the (more plentiful) oil shale seams. By the late 1860s, the industry took off, with a multitude of new companies employing up to 40,000 workers, with a maximum production of 27.5 million barrels of oil in 1913 (Harvie 2005, 4).

To obtain the shale, mines were dug up to 450 metres beneath West Lothian, with the minerals extracted by hand and brought to the surface in hutches (mine carts). The raw shale was crushed and heated to above 500 degrees Celsius in large vessels called retorts. This freed the hydrocarbon vapour to be collected and condensed for refining, with the 'spent' blaes collected from an opening beneath. The refining of the crude oil took place in stills where it was split into paraffin and other valuable fractions, including heavy fuel oils, lubricating oils, sulphuric acid and, later, motor spirit (petrol/gasoline) and detergents.

An especially significant by-product was ammonium sulphate, which, having initially been disposed of as waste in the early years of the industry, was extensively reclaimed to produce fertiliser. This was used in agricultural production across the British Empire, until outcompeted by Chilean nitrate-based fertilisers (Knox 2013, 40). The reuse of such waste material meant that the industry was far more competitive than it would have been if it had only focussed on paraffin production alone – another example of how notions of value and waste can shift rapidly. This waste was not the only link to British imperialism. James Young spent some of his vast profits bankrolling his friend and former classmate David Livingstone's journeys into Eastern Africa.

The exploitation of cheaper free-flowing oil in the USA and Middle East from the 1860s meant the days of the industry were always numbered, given shale oil's more labour-intensive and costly processes. Nonetheless, the West Lothian industry persisted (with government subsidy) until 1962 when the last works were closed (Knox 2013, 41). This simplified summary can, to some extent, be seen as its conventional heritage narrative: a Victorian innovation, capitalistic accumulation, and raw material combined to create a pioneering industry. This is recognised in the Museum of the Scottish Shale Oil Industry (https://www.scottishshale.co.uk/) and publications that celebrate the 'world's first' oil industry (e.g. Kerr 1994, McKay 2012, Knox 2013). This is not to be disrespectful of these important efforts, but rather to suggest that this is only a partial perspective. As artist Kim Wilson puts it, the blaes and bings somehow remain 'untimely'; both representative of the industrial past but yet seeming to offer something more for the present and future (2018, 176–7).

Monumentality

Almost all the products of the shale oil industry exhibit chemically toxic properties that are harmful to life. While the spent blaes themselves contain few remaining hydrocarbons and are valuable habitats (below), the products derived from them in the past are likely to still contaminate nearby areas. Though no data is available on the former oil shale works in the vicinity of the bings discussed here, the redevelopment of similar works at Pumpherston four kilometres southwest, revealed high concentrations of carcinogenic by-products that required substantial remediation (Couper 1995). It, therefore, seems likely that similarly toxic substances remain adjacent to Greendykes and the other bings (see West Lothian Council 2020). Though no recent shale industry-related epidemiological data are available on human communities in West Lothian, a previous study suggested excess mortality and an increased risk of certain types of cancer in former shale workers (Randall et al. 1990, 38–9). While toxic risks to both human and non-human life are of continued concern, a far more obvious reminder of the industry are the millions of cubic metres of the blaes itself. While such bings have sometimes been seen as eyesores, or 'unattractive' features in the landscape (e.g. West Lothian Council 2015), like other post-industrial sites, their lingering monumentality is not easily reducible to a negative or positive categorisation. Instead, they provide a broad range of uses and valuations as habitat, heritage and waste (see also Storm 2014, 6).

The monumentality of the bings is ultimately an accident of waste deposition. They emerged gradually, one hutch at a time, the spent shale still steaming from the retorts, pulled on steep tramways up ever-growing slopes. The steel and wooden structures of some of these tramways can still be found, while the routes they created continue to act as trails for dirt bikers, walkers, and animals (Figure 2.3). Climbing such a bing today requires a surprising level of exertion, with the blaes slipping treacherously beneath you as you ascend. For every 10 barrels of crude oil retorted from the shale, almost seven tonnes of blaes were produced (Harvie and Hobbs 2013). It is nonetheless sometimes hard to imagine that every fragment of blaes you encounter was removed from deep below the surface by someone. Other than the bikers, few people seem to visit these sites. This may be partly due to their perceived isolation and a belief that they are unsafe. While Wilson argues that these monuments can sometimes seem uncanny, for her, as a local, 'these shaley gatherings [nonetheless] signal home' (2018, 177). An appreciation is also shown in guides devoted to 'bing bagging' – summiting the 'peaks' across the district to explore 'terrain that is truly out of this world' (Carron 2020, 5).

The blaes themselves are surprisingly non-toxic; unlike coal tips, they are free from heavy metals and other contaminants and their slightly alkaline ph. make them attractive to a range of species. Harvie's surveys of eight bings revealed 357 plant varieties, including 32 that were rare or threatened. The bings also provide a rich habitat for mammals (including badgers, hares and foxes) and many species of insects and birds (Harvie 2005). Two of the bings in the district are protected primarily for their natural heritage rather than their industrial and cultural associations, while all the bings are recognised as key contributors to West Lothian's biodiversity by the local authority (Harvie 2005).

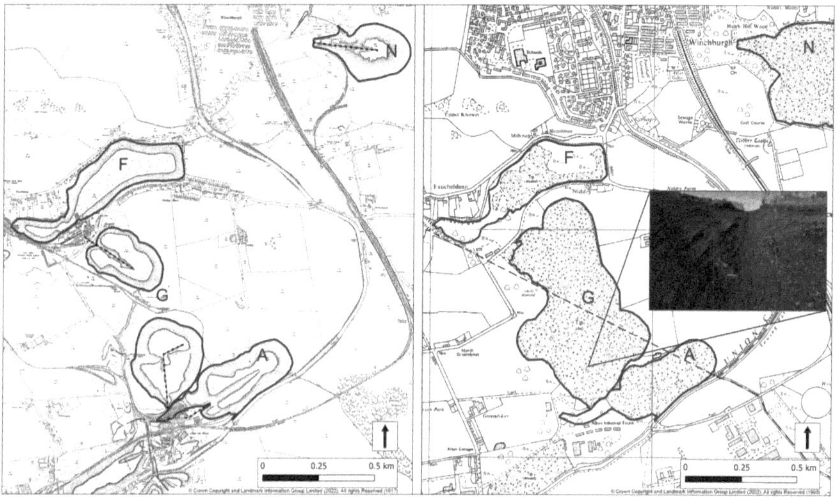

FIGURE 2.3 The four shale bings discussed outlined and labelled by letter on Ordnance Survey maps of 1917 (left) under formation at the peak of the industry, and of 1995 (right), 30 years after their closure and at full extent: Albyn, Greendykes, Faucheldean and Niddry. On the 1917 map, blaes tipping tramways are marked with a thick black dotted line. The inset photograph shows the remnants of such a tramway in-situ on Greendykes in December 2021 (20 cm scale). Note: the maps are reproduced here at the same display scale (1:16454) but were originally created at 1:2500 and 1:10,000 respectively; hence the latter contains less detail. Maps: Crown Copyright and Landmark Information group (2022). All rights reserved (1917, 1995). Photograph by Jonathan Gardner, CC BY-NC 4.0.

The colonisation of bings by this non-human life has made many of them surprisingly green and resemblant of natural hills (e.g. Faucheldean, Figure 2.1). While this habitat is rightly celebrated, Harvie argues that their 'historical importance', 'is in danger of being forgotten' (2005, 21). While Greendykes and Faucheldean are protected from development as Scheduled Ancient Monuments (along with the 'Five Sisters' bing further west), Albyn, Niddry and the others are left unprotected. I shall return to the implications of such protection for toxic heritage later, but though there is the museum and comprehensive (online) 'Shale trail', none of the bings have any on-site interpretation and most remain difficult to access safely.

Revaluation

Both before and after the decline of the shale oil industry, the bings were called 'unsightly' (e.g. Dundee Courier 1938), 'blots on the landscape' (Harvie and Hobbs 2013), associated with social deprivation, and even a source of 'shame'.[2]

The long-standing perception that these heaps exhibit an aesthetic and social toxicity, if not a chemical one, has led to attempts to remove them entirely, with around half of the bings removed or radically altered since 1962.

In the 1950s and 1960s, investors bought up many of the bings for bargain prices to mine their blaes as aggregate. One contractor – William Griffith – was so successful in this business that he became known as the 'King of the Bings'. In a remarkable advert in the *West Lothian Courier* from 1967, entitled 'WE Pave The Way ...', Griffith proudly relates how the company supplied 1.5 million tons (tonnes) to build the M8 motorway and that: 'Shale blaes is the most adaptable, most plentiful and cheapest infilling and bottoming material in Britain' (Figure 2.4). He continues: 'It is a romantic story. The new Scottish Industrial Revolution speeds along on the refuse of the old, on blaes supplied by GRIFFITH. We move mountains!' (William Griffith Ltd 1967). This insatiable demand for material also saw blaes used in the building of the first Forth Road Bridge, the M9 motorway and the Edinburgh City Bypass.

By the advent of the new (North Sea) oil boom of the late 1970s and 1980s, blaes was also to prove 'foundational' to the new industry, forming the base of a major

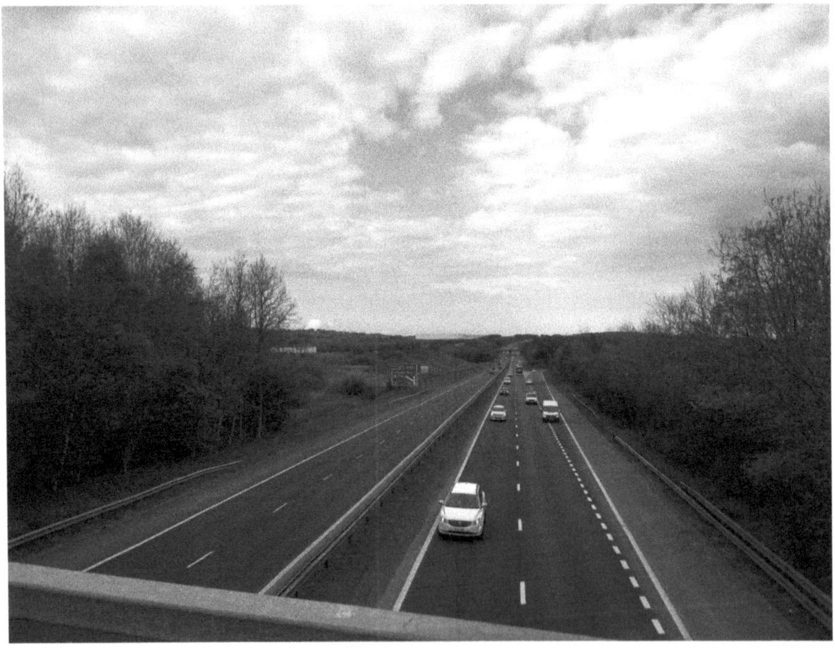

FIGURE 2.4 The M8 motorway, looking down the eastbound carriageway near junction 3a, back towards the outskirts of Livingston. Both the motorway and the town were built on thousands of tons of shale blaes extracted from nearby bings. Photograph by Jonathan Gardner, CC BY-NC 4.0.

new oil terminal at Grangemouth (WLC 1967) – with the shale oil industry recognised at the time as, 'the embryo from which the world-wide oil industry sprung' (WLC 1970). At the same time, large programs of 'slum' clearance and redevelopment saw several New Towns built, including Glenrothes in Fife and Livingston in West Lothian itself, and using blaes as a foundation material (Figure 2.4). Mining of blaes continues today at Niddry Bing and it has recently been used in the approach roads for the new Queensferry Crossing over the River Forth (see the background, Figure 2.1, top right).

Reimagination

While the use of the blaes to rebuild central Scotland was a striking turnaround in fortunes for what was once unwanted waste, this was not its only reimagining. In 1972, as part of plans to deal with the 'problem' of the bings, the Scottish Development Agency (SDA; part of the UK government's Scottish Office) engaged an organisation called the Artist Placement Group (APG) and the artist John Latham.

APG's placements were not artists' residencies in a traditional sense, but, instead, followed a model intended to mix 'business, science, and civil service practices with socially engaged art', predicated on the idea that artists had something to offer outside of art galleries (Richardson 2019, 92). Latham's 1975–6 placement with the SDA looked at numerous issues facing a rapidly deindustrialising Scotland but it was the bings that attracted him most strongly. After six months of research, he produced a 'Feasibility Study',[3] of which Niddry, Faucheldean, Greendykes and Albyn bings formed a major part.

Rather than recommending their removal, Latham proclaimed the four bings as an artwork. Using aerial imagery, he viewed the bings from above as a passenger flying into nearby Edinburgh Airport might, and conceptualised them as a gigantic goddess-like figure that he called the *Niddrie Woman*.[4] Each bing was seen as a body part of this woman: Albyn as head, Greendykes as torso, Faucheldean as limb, with Niddry forming a dismembered heart at some distance from the rest. Latham argued that this *Niddrie Woman* was a form of Duchampian 'ready-made' and that the bings had immense value as 'monuments to the period we live in', as 'process sculptures' outside of traditional (textual or verbal) representations of history and economic understandings of value.[5] Latham felt that the 'unconscious' creation of this landform was unprecedented and, thus, 'presentable as at least the equal of any of the great monuments of history'.[6]

That said, Latham was not above seeing the bings as more conventional monuments to the oil industry itself. In a letter pleading with the aristocratic owner of Niddry bing to save it from mining, Latham noted that the APG had 'prepared an appeal to worldwide oil interests' and that, '[t]he objective will be to establish the *Niddry Woman* as the Oil Industry's natural contribution to the cause of unity and continuity in the human story'.[7] Thus, he seemed to suggest the bings were both heritage *and* art, and invoked a pragmatic heritage preservation argument alongside his more cosmological theorisations to advocate for their protection. As well as suggesting the erection of a gigantic book-like sculpture of his own on Niddry Bing (entitled, *The Handbook of Reason*), he also proposed improved access, heritage trails and plaques. In spite of Latham's representations, Niddry bing was not saved and has been substantially reduced in size by mining that is set to continue to the 2050s. Upon Latham's death in 2006, his ashes were scattered on the bing so it is possible (and perhaps appropriate) that some of them ended up in yet another monumental motorway or building project.

Alternatively, a few atoms may have made it into the works of Kim Wilson – one of several artists I recently interviewed to understand how the bings continue to be reimagined. Wilson's work diverges sharply from Latham's possessive approach to the bings.[8] She describes herself straightforwardly as someone with an 'interest in materials' and a sculptor whose first degree was in archaeology (Interview: Kim Wilson). Her practice takes the form of transient experimentations with substances including wool grease, animal glue, bone ash, sulphur, and, crucially, various hues of oil shale blaes. In hand-mining the blaes from the slopes of Niddry Bing (using her former archaeological trowel), Wilson describes the sometimes treacherous experience as a key part of the work in its connections to stories of its original mining, its geological variability, and her own connections to the landscape as a local.

In her mixing of these materials, Wilson makes bioplastics and moulds them into plaques, cylinders, and foams that are further assembled into installations (Figure 2.5; see also https://www.kimwwilson.com/). Her work's efforts to understand the chemical properties of the blaes (each colour varying subtly in its material affordance) recall the inventiveness of the shale oil industry itself. The transitory nature of these ancient and not-so-ancient materials, mineral and animal, is at the core of her work, reflecting her argument that the blaes are 'untimely', as a 'storied matter' that continues to change in value, to be used and reused, made and remade (Wilson 2018, 179; Interview). In this, her work is more obviously bound up with the complex geosocial itineraries of the blaes than Latham's. She rejects a relationship to such materials where 'everything is subject to us' (Interview); instead, she sees the blaes as generative, laden with possibilities derived from the stories of their human and non-human engagements throughout time, rather than as a stand-in or representative, whether for an ancient or mythological monumentality, or simplified heritage narrative. This approach hints at a means of going beyond simply 'toxic' or 'non-toxic' or 'wasted' and 'valued' binaries in connection to the bings.

Discussion

The bings and their blaes have been subject to continual processes of re-valuation over their 160-odd years on the surface of West Lothian (to say nothing of the shale's ca. 300 million years of existence). This began with their nineteenth-century categorisation and identification as valued raw material and source of profit by Cadell, Young and others, and as contributors to the industrial and agricultural

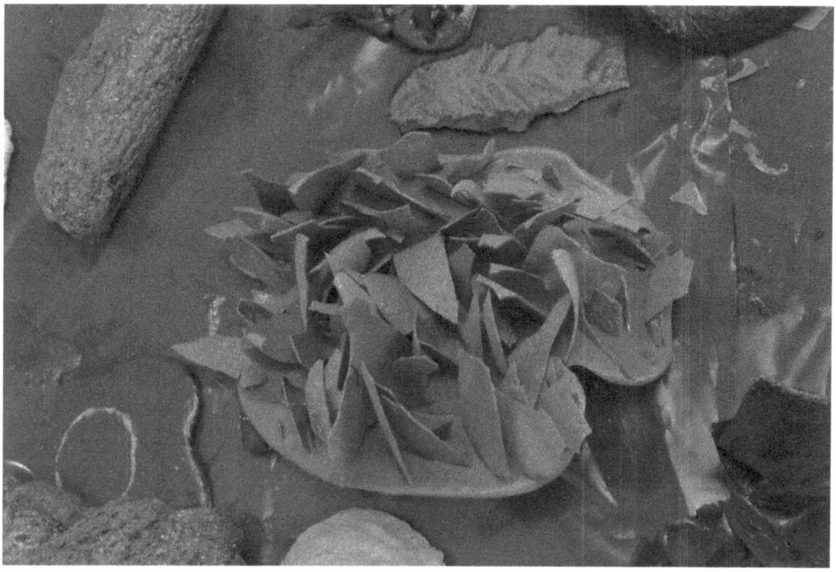

FIGURE 2.5 Experimental forms created by the artist Kim Wilson from red oil shale blaes and animal glue. The other artworks in the image are made from a combination of different coloured blaes, glue, wool grease and other substances. Artwork by Kim Wilson. Photographed with permission by Jonathan Gardner.

development of the UK and the British Empire. It is this contribution and the stories of those who developed the industry that are celebrated in the museum, the aforementioned books, and the listing of (some of) the bings as, 'the foundation of some of the earliest petro-chemical industries' in the world (Historic Environment Scotland 2022). Though originally seen only as an extraneous by-product, the blaes of these four bings, along with the Five Sisters (the other protected bing), are no longer primarily seen as waste, but as natural and cultural heritage. It has been relatively easy to transform some of this inert waste into heritage icons – literally, in the case of West Lothian Council's logo, which features the Five Sisters.[9] The situation may have been different if, like coal tips, the material was poisonous or at risk of collapse (e.g. Fairclough 2021).

Though obvious, it is worth restating that the blaes and their bings are also not simply representations of something else; they have provided a (literal) material platform for human and non-human life. For example, the bings freely offer sites of leisure: each weekend dozens of dirt bikers descend upon the slopes of Greendykes and Albyn. Though contested, their actions in churning up the terrain are said to actually enhance the growing environment for plants. These species are themselves an emergent property of the waste of human industry that benefits from our input but nonetheless escape our control and representations.

The bings also offer a second, more conceptual, terrain where ideas and representations are constructed. Latham's *Niddrie Woman*, as a radical reimagining of what had been seen as 'derelict land' is the most dramatic example, but it is clear the conceptual flexibility of the blaes also reshaped the post-industrial Scottish landscape. This is most clearly seen with its use in building central Scotland's New Towns, motorways and oil refineries, with the simultaneous (pharmakonic) perception of blaes as both worthless waste *and* valuable raw material (e.g. WLC 1964). While these varied material and conceptual opportunities were, in most cases, 'non-toxic', I want to suggest that a third valuation is also possible, one that re-recognises the blaes and bings as toxic (though not necessarily negative) representatives of the contemporary world and its uncertain future.

It is not without significance that James Young kick-started the oil shale industry in 1851, a year often associated with the supposed apogee of the Victorian age in the form of the Great Exhibition (Gardner 2022). Indeed, the Exhibition hosted many displays of hydrocarbons in both raw and processed form (including 'bituminous [oil] shale': Great Exhibition 1851, 141). Coal in particular was heralded as the material that not only made the event itself possible, but also the technological industrial civilisation that it showcased (Gardner 2022, 80–3). While coal (and shale oil) was lauded as a wonder stuff in the nineteenth century, this valorisation was greatly exceeded by reactions to the exploitation of free-flowing oil by the mid-twentieth century. By the 1970s, the massive deposits of the North Sea basin were heralded as a means of transforming the British economy and a means of enriching the quality of life of its population. Crucially, the process started by Young not only kick-started the first large-scale oil industry but also provided the technologies that continue to underpin contemporary oil extraction and processing: the hydrocarbon cracking process patented in 1865 by Young's son (James Young Junior) remains the basis for producing the enormous range of petroleum-based products we use every day (Knox 2013, 88). Thus to some extent, the pioneering narrative of the shale oil industry can now also be connected to a 'heritage' of global heating.

New research is increasingly drawing attention to examples of this heritage in the era of massive climate change (Byrne 2018, Lafrenz Samuels 2016). Morel and Oud Ammerveld identify climate change itself as 'our heritage'. Though acknowledging its potentially 'dark' and apocalyptic nature, bearing in mind debates around the often over-simplified concept of 'negative heritage', they argue that this heritage acts to contextualise and mobilise movements to mitigate the worst effects of industrial societies polluting actions, past and present (Morel and Oud Ammerveld 2021, 274). What role then might the blaes and bings play in such a mobilisation?

Despite their bulky presence outside of Edinburgh alongside major transport infrastructure (the M8 and M9 motorways, the Glasgow-Edinburgh railway, the airport), the bings nonetheless remain oddly invisible. However, I want to suggest that we might subvert Latham's idea of seeing them 'as monuments to the period we live in' and act to re-materialise them through further interpretation, education and engagement as a toxic heritage site. Following Kim Wilson, the blaes and bings could be reimagined as a form of 'untimely' geosocial heritage of our ongoing hydrocarbon addiction, a pharmakonic manifestation of both the benefits we have derived from oil and the devastating consequences it has left behind; vast monuments to extractivism that act as visually and materially unavoidable reminders of what is at stake for our future.

Notes

1 Blaes is both singular and plural. *Blae* is simply the Scots word for dark blue, originally from the Old Norse *blār*. Bing comes from a Norse word for heap.
2 Tate Gallery Archive (TGA) 20042/9/2/10 Feasibility study folder. Part 2, 1.
3 TGA 20042/9/2/10 Feasibility study folder.
4 Latham used both 'Niddrie' and 'Niddry', though the latter is customary.
5 TGA 20042/2/2/13/7 Statement by John Latham, undated, entitled, 'BIRTH OF THE OIL INDUSTRY, IN SCOTLAND […]'.
6 TGA 20042/2/2/13/6 Statement by John Latham, undated, entitled 'Process Sculptures'.
7 TGA 20042/2/2/13/9 Letter from John Latham to Lord Linlithgow, (September/October?) 1980, entitled, 'Your Niddry Bing'.
8 Interviews were approved by the Edinburgh College of Art Ethics Committee (22/6/2021). The artist consented to being identified in research outputs and to the reproduction of images of her studio works.
9 See https://www.westlothian.gov.uk/.

Bibliography

Brown, D., Mah, A., and Walker, G., 2021. The tenacity of trust in petrochemical communities: reckoning with risk on the Fawley Waterside (1997–2019). *Environment and Planning E: Nature and Space* online first, 1–23. 10.1177/25148486211045367

Butler, B., 2011. Heritage as Pharmakon and the Muses as Deconstruction: Problematising Curative Museologies and Heritage Healing. In: S. Dudley, A. Barnes, J. Binnie, J. Petrov, and J. Walklate, eds. *The Thing about Museums: Objects and Experience, Representation and Contestation*. London: Routledge, 354–371.

Byrne, D., 2018. Reclaiming landscape: Coastal reclamations before and during the Anthropocene. In: P. Howard, I. Thompson, E. Waterton, and M. Atha, eds. *The Routledge Companion to Landscape Studies*. London: Routledge, 277–287.

Cadell, H. M., 1894. The Oil Shales of the Scottish Carboniferous System. *The Journal of Geology* 2(3), 243–249. https://www.journals.uchicago.edu/doi/10.1086/606943

Carron, J., 2020. *Bing Bagging: Central Scotland*. No publication place: Amenta Publishing.

Clark, N. and Yusoff, K., 2017. Geosocial Formations and the Anthropocene. *Theory, Culture & Society* 34(2–3), 3–23. 10.1177/0263276416688946

Couper, A. S., 1995. Drumshoreland Redevelopment Area, West Lothian. In: W. J. Van Den Brink, R. Bosman, and F. Arendt, eds. *Contaminated Soil '95*. Dordrecht: Springer, 1437–1446.

Dundee Courier, 1938. Bricks Made from Shale. *Dundee Courier*, 29 January (26413), 7.

Fairclough, S., 2021. Coal tips: Almost 300 in Wales classed as 'high-risk'. *BBC News*, 16 February. Available from https://www.bbc.com/news/uk-wales-56073459 [Accessed 16 February 2021]

Gardner, J., 2022. *A Contemporary Archaeology of London's Mega Events: From the Great Exhibition to London 2012*. London: UCL Press. 10.14324/111.9781787358447

Great Exhibition, 1851. *Official descriptive and illustrated catalogue. Great Exhibition of the Works of Industry of All Nations, 1851. Vol. I.* London: Spicer Brothers. https://archive.org/details/b21495361_0001

Grosz, E., 2011. *Becoming undone Darwinian reflections on life, politics, and art.* Durham, N.C: Duke University Press.

Harrison, R., 2021. Rethinking the futures of heritage and waste in the Anthropocene. In: T. R. Bangstad and Þ. Pétursdóttir, eds. *Heritage Ecologies.* London: Routledge.

Harvie, B., 2005. *West Lothian Biodiversity Action Plan: Oil Shale Bings.* Bathgate: West Lothian Local Biodiversity Plan partnership.

Harvie, B. and Hobbs, R., 2013. Case Study: Shale Bings in Central Scotland: From Ugly Blots on the Landscape to Cultural and Biological Heritage. In: R. Hobbs, E. Higgs, and C. Hall, eds. *Novel Ecosystems: Intervening in the New Ecological World Order.* Oxford: Wiley, 286–289.

Historic Environment Scotland, 2022. Greendykes shale bing, Broxburn (SM6186). Available from: http://portal.historicenvironment.scot/designation/SM6186 [Accessed 5 July 2022].

Joyce, R. and Gillespie, S., eds., 2015. *Things in Motion: Object Itineraries in Anthropological Practice.* Santa Fe, NM: SAR Press.

Kerr, D., 1994. *Shale Oil Scotland: the World's Pioneering Oil Industry.* Edinburgh: Self-published.

Knox, H., 2013. *The Scottish Shale Oil Industry & Mineral Railway Lines.* Oxford: Lightmoor Press.

Lafrenz Samuels, K., 2016. The cadence of climate: Heritage proxies and social change. *Journal of Social Archaeology* 16(2), 142–163. https://journals.sagepub.com/doi/10.1177/1469605316639804

McKay, J., 2012. *Scotland's First Oil Boom: The Scottish Shale-Oil Industry, 1851-1914.* Edinburgh: John Donald.

Meighan, M., 2012. *Scotland's Lost Industries.* Stroud: Amberley.

Mendum, J., 2010. Henry Moubray Cadell: a geological and industrial innovator. *The Edinburgh Geologist* Autumn 2010 (48), 5–14. https://edinburghgeolsoc.org/eg_pdfs/issue48_00_fullissue.pdf

Meskell, L., 2002. Negative Heritage and Past Mastering in Archaeology. *Anthropological Quarterly* 75(3), 557–574. 10.1353/anq.2002.0050.

Moore, S., 2012. Garbage matters: Concepts in new geographies of waste. *Progress in Human Geography* 36(6), 780–799. 10.1177/0309132512437077

Morel, H. and Oud Ammerveld, J., 2021. From Climate Crisis to Climate Action: Exploring the Entanglement of Changing Heritage in the Anthropocene. *The Historic Environment: Policy & Practice* 12(3–4), 271–291. 10.1080/17567505.2021.1957261

Morgan, B., 2016. Fin du Globe: On Decadent Planets. *Victorian Studies* 58(4), 609–635. 10.2979/victorianstudies.58.4.01

Randall, S., Cowie, H., Hurley, J., and Jacobsen, M., 1990. *Studies of the Scottish Shale Oil Areas. Volume 2: Mortality in the Scottish Shale Mining Communities.* Edinburgh: Institute of Occupational Medicine/Department of Energy.

Reno, J., 2018. What is Waste? *Worldwide Waste: Journal of Interdisciplinary Studies* 1(1), 1–10. 10.5334/wwwj.9

Richardson, C., 2012. 'Waste to Monument: John Latham's 'Niddrie Woman (1975-6)'. *Tate Papers* (17). www.tate.org.uk/research/publications/tate-papers/issue-17

Richardson, C., 2019. Monuments to the period we live in. In: D. Cateforis, S. Duval, and S. Steiner, eds. *Hybrid Practices: Art in Collaboration with Science and Technology in the Long 1960s.* Oakland, CA: University of California Press, 91–109.

Rico, T., 2016. *Constructing Destruction: Heritage Narratives in the Tsunami City*. London: Routledge.

Storm, A., 2014. *Post-Industrial Landscape Scars*. Basingstoke: Palgrave Macmillan

Viney, W., 2015. *Waste: A Philosophy of Things*. London: Bloomsbury.

West Lothian Council, 2015. Easter Inch Moss & Seafield Law. Available from: https://www.westlothian.gov.uk/article/34244/Easter-Inch-Moss-Seafield-Law [Accessed 1 Jul 2022].

West Lothian Council, 2020. Supplementary Planning Guidance: Development of land potentially affected by contamination. *westlothian.gov.uk*. Available from: https://geosmartinfo.co.uk/wp-content/uploads/2020/02/West-Lothian-Guidance.pdf [Accessed 21 Jun 2022].

William Griffith Ltd, 1967. [Advertisement] We Pave The Way ... *West Lothian Courier*, 24 February (4900), 13.

Wilson, K., 2018. Untimely Mountains | Entangled Matter. In: C. Kakalis and E. Goetsch, eds. *Mountains, Mobilities and Movement*. London: Palgrave Macmillan, 171–186.

WLC, 1964. THEM THAR BINGS! Big Prices For Waste. *West Lothian Courier*, 12 June (4729), 8.

WLC, 1967. King of the Bings – He Moves Mountains – An Idea Turned Waste into Gold. *West Lothian Courier*, 24 February (4900), 13.

WLC, 1970. New Discovery To Landscape Bings. *West Lothian Courier*, 24 July (5728), 12.

Yusoff, K., 2015. Queer Coal: Genealogies in/of the Blood. *philoSOPHIA* 5(2), 203–229. https://muse.jhu.edu/article/608468

YussoffK., 2020. Geophysics after Life: On the Way to a Political Geology of the Anthropocene. *springerin* 3: Post-Anthropocene. https://www.springerin.at/en/2020/3/geophysik-nach-dem-leben/

3

WHEN TOXIC HERITAGE IS FOREVER: CONFRONTING PFAS CONTAMINATION AND TOXICITY AS LIVED EXPERIENCE

Thomas W. Pearson and Daniel Renfrew

Consumer capitalism ushered in a world of pervasive contamination, leaving a legacy of harm and a future of uncertainty. Few "toxic events" better encapsulate this than *per-* and *poly-fluoroalkyl substances*, or PFAS, a broad category of synthetic chemicals that includes several thousand substances (Renfrew and Pearson 2021). They are used to coat things or make materials that repel water or oil. Such properties have innumerable uses, ranging from clothing and furniture to nonstick cookware, food packaging, and cosmetics. PFAS came into widespread use in the 1950s before significant environmental regulation existed in the United States and have evaded meaningful oversight ever since (Richter, Cordner, and Brown 2021). By the 1970s, the chemicals were turning up in drinking water, fish, and people's bodies, but it wasn't until the 2000s that PFAS became an emerging contaminant of scientific and regulatory concern. Their commercially desirable characteristics also make them a pernicious threat to public health. Some bioaccumulate and move easily with surface and groundwater, through air emissions, or with weather systems. And they resist degrading, some never naturally breaking down, earning the nickname "forever chemicals."

We all embody the legacy and uncertain future of forever chemicals. Biomonitoring studies by the Centers for Disease Control and Prevention have found PFAS in the blood of almost everyone sampled, suggesting ubiquitous human exposure (CDC 2017). They're also toxic, with extremely low levels associated with a wide range of health effects, including testicular and kidney cancer, ulcerative colitis, thyroid disease, pregnancy-induced hypertension, hypercholesterolemia, decreased vaccine response, and developmental problems (ATSDR 2020). In 2016, the Environmental Protection Agency issued a health advisory for two PFAS substances, *perfluorooctane sulfonate* (PFOS) and *perfluorooctanoic acid* (PFOA), recommending a drinking water limit of 70 parts per trillion. Then in 2022, it dramatically lowered its

DOI: 10.4324/9781003365259-6

advisory levels to 0.02 ppt for PFOS and 0.004 ppt for PFOA. At those levels, more than 2,000 communities, serving 43 million people, drink contaminated water (EWG 2022).

Knowledge of PFAS toxicity was secluded for decades within the companies that produce the chemicals (Bilott 2019; Richter, Cordner, and Brown 2018). As a result, PFAS contamination simmered quietly, a slow-motion toxic disaster enabled by flawed regulatory structures and chemical industries that prioritize corporate profit. Today, its frequent discovery forces many communities to confront the hidden threats of once proud industrial pasts, revalued as toxic heritage. To illustrate, we trace three histories of contamination related to PFOS, one of the first PFAS substances commercialized on a mass scale. PFOS was originally in Scotchgard, a fabric coating made by the 3M company, used on iconic brands such as Hush Puppies shoes. PFOS was also a crucial component in the development of aqueous film-forming foams (AFFF) to suppress oil fires. Each of these products, with interconnected origins and histories, reveals how the discovery of PFAS contamination forces a reckoning with and reinterpretation of the past. It also prompts a new orientation towards the future, with toxic heritage signifying a "future-making" (Harrison 2022, 32) or "emergent presence" (Bangstad and Pétursdóttir 2022, 21) that informs the lived experience of PFAS toxicity, characterized by uncertain health impacts and ongoing struggles to achieve long-term solutions to a forever problem (Wollentz et al. 2020).

Wildest Hellcat

The toxic heritage of PFOS begins with an accident. After acquiring a key patent in 1948, 3M started working on commercial applications for the first PFAS, known as fluorocarbon chemicals or fluorochemicals. "Almost every day we turned out a new molecule which had never been on the face of the earth before," recalls J. Donald LaZerte, a 3M chemist who started in 1949 (LaZerte 1989, 25). Management, however, worried about the prospects of fluorochemical research. Fluorocarbons were known, as one journalist wrote, as the "chemical daughters of the wildest hellcat of elements, fluorine, a yellow-green gas that will burn sand, glass, steel wool, asbestos, concrete or even water, and is extremely dangerous to touch or inhale" (Manchester 1959, 120). When tamed, however, the hellcat opened new possibilities. While the violence of fluorine "is born of a desperate yen to combine chemically with everything in sight," a stable carbon-fluorine bond is considered one of the strongest known in the field of chemistry (Manchester 1959, 120).

In 1950, 3M struck a deal to produce PFOA for DuPont, which DuPont then used to manufacture Teflon, a material that had emerged from the World War II Manhattan Project, central to the numerous components, valves, and pumps needed to enrich uranium. Slippery, durable, heat-resistant Teflon coatings were revolutionizing other products, including cookware. Touted as a miracle substance, some imagined fantastic applications, from "100,000-mile tires," "greaseless bearings," and "rustproof steel" to "artificial arteries" (Manchester 1959, 117). But 3M was struggling to commercialize fluorochemical products on its own, until an unexpected breakthrough. Scientists working on a military contract to develop jet fuel hoses spilled an experimental fluorochemical on a canvas shoe and couldn't remove the substance. "A fluorochemical surface is not wet by oil or water," a chemist reported in 1952. The compound surrounded each individual fiber, resulting in a protective barrier. "A textile not wet by oil or water is a highly desirable commodity," explained the report, as recounted by LaZerte (1989, 25). "Therefore, we're in the money!"

To commercialize this discovery, which utilized PFOS, 3M needed to create a market for it (LaZerte 1989, 26). Around 1956, they promoted it to textile manufacturers as Scotchgard, targeting clothing, footwear, furniture, and luggage. As uptake of Scotchgard grew within product manufacturing, 3M worked with retail stores to generate consumer demand, placing hang tags on products with Scotchgard and coordinating advertising. The challenge was in marketing an invisible product, "one that the end-user had to accept on faith alone. Or at least until a cup of coffee was spilled on the sofa" (LaZerte 1989, 26). To help visualize Scotchgard, 3M advertising blended notions of tradition and modernity. A logo featured a castle tower, conveying protection, and a tartan layout, the plaid textile design evoking Scottish tradition. The Scotchgard logo accompanied images of pretreated products and families engaged in the mundane activities of a modern, consumer lifestyle, such as a dinner party or a child's birthday. "Look for the name 'Scotchgard' Stain Repeller on clothes you buy, on furniture, furnishings, even tablecloths," reads one advertisement. "Take your share of a new kind of carefree living" (*Life Magazine*, September 5, 1960).

Scotchgard normalized chemically treated fabrics and unleashed a large-scale human experiment. As early as the 1960s, 3M was collecting internal data from lab studies on the toxicity of fluorocarbons, with indications that PFOS accumulated in the blood of rats (Hayes 2019). Then in 1975, a bombshell landed on LaZerte's desk, who by then directed 3M's commercial chemical division. An internal memo reported on a disturbing call from William Guy, a University of Florida researcher. Guy and his colleague Donald Tayes discovered fluorocarbons in human blood serum (Bilott 2019, 75–76; Lerner 2018b). They contacted 3M and speculated about the source, including products such as Teflon and Scotchgard. Internally, 3M officials expressed alarm, but then adopted an outward posture of naivete. "We plead ignorance," according to the memo author, "a position of scientific curiosity," working with Guy and Tayes in a spirit of cooperation but without leading them to dig too deeply (Crawford 1975).

3M accelerated its studies of fluorocarbon toxicity, documenting accumulation in the blood of its workers in 1976. In 1978, 3M officials speculated that PFOS "should be regarded as toxic," but also asserted that "no substantial risk exists," concluding they had no obligation to report their findings (Prokop 1978; see also Bilott 2019, 172; Lerner 2018b). Some 3M scientists called for urgent health studies (Case 1979), raising concerns about immunosuppressive effects in 1983, and associating worker exposure with elevated cancer rates in 1989 (Lerner 2018b; 2020). Inside 3M, health studies of Scotchgard increased in the 1990s, but it would be another decade before action was taken to reduce human exposure (Lerner 2018b).

In 1998, 3M finally divulged its studies to the EPA (Faber 2020). Still frustrated with its lack of urgency, Richard Purdy, a 3M environmental scientist, criticized the company's handling of PFOS in a resignation letter, describing it as "the most insidious pollutant since PCB" (Purdy 1999). Then in 2000, under EPA pressure, 3M surprisingly announced a market phaseout of Scotchgard (Barboza 2000). Three years later, however, 3M reformulated the ingredients, replacing PFOS with an alternative, *perfluorobutane sulfonic acid* (PFBS). In 2006, the EPA negotiated a voluntary phaseout of PFOA and PFOS, completed in 2015. But efforts to regulate the chemicals continued into the 2020s, and management of widespread contamination will continue indefinitely.

"Today," boasted LaZerte in 1989, "our 'hellcat' can be found on everything … You can also capture this 'hellcat' in aerosol cans in your supermarket for do-it-yourself applications" (LaZerte 1989, 27). Indeed, the PFAS hellcat is everywhere. Even in the shadow of its corporate headquarters in Minnesota, 3M has contaminated the drinking water of hundreds of thousands of people. From the 1950s to the 1970s, it dumped waste from its factory in Cottage Grove at four regional dump sites. Minnesota sued 3M in 2010 and sought $5 billion for damaging the state's natural resources with permanent groundwater contamination stretching some 150 square miles. Just before the trial in 2018, 3M settled for $850 million, and efforts to secure long-term drinking water solutions continue.

Lost Soles

Beyond the birthplace of PFOS, contamination follows the routes of industrial development and along the commodity chains of modern consumer life. Many manufacturers incorporated Scotchgard into their production processes, but the story of Hush Puppies shoes aptly captures how the discovery of PFAS contamination involves revaluing a once proud industrial heritage as newly toxic. A leather suede shoe brand launched in 1958 out of Rockford, Michigan, Hush Puppies utilized Scotchgard, and went on to help define the counterculture trends of the 1960s, its branding as "casual" and "relaxed" synching with 3M's "carefree living" theme.

Hush Puppies fueled the growth of Wolverine Worldwide, initially founded in Rockford in 1908. Much of the original factory, built along the Rogue River, was demolished in 2010, but Wolverine's corporate headquarters remains. Now a city of 6,000 people, Rockford's economic well-being once hinged on the fortunes of Wolverine, its presence shaping daily life. Raw pigskins would arrive by rail for processing into suede, and a disagreeable odor, like chemically infused, rancid bacon, wafted over the city. It was variously described as "the Rockford stink" or "the smell of money" (Gardner 2019). Today, part of the original factory has been repurposed into a shopping store, allowing consumers to nostalgically experience industrial heritage as they buy shoes from a now multinational corporation. For some residents, however, Rockford's postindustrial landscape has brought them illness and trauma (Trepal and Hurstad 2019, 235).

Before outsourcing manufacturing, Wolverine produced Hush Puppies in Rockford, but also tons of waste. Its wastewater discharged into the Rogue River, while truck drivers dumped animal skin remnants, leather debris, sludge laced with Scotchgard, and metal drums on land outside of town (Ellison 2019). Sludge was even spread on local farm fields, presumably intended as fertilizer (Ellison 2017d). Into the 2010s, twisted strips of leather littered the banks of the Rogue River. Entangled with plants and tree roots, the leather scraps snaked out of the ground like Medusa's hair and amassed on cattail-covered mudflats, an area dubbed the "Island of Lost Soles" (Ellison 2017a).

Contamination incubated for decades. 3M warned Wolverine officials about the hazards of Scotchgard in the late 1990s, but Wolverine did nothing (Ellison 2017c). Other signs went overlooked, regarded as curious features of a postindustrial landscape, even as housing development expanded around the small city. Residents encountered leather strips along the riverbank, or in even their yards. Hiking in the woods, one might come across a rusted barrel poking out of a hillside, or a shoe sole imprinted with the Hush Puppies logo. Such artifacts formed the unremarkable legacy of a company town, like the previous foul odor that reminded of a steady paycheck. Residents lacked a framework for perceiving the relics as anything else.

When the shoe factory was slated for demolition in 2010, some residents questioned the environmental impact of the operation. Wolverine assured the community there was "no known contamination on the property," but some residents organized as the Concerned Citizens for Responsible Remediation (CCRR) and petitioned for an EPA assessment, which found numerous hazardous substances (Ellison 2019). Wolverine, however, with the support of Rockford city officials, escaped listing the site on the federal Superfund list, with the city manager asserting that "there are no conditions at the property that present a health threat to the public" (Ellison 2019).

In 2011, a scientist for what was then the Michigan Department of Environmental Quality (DEQ) sampled for PFAS contamination at the former Wurtsmith Airforce Base in Oscoda (Ellison 2018b). His calls for a wider-scale sampling of drinking water went ignored but prompted the CCRR to start looking for signs of PFAS pollution from Scotchgard. The group petitioned to test fish from the Rogue River, with results showing elevated PFAS levels near the former tannery (Ellison 2017a). Nationwide EPA monitoring of unregulated pollutants in 2013 found PFAS in the municipal drinking water of a nearby township, and the contamination was traced to a landfill previously used by Wolverine. By 2016, the EPA had issued its first drinking water advisory for PFOS and PFOA, and Michigan subsequently issued its own advisory in Oscoda. With awareness of PFAS growing, the CCRR collected evidence of historic Scotchgard use by Wolverine and spoke to retired truck drivers to identify former dump sites. In early 2017, they presented their findings to the DEQ, showing decades of Scotchgard use and waste disposal at forgotten dump sites, which prompted testing that occurred later in the year, triggering the discovery of widespread PFAS contamination (Ellison 2019).

Sandy Wynn-Stelt had lived across from Wolverine's House Street Dump site for 24 years when she learned that her well was highly contaminated. Her husband died the year before at age 61 from liver cancer. She herself suffers from thyroid disorders. The discovery of contamination was devastating. The recent loss of her husband was compounded by a shattering of her assumptions about the place she lives and the surrounding environment. "You lose your husband and it's the worst thing on Earth," she said in 2017. "But you slowly get back. You slowly kind of come back to life … I remember in June this year thinking 'I feel like I'm getting my joy back.' And in July, I get these government people walking up my driveway and saying, 'we think you've got poisoned groundwater.' And it's just brought it all back." Her home, once a grounding source of order in her life, became a threat to her health and economic security. "I have a property now that is sitting next to a toxic waste dump, so my entire life savings is probably down the drain," she said (Ellison 2017b).

Her neighbor's water wasn't tested until a few months later. By then, Seth McNaughton's son was almost two, an age when every day of exposure has a compounding impact. "In his short life," he said, "he would have had so much less water" contaminated with PFAS, if they had known earlier. "He would have a lot less in his blood" (Ellison 2018a). Like other exposure experiences, they now contend with immense stress and uncertainty about the future (Edelstein 2004). Health outcomes may also take years or decades to register. "The biggest thing," said McNaughton (MLive 2018), "is like, the fear and worry of what might happen. You know, that's really where we're at now, just dealing with that and just moving on with our life. We're trying not to worry about it too much because then it will just make it worse. So, you just live like it's not happening."

Black Plumes

Beyond Scotchgard, the 1960s saw another turning point for 3M's fluorocarbon research, which would become a major source of PFAS contamination worldwide. In 1966, the U.S. Navy patented a process for using foams to suppress oil fires, relying on an AFFF product developed by 3M with PFOS as a key component (ITRC 2021, 52). After a fire on the USS Forrestal in 1967 killed 134 sailors, the Navy began requiring all its ships to carry 3M's AFFF (NSRDC 1978). By the 1970s, the Department of Defense required AFFF at military installations (Lerner 2018a), and it was soon adopted at civilian airports and other industrial facilities. While 3M no longer produces AFFF, it supplied the military until around 2001. When 3M phased out PFOS and left the AFFF market, other companies raced to fill the lucrative void with alternative PFAS-containing foams (Bond 2021; ITRC 2021, 52–53; Lerner 2018a).

Over the decades, AFFF has been used in response to fires, but also during routine testing or training, indiscriminately released into the environment. Today, numerous communities that host airports, military installations, or industries wrestle with the toxic legacy of AFFF. One such community, Peshtigo, Wisconsin, is a rural township that surrounds Marinette, a small city on the mouth of Menominee River as it empties into Lake Michigan. Paper mills, a metal casting foundry, and a shipbuilding plant crowd the industrial riverfront, joined by Johnson Controls International (JCI) and ChemDesign, sibling companies that sell AFFF firefighting foams, a local industry that began with the Ansul chemical company. Founded in 1915 to produce refrigerants, Ansul expanded into fire extinguishers and fire suppression systems in 1939 and soon established a firefighting school. By the 1960s, Ansul expanded into producing specialized herbicides and pesticides, including Agent Orange during the Vietnam War. Ansul began testing 3M's foam concentrates around 1962 at its outdoor training grounds, today called the Fire Technology Training Center, located on the boundary with Peshtigo. It distributed 3M's AFFF and developed its own product around 1973 (Mator 2018). Ansul was acquired by Tyco International in 1990 and renamed Tyco Fire Products, and Tyco was acquired by JCI in 2016.

The Fire Technology Center is surrounded by acres of wooded land. While it feels isolated, just beyond the site are the high school, shopping centers, neighborhoods, a hospital, and the town of Peshtigo. Every summer, people from around the country visit the center to work with AFFF, extinguishing live fires at the outdoor training ground. "For forty years, there'd be black plumes comin' up every day" during the summer, recalled John Kowalski in an interview in December 2018. Over the decades, foams seeped into the ground or were flushed into the sewer system. The city of Marinette draws water from Lake Michigan, considered safe. Residents in Peshtigo, however, including John and his wife Ruth, rely on groundwater, with some wells contaminated by PFAS. Tyco has accepted responsibility for an area near its property but denies blame for wells contaminated elsewhere. In addition, biosolids from Marinette's wastewater treatment plant, tainted with PFAS, were spread on farm fields for years, distributing contamination regionally. The Department of Natural Resources has spent years trying to determine the extent of contamination and hold Tyco accountable, an ongoing effort.

Residents like John and Ruth were informed about PFAS in December of 2017, four years after Tyco first detected groundwater contamination. The discovery was disorienting. Ruth had recently been treated for thyroid disease, and John for prostate cancer. Memories of their children's inexplicable health issues while growing up came flooding back, and then they pondered their grandchildren's illnesses and developmental challenges. Ruth's mother had had thyroid cancer, and now her granddaughter has thyroid disease. "I'm just shocked by all the cancer around here," John explained.

Like in Rockford, MI, exposure to PFAS emerged quickly as a "perceptual framework" for reinterpreting daily life in relation to a newly conceived toxic heritage (Edelstein 2004, 75). Knowledge of exposure, however, does not bring clarity, but heightened stress and uncertainty about unknown consequences. "One house can be contaminated, the next one a non-detect," said John. Facing complex histories of industrial pollution and potential exposures, uncertainty fuels a sense of insecurity. Their home, which belonged to John's parents and has been in their family for decades, was once a place of refuge, several acres of land with apple trees, gardens, and a pond. Now they fear drinking or cooking with tap water. That sense of loss extends further, with PFAS contamination transforming how some residents relate to places they have long known. "For years," explained one woman in a May 2019 social media post, "the plumes I watched rise above the trees behind my grandparents' woods brought me a sense of pride. Pride because all of the fire extinguishers my parochial grade school in Chicago used were from Ansul … Now, it is with shame that I remember how I told all my friends that I knew where that was made."

This new understanding of local toxic heritage may reveal multiple, interacting potential exposures, with past industrial activities, previously forgotten or ignored, now representing a threat. Ruth and John wonder if the paper mills made people sick and they speculate about Ansul's decades of producing chemical herbicides. The company was still working to clean up arsenic contamination along the Menominee River into the 2010s. "They probably tested Agent Orange on our beaches," John sneered. They recall how their children would play in the drainage ditches and cricks that run past the Fire Technology Center, and swim in a small pond located nearby. They would notice foam flowing with the water through the ditches to Lake Michigan or washing up on a local beach. "I never really thought about where it was coming from," says John, so it went ignored. Now, however, the surrounding environment, once a benign backdrop to rural life, is perceived as malevolent (Edelstein 2004, 84). "I wouldn't eat an apple off that tree," remarked Ruth during a December 2018 interview, pointing to the backyard. Contaminated ditches currently have warning signs posted, and residents document foam sightings.

Others are similarly reinterpreting their memories of Ansul as they perceive the rural postindustrial landscape of Peshtigo and Marinette through the new experience of PFAS toxicity. As Annie Boyle Davis recounted, "When we were little kids, we would say to Mom, 'Mommy, it smells like Ansul.' It was just this horrible smell and we just called it 'Ansul'. It was probably the way the wind was blowing and what they were making. One day we went to school, I want to say it was 1970, and all the foliage around the school—the trees, the leaves, the grass—were dead" (Bence 2019). Another area resident recalls the same period—a summer when the trees in their yard had no leaves. At a public listening session hosted by the Wisconsin DNR in December 2019, she said "we had continuous bloody noses, sore throats, blood coming out of ears. The stench was so bad when the wind blew across the river that we had to keep our doors closed. We didn't know what was going on. We were told not to make any waves, because the chemical company could move away and take all the jobs with them." Now, she says, "I believe it was due to the chemicals we were breathing. The trees didn't lie."

Conclusion

Over the twentieth century, modern life became infused with forever chemicals. PFAS were meant to make us into carefree consumers free of messiness and worry, and in the case of AFFF, free from potentially deadly hazards. What we find instead is a simultaneously latent and emerging toxic heritage. The stunning range of uses for PFAS have set the stage for the continuous discovery of unanticipated contamination sources. Confronting PFAS contamination entails reconsidering personal and collective memories that bind people to the places they live and questioning the industries on which their communities were built. Plumes of smoke in Marinette and Peshtigo, or strips of old leather in Rockford, once taken for granted, have come to signal a new hazard.

As communities discover PFAS contamination, people are forced to reckon not only with the past but also with their own embodiment of toxicity. Industrial heritage, once a source of local pride and well-being, is newly experienced as a toxic heritage. This is not merely a question of what happened long ago, however, a postindustrial phenomenon where deindustrialized, neglected, and often racialized urban landscapes unearth toxic timebombs from a bygone era (Mah 2012). With PFAS, the past colonizes a forever unfolding present with chemicals that don't break down, are pervasive in the environment and continue to be widely used. The revaluing of industrial history as toxic heritage informs the perceptual frameworks through which PFAS contamination becomes intelligible, through which people make sense of toxicity as lived experience, and, ultimately, through which some challenge the systems that created this crisis.

To the extent that some victims of contamination fight for accountability, this is also a conflict over the future, over the uncertain consequences for those exposed, and over the management of contamination that is difficult to contain and will persist indefinitely. As one resident in the Peshtigo area said in an April 2021 interview, "they've poisoned our water with a forever chemical, so we want a forever solution."

Bibliography

ATSDR. 2020. "Potential Health Effects of PFAS Chemicals." Agency for Toxic Substances and Disease Registry, U.S. Department of Health and Human Services, June 24. https://www.atsdr.cdc.gov/pfas/health-effects/index.html

Bangstad, Torgeir Rinke and þóra Pétursdóttir. 2022. "An ecological approach to heritage." In *Heritage Ecologies*, edited by Torgeir Rinke Bangstad and þóra Pétursdóttir, 3–27. London: Routledge.

Barboza, David. 2000. "E.P.A. Says It Pressed 3M for Action on Scotchgard Chemical." *The New York Times*, May 19. https://www.nytimes.com/2000/05/19/business/epa-says-it-pressed-3m-for-action-on-scotchgard-chemical.html

Bence, Susan. 2019. "Marinette Residents Want to Get PFAS Chemicals Under Control." *WUWM 89.7*, August 29. https://www.wuwm.com/post/marinette-residents-want-get-pfas-chemicals-under-control

Bilott, Robert. 2019. *Exposure: Poisoned Water, Corporate Greed, and One Lawyer's Twenty-Year Battle against DuPont.* New York: Atria Books.
Bond, David. 2021. "The US Military Is Poisoning Communities across the US with Toxic Chemicals." *The Guardian*, March 25. https://www.theguardian.com/commentisfree/2021/mar/25/us-military-toxic-chemicals-us-states
Case, M. T. 1979. "Fluorochemical Chronic Toxicity, 3M Interoffice Correspondence." Exhibit 1212, State of MN v. 3M Co., Court File No. 27-CV-10-28862. https://www.documentcloud.org/documents/4609339-1979-Paramont-to-Do-Carcinogenicty-Testing.html
CDC. 2017. "Per- and Polyfluorinated Substances (PFAS) Factsheet." Center for Disease Control and Prevention. https://www.cdc.gov/biomonitoring/PFAS_FactSheet.html
Crawford, G. H. 1975. "Confidential Record of a Telephone Conversation, 3M Interoffice Correspondence, Subject: Fluorocarbons in Human Blood Plasma." Exhibit 1118, State of MN v. 3M Co., Court File No. 27-CV-10-28862. https://www.ag.state.mn.us/Office/Cases/3M/docs/PTX/PTX1118.pdf
Edelstein, Michael R. 2004. *Contaminated Communities: Coping with Residential Toxic Exposure.* Boulder, CO: Westview Press.
Ellison, Garret. 2017a. "Scotchgard Chemical Ails Fish Where Tannery Scraps Litter River." *Mlive.com*, August 23. https://www.mlive.com/news/2017/08/wolverine_tannery_pfos.html
Ellison, Garret. 2017b. "Cancer, Thyroid Problems Plague Wolverine Dump Neighbors." *Mlive.Com,* October 1. https://www.mlive.com/news/grand-rapids/2017/10/house_street_dump_neighbors.html
Ellison, Garret. 2017c. "3M Document Shows Wolverine Knew about PFAS in 1999." *Mlive.com*, November 4. https://www.mlive.com/news/grand-rapids/2017/11/3m_scotchgard_wolverine_pfas.html
Ellison, Garret. 2017d. "More than 100 Homes in New PFAS Testing Area near Rockford." *Mlive.com*, November 14. https://www.mlive.com/news/grand-rapids/2017/11/sludge_farm_wellington_ridge.html
Ellison, Garret. 2018a. "DEQ Ignored Crucial Warning on Wolverine Dump Plume." *Mlive.com*, March 12, 2018. https://www.mlive.com/news/grand-rapids/2018/03/pfas_deq_odonnell_www.html
Ellison, Garret. 2018b. "Major Warning about Michigan PFAS Crisis Came 6 Years Ago." *Mlive.com*, July 10. https://www.mlive.com/news/2018/07/meq_pfas_delaney_2012_report.html
Ellison, Garret. 2019. "How Citizen Sleuths Cracked the Wolverine Tannery Pollution Case." *Mlive.com*, June 12. https://www.mlive.com/news/2019/06/how-citizen-sleuths-cracked-the-wolverine-tannery-pollution-case.html
EWG. 2022. "More than 2,000 Communities Have Drinking Water with 'Forever Chemicals' above New EPA Levels." *Environmental Working Group*, June 15. https://www.ewg.org/news-insights/news-release/2022/06/more-2000-communities-have-drinking-water-forever-chemicals
Faber, Scott. 2020. "For 20-plus Years, EPA Has Failed to Regulate 'Forever Chemicals.'" *Environmental Working Group*, January. https://www.ewg.org/research/20-plus-years-epa-has-failed-regulate-forever-chemicals
Gardner, Paula. 2019. "When the Biggest Company in Town Poisons the Water." *Mlive*, June 11. https://www.mlive.com/news/2019/06/when-the-biggest-company-in-town-poisons-the-water.html

Hayes, Jared. 2019. "For Decades, Polluters Knew PFAS Chemicals Were Dangerous but Hid Risks from Public." *Environmental Working Group*, August 29. https://www.ewg.org/research/decades-polluters-knew-pfas-chemicals-were-dangerous-hid-risks-public

Harrison, Rodney. 2022. "Legacies: Rethinking the futures of heritage and waste in the Anthropocene." In *Heritage Ecologies*, edited by Torgeir Rinke Bangstad and Þóra Pétursdóttir, 31–48. London: Routledge.

ITRC. 2021. "Per- and Polyfluroalkyl Substances Technical and Regulatory Guidance." Interstate Technology and Regulatory Council. https://pfas-1.itrcweb.org/wp-content/uploads/2022/03/PFAS-Full-PDF-December-2021-Update.pdf

LaZerte, J. Donald. 1989. "3M'S Scotchgard Brand Fabric Protector." *Research Technology Management* 32 (2): 25–27.

Lerner, Sharon. 2018a. "The U.S. Military Is Spending Millions to Replace Toxic Firefighting Foam with Toxic Firefighting Foam." *The Intercept*, February 10. https://theintercept.com/2018/02/10/firefighting-foam-afff-pfos-pfoa-epa/

Lerner, Sharon. 2018b. "3M Knew About the Dangers of PFOA and PFOS Decades Ago, Internal Documents Show." *The Intercept*, July 31. https://theintercept.com/2018/07/31/3m-pfas-minnesota-pfoa-pfos/

Lerner, Sharon. 2020. "The Battle for Decatur: PFAS Contamination Divides an Alabama Town." *The Intercept*, August 23. https://theintercept.com/2020/08/23/pfas-3m-decatur-alabama/

Mah, Alice. 2012. *Industrial Ruination, Community and Place: Landscapes and Legacies of Urban Decline*. Toronto: University of Toronto Press.

Manchester, Harland. 1959. "Amazing Fluorocarbons Promise Stainproof Clothing." *Popular Science*, January.

Mator, Richard. 2018. "Tyco Fire Products, LP, Response to WDNR, January 16." BRRTS Activity No. 02-38-580695. Bureau for Remediation and Redevelopment Tracking System (BRRTS), Wisconsin Department of Natural Resources.

MLive. 2018. *MDEQ Official Watches Video of a Father Angry over PFAS Contamination*. https://www.youtube.com/watch?v=aD8RoZkF3SA

NSRDC. 1978. "Candidate Environmental Impact Statement on Discharging Firefighting System Aqueous Film Forming Foam (AFFF) into Harbor Waters During Tests of Machinery Space Fire-Fighting Foam Systems Aboard U.S. Navy Ships." Naval Ship Research and Development Center, U.S. Department of the Navy. https://www.documentcloud.org/documents/4343203-US00006835.html

Prokop, R. A. 1978. "Meeting Minutes - Review of Animal Studies, 3M Interoffice Correspondence." Exhibit 1174, State of Minnesota V. 3M Co., Court File No. 27-CV-10-28862. https://static.ewg.org/reports/2019/pfa-timeline/1978_Rats-and-Monkeys-Memo.pdf?_ga=2.55161198.455752248.1651760686-1458525737.1651760686

Purdy, Richard. 1999. "Letter from R. Purdy to 3M Re: Resignation." Exhibit 1001, State of Minnesota v. 3M Co., Court File No. 27-CV-10-28862. https://www.ag.state.mn.us/Office/Cases/3M/docs/PTX/PTX1001.pdf

Renfrew, Daniel and Thomas W. Pearson. 2021. "The Social Life of the 'Forever Chemical': PFAS Pollution Legacies and Toxic Events." *Environment and Society* 12 (1): 146–163. 10.3167/ares.2021.120109.

Richter, Lauren, Alissa Cordner, and Phil Brown. 2018. "Non-Stick Science: Sixty Years of Research and (in)Action on Flourinated Compounds." *Social Studies of Science* 48 (5): 691–714.

Richter, Lauren, Alissa Cordner, and Phil Brown. 2021. "Producing Ignorance Through Regulatory Structure: The Case of Per- and Polyfluoroalkyl Substances (PFAS)." *Sociological Perspectives* 64 (4): 631–656. 10.1177/0731121420964827.

Trepal, Dan and Kaeleigh Herstad. 2019. "Postindustrial Landscapes, Communities, and Heritage." *Journal of Community Archaeology and Heritage* 6(4): 235–237. doi:10.1080/20518196.2019.1670394.

Wollentz, Gustav, Sarah May, Cornelius Holtorf, and Anders Högberg. 2020. "Toxic heritage: Uncertain and unsafe." In *Heritage Futures*, 294–312. London: University College London.

4

PLASTICITY AND TIME: USING THE STRESS-STRAIN CURVE AS A FRAMEWORK FOR INVESTIGATING THE WICKED PROBLEMS OF MARINE POLLUTION AND CLIMATE CHANGE

John Schofield and Celmara Pocock

Introduction

Plastics have very rapidly become the characteristic material of the current Plastic Age, just as stone and metal artefacts characterised and became the type fossils of earlier periods. Plastics have utility and value as everyday items. As such they have become a part of the archaeological record (e.g. Mytum and Meek 2021) and a new category of cultural heritage[1]. However, plastics also present a significant threat to planetary health, forming a particular, disruptive and potentially resilient category of what has become known as toxic heritage (after e.g. Wollentz et al. 2020), the legacies of which will likely extend long into the future, itself now the subject of heritage attention (e.g. Holtorf and Högberg 2021).

In this chapter, we present a temporal framework within which to assess this particular category of toxic heritage: the stress-strain curve (e.g. Beer et al. 2020). This curve recognises that materials pass through a phase of elasticity (in which remediation can bring things back to their 'original' form) towards plasticity (whereby things retain an aspect of their original form but are forever changed) leading ultimately to fracture (where the thing becomes irretrievably broken). We propose that this curve can apply to plastics in two distinct but related ways. First, it can apply to the changing form of plastics, by which they may start as useful and everyday, perhaps even becoming formally recognised as heritage items, to eventually and typically breaking down into potentially dangerous micro- and nano-plastics, becoming at best *in*significant in heritage terms (after Ireland et al. 2020), and at worst destructive. Second, it can apply to the ecosystems that are impacted by plastics: if the problem of plastic pollution is caught early enough, then those ecosystems may be relatively unaffected (elasticity); but if the problem is left untreated, ecosystems might break down altogether (fracture).

DOI: 10.4324/9781003365259-7

We position this theoretical argument alongside studies that have used archaeological methods within the wider framing of heritage research to address the challenges of plastics and waste (e.g. Ross and Angel 2020). As outlined later in the chapter, ongoing work in Galápagos (Ecuador) has demonstrated how archaeology (in the form of object itineraries) can play an important role alongside oceanographic studies in determining the origins of plastic waste in remote locations (van Sebille et al. 2019; Schofield et al. 2020). Archaeology has also been used to help understand (and potentially, ultimately change) people's behaviours, for example through highlighting the distributions of COVID waste on local and international scales (e.g. Magnani et al. 2022 and Schofield et al. 2021 respectively).

The 'Wicked Problem' of Plastic Pollution

It is hard to imagine a world without plastics, and yet such a world existed until only around seventy years ago. It is true that plastics have earlier origins, but their widespread use only began after the conclusion of the Second World War. From that point on, as Geyer et al. (2017) state, 'the ensuing rapid growth in plastics production is extraordinary, surpassing most other man-made [sic] materials' (see also MacLeod et al. 2021). Currently, according to Geyer et al., some 12% of plastic waste globally is incinerated (itself a major environmental and health problem, e.g. Ágnes and Rajmund 2016) and 9% recycled, with the majority being either disposed into landfills (e.g. Rathje 1992) or being carelessly discarded or lost (including from landfill) to enter the environment. Where plastics end up in the environment they present a particular problem, both for that environment and for its inhabitants (human and non-human).

For example, plastics pollute chemically, 'when added chemicals (monomers and plasticizers) escape plastics and interact with bodies and ecosystems' (after Liboiron 2016, 92). Plastic pollution can jeopardise or threaten vital resources that are the basis of livelihoods and cultural knowledge and practices, degrade natural attributes that are the basis of tourism income, and impact aesthetic qualities that underpin social and cultural attachments to places. Plastics thereby compromise the heritage values of diverse environments with implications for the well-being of local peoples.

In terms of dangers, within the marine environment the majority of recorded pollution has involved entanglement and the ingestion of plastics by marine life. For example, of the 340 publications reviewed by Gall and Thompson (2015), 292 reported ingestion or entanglement involving marine debris and organisms, with plastics by far the most common form of debris (at 92%). Further analysis showed that direct harm or death is a significantly more common consequence of entanglement (at 79%) than of ingestion (at 4%).

Like the plastics themselves, the problem is ubiquitous: plastic debris has been found in every ocean basin as well as in freshwater and terrestrial habitats across the globe. It has now also been found in both animal and human bodies, including in people's lungs (Jenner et al. 2022).

Plastic pollution, alongside the interrelated problem of climate change (see Ford et al. 2022; Lavers et al. 2022,) and health (Romanello et al. 2022), has been referred to as a wicked problem (after Rittel and Webber 1973; Head 2022; see also Schofield in prep.), being a particular type of problem which is, 'complex, intractable, open-ended, unpredictable' (Alford and Head 2017, 397). This particular cluster of wicked problems (plastic pollution, climate change and health) has also been defined as 'super wicked' (after Lazarus 2009; Levin et al. 2012), because there are additional factors at play, including that:

1 time is running out;
2 there is no central authority, or only weak authority, to manage the problem; and that
3 the same actors seeking to solve the problem continue to directly contribute to it.

As Peters (2017) states, the time element is critical where irreversible harm will be done if significant policy interventions are not made, and quickly. Peters (2017, 389) also points out that public sector decision-making is not good at dealing with long-term challenges, especially in democratic regimes where changes in government often also mean changes in policy.

It is widely agreed that wicked problems require creative solutions, sometimes referred to as messy or 'clumsy' solutions (e.g. Grint 2010, 24). This chapter suggests how such creativity, in the form of an archaeological-heritage perspective, might help address the particular wicked problem of marine pollution alongside the related problem of climate change.

The Archaeology of Plastics

Plastics are now a category of cultural heritage: a highly visual and impactful component of an ever-changing and evolving landscape of entangled nature and culture. Plastics have therefore rapidly become type-fossils of a Plastic Age, the definition of which is not new. Over a decade ago, Thompson et al. (2009) referred to 'our plastic age' in a landmark publication for the Royal Society, while Brandon et al. (2019) note how the, 'increase in plastic deposition in the post–World War II years can be used as a geological proxy for the Great Acceleration of the Anthropocene in the sedimentary record' (see also Mytum and Meek 2021). Edgeworth et al. (2022) gave this argument more nuance, focusing on the precise temporal parameters for viewing this category of material within the wider context of debates around the Anthropocene. Having established that 'PET (polyethylene terephthalate) bottles were first produced in 1973, and so date what might be termed the mid- to late Anthropocene', they describe how,

> the patterns into which plastics have been shaped infinitely multiply the dating possibilities. The plastic credit card and the Bic Cristal ballpoint pen, for instance,

> both first produced in 1950, may be regarded as zone technofossils for the Anthropocene, while compact discs (surprisingly widely dispersed as trash) only date from 1982. The technostratigraphic possibilities are almost endless, as technodiversity (if measured as the number of specific kinds of technofossil) is now far greater than biodiversity, measured as the number of species.

Beyond dating archaeological sites, these plastics are also indicative of contemporary behaviours. One difference between this contemporary archaeology (and see also Zalasiewicz et al. 2017 in this context) and the study of earlier periods is that archaeologists are not looking back on a deeper past but engaging with the period in which they live, critiquing the very world they shape through their everyday practices (Graves-Brown 2000, 1). Through a critical lens, archaeology can thus become activist as a relevant, engaging and informative mode of enquiry.

The Stress-Strain Curve

The Stress-Strain Curve (Figure 4.1) is well known in engineering and materials science and has been applied to virtually every material, from plastic and glass to concrete (see Beer et al. 2020). As stated earlier, the stress-strain curve is notable for its recognition of the concept of plasticity: that materials pass through a phase of elasticity (in which remediation can bring things back to their 'original' form) towards plasticity (whereby things retain an aspect of their original form, but are forever changed) leading ultimately to fracture (where things become irretrievably broken). Much has been written about stress and strain within plastics and polymers

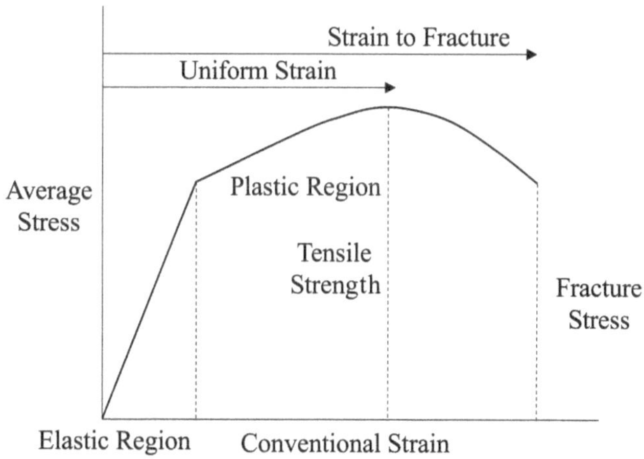

FIGURE 4.1 The Stress-Strain Curve. (Available open source via Wikimedia Commons, https://upload.wikimedia.org/wikipedia/commons/9/98/Stress_Strain.JPG).

(e.g. Williams and Ford 1964; Haward 1987), recognising, for example, that nearly all have a degree of elasticity, particularly during their (often very short) use-life. Unless plastics are conserved in some way, elasticity is gradually reduced as plasticity takes over. In this phase, the plastics will become less flexible, more embrittled. And ultimately the plastic items fracture, creating micro- and nanoplastics that can have a detrimental impact on human and animal health. In the case of object itineraries, discussed below, this curve provides a temporal framework, much as the ageing process does for human and non-human beings.

We suggest that this curve provides a useful model (and terminology) for understanding how plastic becomes toxic heritage, and simultaneously threatens existing heritage values. Here we explore this particular application through two World Heritage Listed archipelagos whose Outstanding Universal Values are impacted by the related wicked problems of plastics pollution and climate change: Galápagos (Ecuador) and the Great Barrier Reef (Australia). (See Table 4.1 for a summary of the key issues).

Within Galápagos, the problem of plastic pollution and its impact on iconic and endemic species was recognised early by those responsible for managing the island's resources. Recalling Grint's (2010) definition of clumsy solutions to wicked problems, specialist advice was sought from outside the remote archipelago, involving scientists from a diverse range of specialisms with the intention that these specialists work alongside locally based scientists and local communities. This degree of trans-disciplinary collaboration was considered essential if solutions were to be ethical, sustainable and have local support. One part of this collaborative effort involved a Science to Solutions workshop and research programme coordinated initially by the Galapagos Conservation Trust (based in London), itself involving scientists from around the world, including the United Kingdom, Europe, Australia and South America.[2] This collaborative network included archaeology, with its focus on material culture bringing a temporal perspective to the problem (Schofield et al. 2020).

This collaborative and transdisciplinary initiative to address the fast-emerging problem of plastic pollution in Galápagos, hoped to catch the problem early enough to ensure the process of change did not extend beyond the elasticity phase in which remediation can bring things back close to an 'original' form. By appropriate action,

TABLE 4.1 The key issues related to the two case studies of Galápagos and the Great Barrier Reef

Wicked Problem	Example	Current emphasis in solution	Stress-Strain Stage	Aim	URL for OUV summary
Plastic pollution	Galápagos	Behaviours	Elasticity → Plasticity	Reversal	https://whc.unesco.org/en/list/1/
Climate change	Great Barrier Reef	Technical	Plasticity → Fracture	Mitigation	https://whc.unesco.org/en/list/154/

it is argued, the problem can be stopped in its tracks, or at least slowed, provided the efforts are then maintained to ensure it does not, once again, get out of hand. This is where the local community becomes so important. Without local support, holding the line and even continuing to reduce the scale of the problem would likely prove impossible.

Part of the clumsy solution to plastic pollution in Galápagos involved using object itineraries (versions of which are also referred to as narratives or biographies), which have long formed a part of material cultural research (e.g. Joyce and Gillespie 2015; after Kopytoff 1986). In the Science to Solutions workshop, the team used object itineraries to investigate a plastic assemblage from a remote beach to help understand the source of the plastic waste accumulating there (Figure 4.2). This was achieved through building stories around each plastic item, one element of which was to use its condition (alongside information from labelling and biofouling) as an indicator of both how long it had first been at sea, and then how long it had been exposed on the beach - identifying its natural and its cultural transformations, in Schiffer's (1976) terms. Through this work, and by combining the evidence for longevity with evidence of the objects' origins including through oceanographic modelling (specifically, van Sebille et al. 2019), Galápagos plastics have predominantly been sourced to the Pacific coast of South America (specifically northern

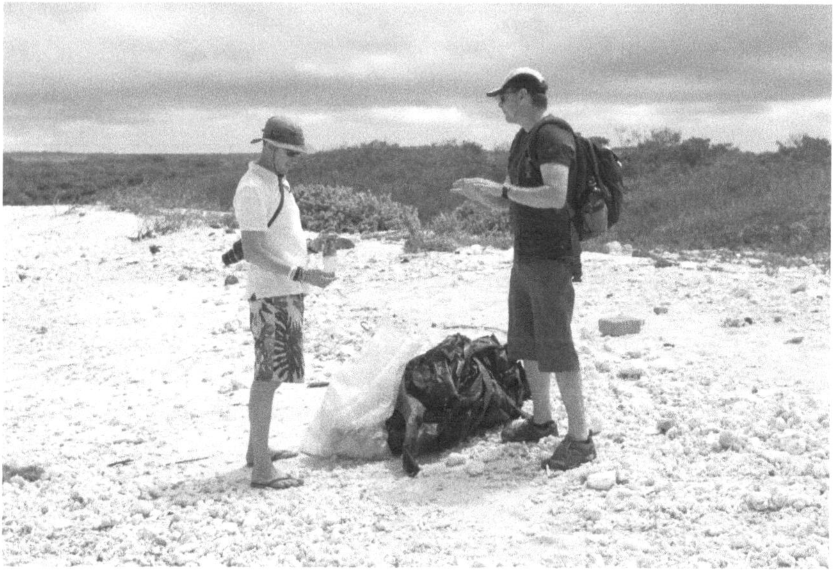

FIGURE 4.2 Members of the Science to Solutions workshop team collecting a representative sample of plastic objects from a remote beach in Galápagos, in 2018. The objects collected were then subject to 'object itinerary' investigations to help better understand where the items were coming from. (Image: John Schofield).

Peru and southern Ecuador) and to the International fishing fleet, which operates legally immediately beyond the exclusion zone around Galápagos.

In general terms, and relating to the stress-strain curve, plastic objects remain low on the curve to begin with, in the elasticity phase. In this phase, they may sometimes be recognised as valued heritage objects[3]. Perhaps during their use phase, but more usually after discard, the objects enter the plasticity phase. In heritage terms, we can recognise this as a time of vulnerability for plastics where they might hold heritage significance and be conserved, or more likely discarded and at risk of becoming toxic heritage, both as objects lacking integrity and as plastics that fracture, as the objects become brittle and break down into smaller parts. This latter part of the process often coincides with environmental degradation. In this way the object itineraries and the landscape history are enmeshed; the trajectory of one (the environment) being closely entwined with that of the other (plastics).

The impacts of climate change are widespread and catastrophic at the Great Barrier Reef, where a confluence of factors wreaks untold damage on individual reefs and undermines the integrity of the system. The increased magnitude and frequency of extreme weather events cause both immediate physical damage (e.g. uprooted corals), and conditions of vulnerability (e.g. higher nutrient and pollution runoff into reef waters). These changed conditions lead to imbalances, notably in increasing numbers of the naturally occurring Crown-of-Thorns Starfish (COTS). COTS have a voracious appetite for coral polyps, and in large numbers, they ravage swathes of reefs. While reefs can recover between outbreaks of COTS, this takes a couple of decades. But as cyclones repeatedly batter the Great Barrier Reef and the time between COTS outbreaks is reduced, there is insufficient time for such recovery (Westcott et al. 2020). These immediate effects of climate change are amplified by the most compelling of all problems facing the Reef: coral bleaching (or loss of symbionts), which is a response to rising sea temperature and increased UV radiation (Figure 4.3). Cheung et al (2021) describe coral bleaching as 'one of the most striking manifestations of marine heatwaves' that can 'cause mass coral mortality over thousands of hectares within a few months'. These events are also becoming more intense and frequent, with three mass events in just five years (Cheung et al. 2021). And like the COTS outbreaks, the cumulative nature of these impacts could be a tipping point, with Cheung et al. suggesting that 'some reefs will become untenable under repeated stress' (Cheung et al. 2021).

In other words, parts of the Reef remain in the elasticity phase but are likely to degrade due to climate change. Other parts have already reached the plasticity phase at which point change is typically irreparable. It is also possible, if not likely, that areas of the Reef have already fractured, in those places where ecosystems have broken down altogether. Where plasticity and fracture have already been reached, it is hard to see a way back. As Ainsworth et al. (2016, 338) have stated, this bleaching of the coral reefs relates directly to climate change, a wicked problem for which solutions are near impossible: 'We find that near-future increases in local

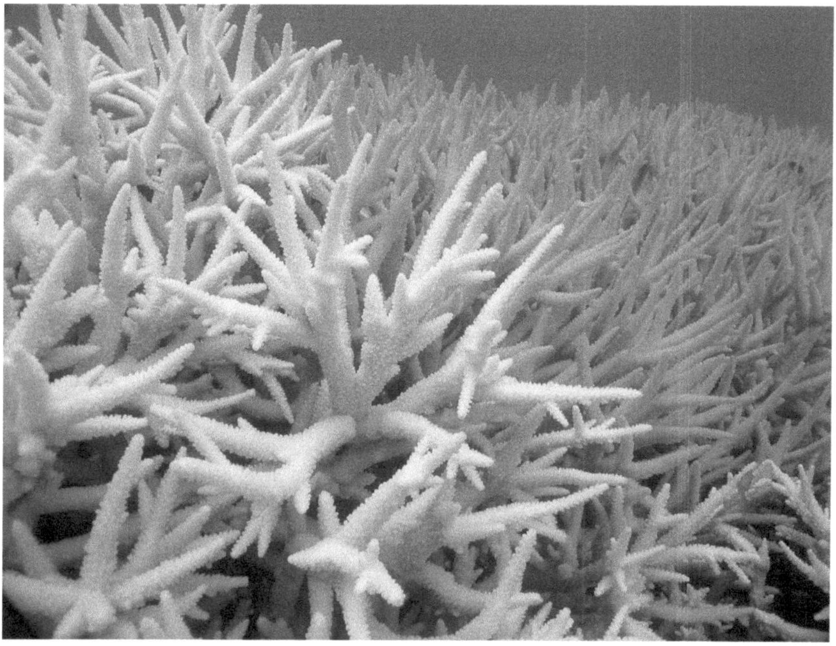

FIGURE 4.3 Bleached corals at Keppel Island, Great Barrier Reef. (Photograph by P. Marshall. Copyright Commonwealth of Australia - GBRMPA).

temperature of as little as 0.5°C result in [the Reef's] protective mechanism being lost, which may increase the rate of degradation'.

Indeed, in the face of what appear to be irreversible impacts, and taking the opposite approach to that being developed in Galápagos, conservation of the Great Barrier Reef has shifted away from efforts to change human behaviours to one focused on finding technical solutions to artificially build and sustain the Reef. These experimental methods include engineering more heat-tolerant species, adding chemical sunshade to the water, and transplanting corals to bleach-affected reefs (Great Barrier Reef Foundation 2017, Suggett et al. 2019). However creative and life-saving these efforts appear, they do not address the stress-strain context produced by exponentially rising sea temperatures and increasingly frequent extreme weather events which have destructive consequences for the Great Barrier Reef (Pocock 2021).

Conclusion

As this chapter has shown, the more nebulous and intractable issue of climate change has pushed the Great Barrier Reef to the point of fracture, while interventions to address the more immediately visible problem of plastic pollution in

Galápagos are helping to secure a position within the elasticity part of the stress-strain curve, whereby things can still revert to an 'original state'. These differences can be viewed within the context of recent arguments that suggest attention to plastic pollution is a distraction from the more pressing and significant issue of climate change (e.g. Stafford and Jones 2019, Jones and Stafford 2021), claims which are rebutted both by arguments that suggest one global issue should not have priority over another in the current planetary crisis (Avery-Gomm et al. 2019), and by evidence that plastic contributes directly to climate change including through the production of Greenhouse Gases (GHGs) through every stage of their lifecycle; from extraction, through production and consumption, to End of Life (Walker et al. 2021, Ford et al. 2022, Bauer et al. 2022). Others point out that far from diverting resources from climate change, plastic pollution has received far less research attention and funding (Lavers et al. 2022).

One of the advantages of plastic pollution over climate change in attracting public support is its immediacy; people can see the pollution and literally pick it up. Yet while one of these wicked problems may be more tangible than the other, both are cumulative and require urgent action beyond such expedient action. While these time pressures are particularly difficult for people to comprehend or to act upon, the temporal framework presented here offers a different way of thinking. Rather than focusing on immediate risk, threat and priorities, we suggest using the stress-strain curve to determine where we are on the trajectory of change, and what can still be realistically achieved. Additionally, object itineraries make this sense of time more explicit and enable communities to understand how time is implicated in the emergence of toxic heritage. By making time explicit through the stress-strain curve, this framework can be used to help understand interconnected trajectories that on the one hand risk transforming plastics into toxic heritage, and on the other recognise how the creation of toxic heritage becomes its own source of stress-strain in ecological systems that underpin World Heritage values. These heritage approaches can also be invoked to help build temporal appreciation of other wicked problems including less visible, environmental challenges, such as climate change.

Notes

1 The Getty Conservation Institute, for example, has a long-term project on the conservation of plastics - https://www.getty.edu/conservation/our_projects/science/plastics/

2 The workshop from 2018 is summarised in Schofield et al. 2020; Pacific Plastics: Science to Solutions is now a GCRF-funded project centred on Exeter University - https://www.pacificplasticssciencetosolutions.com/partners/

3 A child's shoe (one of the Galápagos objects investigated by the team) may have been worn by several children at different times, passing down from one sibling to the next. This is an example of an object that may have been valued as a heritage item (described in Schofield et al. 2020).

References

Ágnes, N. and Rajmund, K. 2016. The environmental impact of plastic waste incineration. *AARMS–Academic and Applied Research in Military and Public Management Science* 15(3), 231–237.

Ainsworth, T. D., Heron, S. F., Ortiz, J. C., Mumby, P. J., Grech, A., Ogawa, D., Eakin, C. M. and Leggat, W. 2016. Climate change disables coral bleaching protection on the Great Barrier Reef. *Science* 352(6283), 338–342. 10.1126/science.aac7125

Alford, J. and Head, B. W. 2017. Wicked and less wicked problems: a typology and a contingency framework. *Policy and Society* 26(3), 397–413.

Avery-Gomm, S., Walker, T. R., Mallory, M. L. and Provencher, J. F. 2019. There is nothing convenient about plastic pollution. Rejoinder to Stafford and Jones Viewpoint–Ocean plastic pollution: A convenient but distracting truth? *Marine Policy* 106, 103552.

Bauer, F., Nielsen, T. D., Nilsson, L. J., Palm, E., Ericsson, K., Fråne, A. and Cullen, J. 2022. Plastics and climate change breaking carbon lock-ins through three mitigation pathways. *One Earth* 5(4), 361–376.

Beer, F., Johnston, R., DeWolf, J., Mazurek, D. and Sanghi, S. 2020. *Mechanics of Materials (eighth edition)*. New York: McGraw-Hill.

Brandon, J. A., Jones, W. and Ohman, M. D. 2019. Multidecadal increase in plastic particles in coastal ocean sediments. *Science Advances* 5(9). 10.1126/sciadv.aax0587

Cheung, M. W. M., Hock, K., Skirving, W. and Mumby, P. J. 2021. Cumulative bleaching undermines systemic resilience of the Great Barrier Reef. *Current Biology* 31(23), 5385–5392.e4. 10.1016/j.cub.2021.09.078

Edgeworth, M., Haff, P. K., Ivar do Sul, J. A., Richter, D. and Zalasiewicz, J. 2022. The Technofossil Record: Where Archaeology and Paleontology Meet. *Anthropocene Curriculum* (22 April 2022). https://www.anthropocene-curriculum.org/contribution/the-technofossil-record-where-archaeology-and-paleontology-meet

Ford, H. V., Jones, N. H., Davies, A. J., Godley, B. J., Jambeck, J. R., Napper, I. E., Suckling, C. S., Williams, G. J., Woodall, L. C. and Koldeway, H. J. 2022. The fundamental links between climate change and marine plastic pollution. *Science of the Total Environment* 806, 150392. 10.1016/j.scitotenv.2021.150392

Gall, S. C. and Thompson, R. C. 2015. The impact of debris on marine life. *Marine Pollution Bulletin* 92, 170–179.

Geyer, R., Jambeck, J. R. and Law, K. L. 2017. Production, use, and fate of all plastics ever made. *Science Advances 3*(7), e1700782. Available at https://advances.sciencemag.org/content/advances/3/7/e1700782.full.pdf.

Graves-Brown, P. (ed.) 2000. *Matter, Materiality and Modern Culture*. London: Routledge.

Great Barrier Reef Foundation. 2017. Resilient Reefs Successfully Adapting to Climate Change. Retrieved from https://www.barrierreef.org/uploads/Final%20report%20Resilient%20Reefs%20Successfully%20Adapting%20to%20Climate%20Change%20program.pdf

Grint, K. 2010. *Leadership: A Very Short Introduction*. Oxford: Oxford University Press.

Haward, R. N. 1987. The application of a simplified model for the stress-strain curves of polymers. *Polymer* 28(9), 1485–1488. 10.1016/0032-3861(87)90347-8

Head, B. 2022. *Wicked Problems in Public Policy: Understanding and Responding to Complex Challenges*. London: Palgrave Macmillan. Open Access at https://www.ippapublicpolicy.org/book/wicked-problems-in-public-policy/34

Holtorf, C. and Högberg, A. (eds) 2021. *Cultural Heritage and the Future*. London: Routledge.

Ireland, T., Brown, S. and Schofield, J. 2020. Situating (in)significance. *International Journal of Heritage Studies* 26(9), 826–844. 10.1080/13527258.2020.1755882

Jenner, L. C., Rotchell, J. M., Bennet, R. T., Cowen, M., Tentzeris, V. and Sadofsky, L. R. 2022. Detection of microplastics in human lung tissue using μFTIR Spectroscopy. *Science of the Total Environment* 831. 10.1016/j.scitotenv.2022.154907

Jones, E. A. and Stafford, R., 2021. Neoliberalism and the Environment: Are We Aware of Appropriate Action to Save the Planet and Do We Think We Are Doing Enough? *Earth* 2(2), 331–339.

Joyce, R. A. and Gillespie, S. D. (eds) 2015. *Things in Motion: Object Itineraries in Anthropological Practice*. Santa Fe: SAR Press.

Kopytoff, I. 1986. The Cultural Biography of Things: Commoditization as process. In Appadurai, A. (Ed), *The Social Life of Things: Commodities in Cultural Perspective*, 64–91. Cambridge: Cambridge University Press.

Lavers, J. L., Bond, A. L. and Rolsky, C. 2022. Far from a distraction: Plastic pollution and the planetary emergency. *Biological Conservation* 272, 109655. 10.1016/j.biocon.2022.109655

Lazarus, R. J. 2009. Super-wicked problems and climate change: Restraining the present to liberate the future. *Cornell Law Review* 94, 1153–1233.

Levin, K., Cashore, B., Bernstein, S. and Auld, G. 2012. Overcoming the tragedy of super wicked problems: Constraining our future selves to ameliorate global climate change. *Policy Sciences* 45, 121–152.

Liboiron, M. 2016. Redefining pollution and action: The matter of plastics. *Journal of Material Culture* 21(1), 87–110.

MacLeod, M., Arp, H. P. H., Tekman, M. B. and Jahnke, A. 2021. The global threat from plastic pollution. *Science* 373(6550), 61–65. 10.1126/science.abg5433

Magnani, M., Magnani, N., Venovcevs, A. and Farstadvoll, S. 2022. A contemporary archaeology of pandemic. *Journal of Social Archaeology* 22(1), 48–81. 10.1177/14696053211043430

Mytum, H. and Meek, J. 2021. The Iron Age in the Plastic Age: Anthropocene Signatures at Castell Henllys. *Antiquity* 95(379), 198–214.

Peters, B. G. 2017. What is so wicked about wicked problems? A conceptual analysis and a research program. *Policy and Society* 36(3), 385–396.

Pocock, C. 2021. Great Barrier Reef World Heritage: Nature in Danger. *Queensland Review* 28(2), 118–129.

Rathje, W., Hughes, W., Wilson, D., Tani, M., Archer, G., Hunt, R. and Jones, T. 1992. The Archaeology of Contemporary Landfills. *American Antiquity* 57(3), 437–447.

Rittel, H. W. J. and Webber, M. M. 1973. Dilemmas in a general Theory of Planning. *Policy Sciences* 4(2), 155–169.

Romanello, M., Di Napoli, C., Drummond, P. et al. 2022. The 2022 report of the Lancet Countdown on health and climate change: health at the mercy of fossil fuels. *The Lancet*. 10.1016/S0140-6736(22)01540-9.

Ross, S. and Angel, V. 2020. Heritage and Waste: Introduction. *Journal of Cultural Heritage Management and Sustainable Development* 10(1), 1–4.

Schiffer, M. M. 1976. *Behavioral Archaeology*. New York: Academic Press.

Schofield, J., Wyles, K., Doherty, S., Donnelly, A., Jones, J. and Porter, A. 2020. Object Narratives as a Methodology for Mitigating Marine Plastic Pollution: a New Multidisciplinary Approach, and a Case Study from Galápagos. *Antiquity* 94, 228–244.

Schofield, J., Praet, E., Townsend, K. and Vince, J. 2021. 'COVID Waste' and Social Media as Method: An Archaeology of PPE and its Contribution to Policy. *Antiquity* 95 (380), 435–449.

Stafford, R. and Jones, P. J. S. 2019. Viewpoint–Ocean plastic pollution: A convenient but distracting truth? *Marine Policy* 103, 187–191.

Suggett, D. J., Camp, E. F., Edmondson, J., Boström-Einarsson, L., Ramler, V., Lohr, K. and Patterson, J. T. 2019. Optimizing return-on-effort for coral nursery and outplanting practices to aid restoration of the Great Barrier Reef. *Restoration Ecology* 27(3), 683–693.

Thompson, R. C., Swan, S. H., Moore C. J. and vom Saal, F. S. 2009. Our plastic age. *Philosophical Transactions of the Royal Society* B364, 1973–1976. https://doi.org/10.1098/rstb.2009.0054

van Sebille, E., Delandmeter, P., Schofield, J., Hardesty, B., Jones, J. and Donnelly, A., 2019. Basin-scale sources and pathways of microplastic that end up in the Galápagos Archipelago. *Ocean Science* 15, 1341–1349.

Walker, T. R. and McKay, D. C. 2021. Comment on 'five misperceptions surrounding the environmental impacts of single-use plastic'. *Environmental Science & Technology* 55(2), 1339–1340.

Westcott, D. A., Fletcher, C. S., Kroon, F. J., Babcock, R. C., Plagányi, E. E., Pratchett, M. S. and Bonin, M. C. 2020. Relative efficacy of three approaches to mitigate Crown-of-Thorns Starfish outbreaks on Australia's Great Barrier Reef. *Scientific Reports* 10(1), 1–12.

Williams, J. G. and Ford, H. 1964. Stress-Strain Relationships for Some Unreinforced Plastics. *Journal of Mechanical Engineering Science* 6(4), 405–417. 10.1243/JMES_JOUR_1964_006_055_02

Wollentz, G., May, S., Holtorf, C. and Högberg, A. 2020. Toxic heritage: Uncertain and unsafe. In Harrison, R., DeSilvey, C., Holtorf, C., MacDonald, S., Bartolini, N., Breithoff, E., Fredheim, H., Lyons, A., May, S., Morgan, J. and Penrose, S. (Eds.), *Heritage Futures: Comparative Approaches to Natural and Cultural Heritage Practices*, 294–312. London: UCL Press.

Zalasiewicz, J., Williams, M., Waters, C., Barnosky, A. D., Palmesino, J., Rönnskog, A-S., Edgeworth, M., Neal, C., Cearreta, A., Ellis, E., Grinevald, J., Haff, P., Ivar do Sul, J. A., Jeandel, C., Leinfelder, R., McNeill, J. R., Odada, E., Oreskes, N., Price, S. J., Revkin, A., Steffen, W., Summerhayes, C., Vidas, D., Wing, S. and Wolfe, A. P. 2017. Scale and Diversity of the Physical Technosphere: A Geological Perspective. *The Anthropocene Review* 4, 9–22. 10.1177/2053019616677743

SECTION 2

INTRODUCTION: THE POLITICS OF TOXIC HERITAGE

Elizabeth Kryder-Reid and Sarah May

This section considers the issues of policy, stakeholders, politics, and the fields of power in which the toxic heritage operates. Heritage is always entangled with notions of value, often involving conflicting and contested frameworks of meaning. For toxic heritage, this arena of contestation and meaning-making is particularly fraught because of the implications of culpability and accountability for past polluters, as well as the opportunities for resistance, advocacy, and other political actions in which extractivism is named and challenged. The evidence presented in this section challenges traditional heritage narratives that often sidestep accountability through othering. Political readings probe the entanglements of heritage and toxicity in cases such as Stamatopoulou-Robbins' study of waste wielded as an instrument of oppression in occupied Palestine and May's study of heritage-led regeneration wielded as a tool to stimulate economic and social development in Swansea's post-industrial landscapes. Papoli-Yazdi's garbology study in Tehran demonstrates the ways in which heritage practices of designation and valourisation can intersect with inequality to both increase the disproportional impacts of the toxicity of waste and minimize the benefits of historic preservation policies in marginalized communities.

This section also highlights the flows of toxic heritage along global political contours. Naming toxic heritage as a global phenomenon locates it as a set of political acts reaching across borders. For example, the toxic heritage of war and military activities is inextricable from its geopolitical contexts, and its residues are similarly complicated, whether from former training grounds in Indigenous territories (Rankin et al.) or underwater shipwrecks (Carter et al.). "Biopolitical heritage" visited upon marginalized communities and remote locations can be implicated in the politics of memory (Little and Akese) and the strategic invisibility akin to forgetting (Weinberg and Figueroa). These cases offer counterpoints to the valourised industrial heritage of northern European nations that typically disavows connections with toxicity experienced elsewhere.

DOI: 10.4324/9781003365259-8

5

HERITAGE-LED REGENERATION AND THE SANITISATION OF MEMORY IN THE LOWER SWANSEA VALLEY

Sarah May

Introduction

When the Copper industry in Swansea came crumbling to a close in the 1960s, the Lower Swansea Valley was one of the most contaminated landscapes in the world: Slag heaps, ruined buildings, a hundred years of intense air pollution, the water table full of heavy metals. The urban population which had been built on the strength of the industry had few opportunities for employment. The families which ran the companies withdrew and took their capital with them. The city, its government, citizens, remaining industries, and universities were left to recreate a healthy landscape – a long process of decontamination.

While decontamination typically involves the dilution, containment, or removal of toxic materials, it is also a process of managing relations between the past and the future, as such it is a heritage practice (Wollentz et al. 2020). Throughout this period of decontamination, cultural heritage has played different roles in establishing a liveable city. As the process is ongoing, we can expect cultural heritage to continue to take on different roles. In this chapter, I will use academic, policy and public-facing discourse to define four phases of decontamination and the different types of heritage narratives associated with them. I will then consider how these narratives contribute to the ongoing decontamination and political processes at play and how they are expressed in the contemporary landscape.

Chronology of post-industrial Swansea

The toxic impacts of metal processing in the Lower Swansea Valley were re-cognised even in the heyday of the industry:

DOI: 10.4324/9781003365259-9

> By the 182Os, the sulphurous smoke from the Lower Swansea Valley copperworks choked the town and surrounding area. A number of court cases were brought against the copper companies by local farmers and residents. They contained vivid testimonies of the effect of the smoke on cattle, horses and the growth of crops. John Henry Vivian took a personal interest in the issue of copper smoke, and funded extensive research into alleviating it, enlisting the help of other scientists such as Michael Faraday.
>
> (Goskar 2011, 17)

In the late 1950s, Swansea was struggling with the legacy of nearly two centuries of intensive industrial exploitation, notably copper production, fuelled by the ready availability of coal (Evans and Miskell 2020). In the 1960s the 'Lower Swansea Valley Project' (LSVP) was established to tackle the immense challenge of air, land and water contamination. Since there was no UK legislation surrounding the disposal or management of toxic waste until 1972 (Salvo 1989) Swansea Council, the local government body, was on its own in how to manage this process and it was framed as research with the University. In 1979 just before the last rolling mill closed, the project came to a close as well with a conference which concluded the 'foundations of reclamation have been laid and the challenge lies in the successful planning and development of the valley' (Bromley and Humphrys 1979) (Figure 5.1).

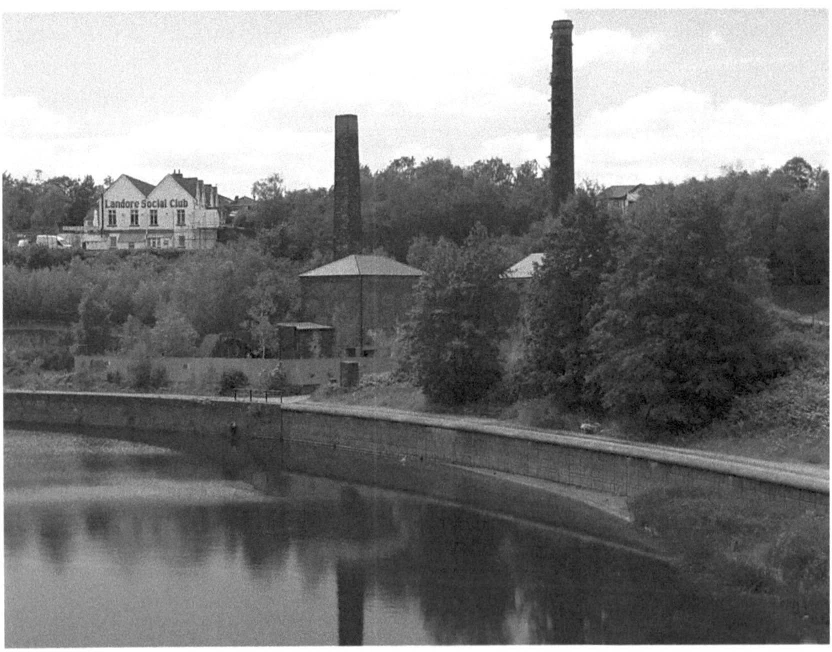

FIGURE 5.1 The contemporary landscape of the copperworks shows little trace of dereliction. The two remaining chimneys are managed as heritage. Attribution: ©Sarah May, 2022.

The LSVP did not frame the problem as contamination, it talked of dereliction. This term is interesting because it draws attention to abandonment, rather than contamination, and because it carries a secondary meaning of a failure to fulfil obligations. The problem was framed as a loss of care and control, the breakdown of a functioning system, rather than the consequences of that system. The primary concern was with how the valley was perceived by residents, visitors and potential investors. The early research on 'human ecology' had a very strong emphasis on the visual, how the 'ruins' and waste were seen and how they formed part of daily life. The LSVP report on Human Ecology even comments 'If the house is turned so that the bad bits can't be seen all the time, it can be more easily forgotten' (Stacey 1962, 122).

The dumps of slag and other industrial waste, known as tips, were definitely an asset as well as waste and using them as an asset – for hardcore more than anything more sophisticated – was important. The main focus of the research on their management related to controlling dust and the visual impact of 'nibbling' – small-scale, ad hoc excavation for reuse of the material (Holt 1966). The response of the LSVP was to use the tips to level the valley and construct embankments on the floodplain. Till deposits from elsewhere were then spread over these areas to 'act as an impermeable seal to the contaminants and as a growing medium for plants and trees' (Waters et al. 2005).

Nonetheless, 20 years after the completion of the LSVP strongly acid soils (pH 4.5) and very high metal amounts, which had been originally recorded in the 1960s (Davies 1966) were still a concern (Bridges and Morgan 1999). A further research project to monitor the problem concluded that the soil contamination in the Swansea region penetrates deeper than usual for metal contamination, this could be because of the acidity of the soil, 'Alternatively, substantial mixing of waste materials with the soils may have occurred' (Waters et al. 2005).

Despite the perception that the toxicity was remediated by the LSVP by the 1970s, soil chemists still monitor conditions and a 2021 study indicates soils in the valley 'contain elevated level of metals, enough to cause direct or indirect effects on human health' (Schilling et al. 2021).

The driver of the LSVP was, of course, the loss of industry and the need for economic recovery. Beyond the physical and chemical aspects of the valley, deindustrialisation was associated with high rates of unemployment, low educational attainment, low average wages and low house prices, as well as poor transport connections, isolation and limited social mobility (Beatty, Fothergill, and Powell 2007; Walkerdine and Jimenez 2012). The social focus of LSVP was planning for how government policy could support economic development. This was the importance and the focus of the 1979 conference which drew the project to a close (Bromley and Humphrys 1979). 1979 also saw the end of the Labour government led by James Callaghan. The Conservative government led by Margaret Thatcher, which replaced it, turned its back on the kinds of economic planning envisioned by the LSVP.

By 1981 the Lower Swansea Valley had been designated the UK's first Enterprise Zone (Evans 1981). These areas where wider planning and taxation policy was in abeyance were 'arguably the neoliberal city's purest policy expression' (Wetherell 2016). By 1983 fifteen Enterprise Zones had been announced and the policy had been adopted by the Reagan administration. The combination of Swansea Council and University College Swansea (as Swansea University was then known) monitoring the Enterprise Zone meant that the economic impact of the Zone was understood in better detail and for longer. So that while the national scheme was declared a success by 1984, the Swansea Enterprise Zone was shown to have been only functioning because of significant government interventions (Bromley and Thomas 1988).

This approach to 'stimulating growth' through deregulation is back in favour at the time of writing in the UK, but by the early 1990s, the 10-year life span of the policy had ended. The Conservative government of John Major moved back towards planning and regeneration based on diversification and the service economy. When Tony Blair led Labour back into government in 1997 Swansea was still looking for a model of post-industrial economic development. Devolution and the creation of the National Assembly of Wales in 1998 established both greater autonomy and greater responsibility for economic development in Wales, and tourism was central to the plans of the new millennium with both policy and funding to drive this new model (Dicks 2019, Price 2021). Heritage became a key component in this economic model, often referred to as heritage-led regeneration (Jones and Munday 2001).

While economic development and heritage had complementary goals, they had contrasting needs in the development of the former industrial landscape. Flynn discusses the process in the Lower Swansea Valley where tourism and recreation are seen as the innovative industries that heritage can support (Flynn 2016). While the Enterprise Zone needed only unregulated space for its economic model, tourism and recreation require pleasant vistas. As Price points out, 'To generate demand, sites must translate the meaning of industrial sites and processes in an entertaining way' (Price 2021, 437). In practice, this translates to interpretation plans which make no reference to the harms of industry only the narratives of innovation and 'entrepreneurial flair' (Cadw 2011 quoted in Price 2021, 446). Similar processes of economic development have been described in Michigan's Mesabi Range in the US where:

> the process of reclamation and remediation involved cleaning up the extensive environmental legacies left over from a century of mining, including the widespread removal of abandoned or derelict mine buildings, the revegetation of mine waste piles, and the stocking of abandoned and flooded mine pits with fish favored by anglers.
>
> (Baeten 2020, 90)

A primary aspect of the use of heritage to drive tourism as an industry is the idea of distinctiveness – the past as something which creates a unique brand. Although industrial heritage is seen as an antidote to 'outdated' visions of Wales associated with 'castles, costumers, and coracles' (WTB 2000 p. 81 (Quoted in Price 2021), the details of that past can be less important than the branding. This is evident in the place branding of the Penderyn Distillery, which uses the heritage of Wales as part of a distinctive offering for their relatively new whisky which is developing a position in the international market (Morrish et al. 2011). The Distillery was originally based in the village of Penderyn but they have agreed to develop a visitor centre, with a distillery tour, in Swansea at the Hafod-Morfa copperworks, the site of the few remaining standing buildings associated with the industry.

> The work has been made possible thanks to a £3.75m grant from The National Lottery Heritage Fund in Wales, which Swansea Council bid with partners Swansea University and Penderyn Whisky. Construction began in summer 2020 and is part of the council's £1bn regeneration programme the authority hopes will lead the way out of pandemic.
>
> (Williams 2022)

It is yet to be seen whether the heritage of copper will be toxic for the Penderyn brand, or whether the heritage will be reclaimed through association with a successful tourism and luxury industry (Figure 5.2).

Politics of heritage projects

As Dicks has argued, industrial heritage has been central to the national heritage narrative of Wales though this has not always been an easy heroic narrative because inter-regional competition pulled against national narratives. Discussing the debates in Welsh history writing traditions about the relationship between rural and urban nationalisms she observes 'Practices of industrial heritage-making in Wales developed in dialogue with these debates, and, by erecting material forms of the past, helped to concretise them' (Dicks 2019, 73). These politics led to the creation of the Waterfront Museum – a national museum of industrial heritage in Swansea, which is not focussed on the industrial heritage of Swansea, but on the national and even global narrative of innovation. Of course, this framing also draws attention away from the local consequences of industrialisation and the toxic legacies it leaves behind. The museum has a very strong community engagement programme, but this is rarely focussed on local industrial legacies.

Throughout the process of redeveloping post-industrial Swansea, the idea and value of 'the past' has shifted, and those changes are revealing the politics of heritage. In the 1960s and 1970s for LSVP, the past was something to be rid of. When discussing clearance in the valley, some residents were concerned that clearance would include their houses, but only one person expressed concern for the historic buildings (Stacey 1962, 154). When discussing the value of the project and local acceptance of it, the Territorial Army 'knocking over ruins' was listed as one of the things that had improved public relations (Stacey 1966 185). This demolition

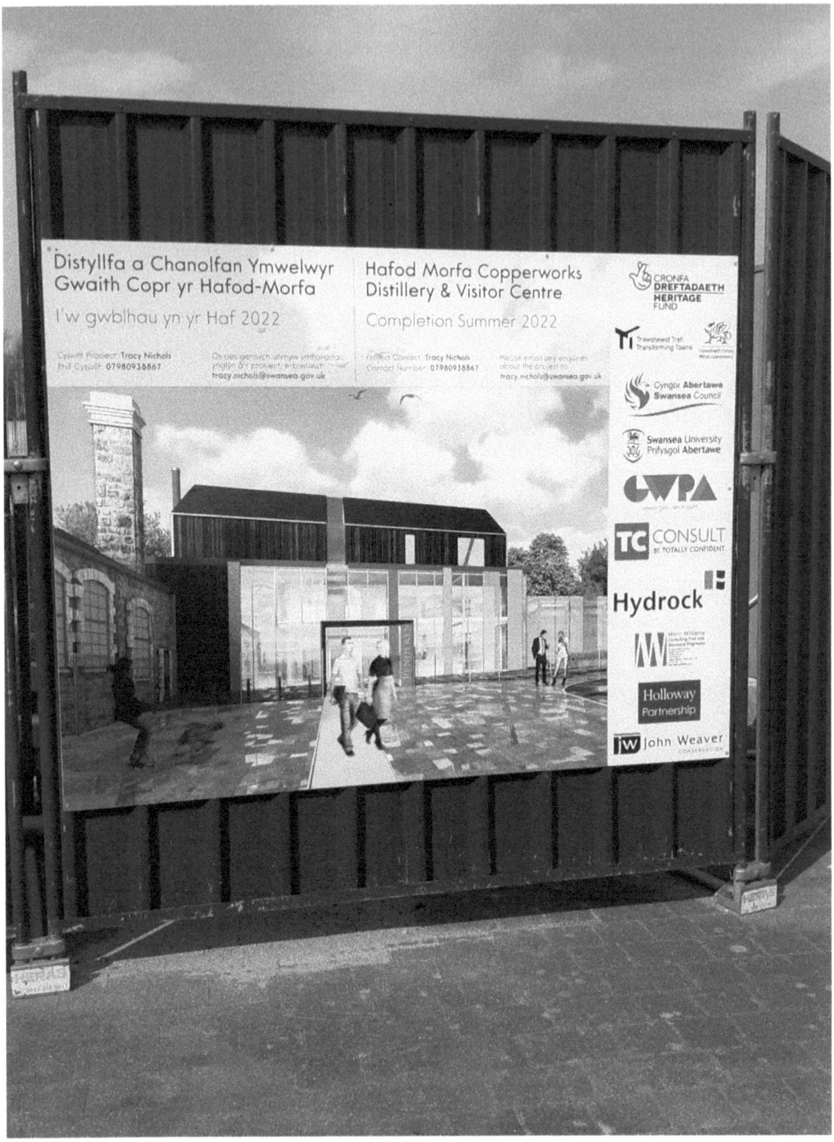

FIGURE 5.2 The artist's impression of the new distillery visitor centre imagines a clean and fresh future, supported by a range of funders. ©Sarah May, 2022.

prepared both the literal and the metaphorical ground for the Enterprise Zone, in the plans for which the earlier use of the land was not mentioned at all.

With the new millennium, however, heritage-led regeneration required 'material forms of the past' which would support the economy. The construction of these material forms of the past required projects in heritage research, management and interpretation. In the 1990s to 2000s 'entrepreneurial governance' emphasised partnership and regions bidding for central cultural funds in a similar fashion to other economic supports. In the late 1990s, EU structural funds were introduced including 'Objective 1' funding for disadvantaged areas. So the negative consequences of deindustrialisation were recognised in heritage funding, if not in the narratives it produced.

These funds encouraged bidding partnerships. By 2013, a new partnership between the Swansea Council and the University envisioned a 'long-term plan … to create a vibrant, multi-purpose place for work, education, leisure, and commercial activity in and around this hugely important industrial heritage site and ensure that the site plays a central role in Swansea's future as it has in the city's past' (Welsh Copper n.d.).

These bidding partnerships also connected groups with diverse relationships with the site: the University of Swansea, local and national government bodies, 'Friends' groups and volunteers. The politics of partnership can blur the responsibilities between groups. Despite the large figures mentioned in the story of Penderyn Whisky described above, most of this work has been done by a series of small to medium projects with funds well under £250,000. These projects combine historical research, community engagement (especially focussed on schoolchildren), archaeological investigation and artistic practice (e.g. CHART n.d.). The nature of the funding produces projects which do more to share the existing narratives than to produce new ones.

Heritage engagement with Swansea copper has been almost entirely focussed on the time of the functioning industry. Although the interpretation has included information about the toxicity of the industry – its reclamation and ongoing legacy are barely a footnote. Writing in 2011, Goskar says 'More than 20 years of tip clearance, demolition and tree-planting has resulted in the return of a tranquil valley and clean River Tawe. The last Swansea works shut in 1981, and now only vestiges of the area's global copper heritage are left'. (Goskar, 2011, 17). Other authors have described the post-industrial period as abandonment and erasure. Writing in 2013 heritage professionals working on the Hafod Morfa copperworks in the heart of the valley declared that it had been 'completely abandoned and left exposed to vandals and the elements for over 30 years' (Betsworth et al. 2014). They saw their work as recovering a lost past for community benefit. Rhodri Morgan described the closure of the Welsh Industrial and Mining Museum as an erasure 'It extinguishes the memory of what made Wales such a powerful force in the industrialisation of the world … not only have railways, mines and iron works been closed in Wales, but we have now closed the museum that commemorates them' (*Independent* 1 June 1998 quoted in Dicks 2019) The heritagisation of the site, in valorising the industrial past through the now well-established tropes of pride and community may erase even the memory of the toxic legacy (Figure 5.3).

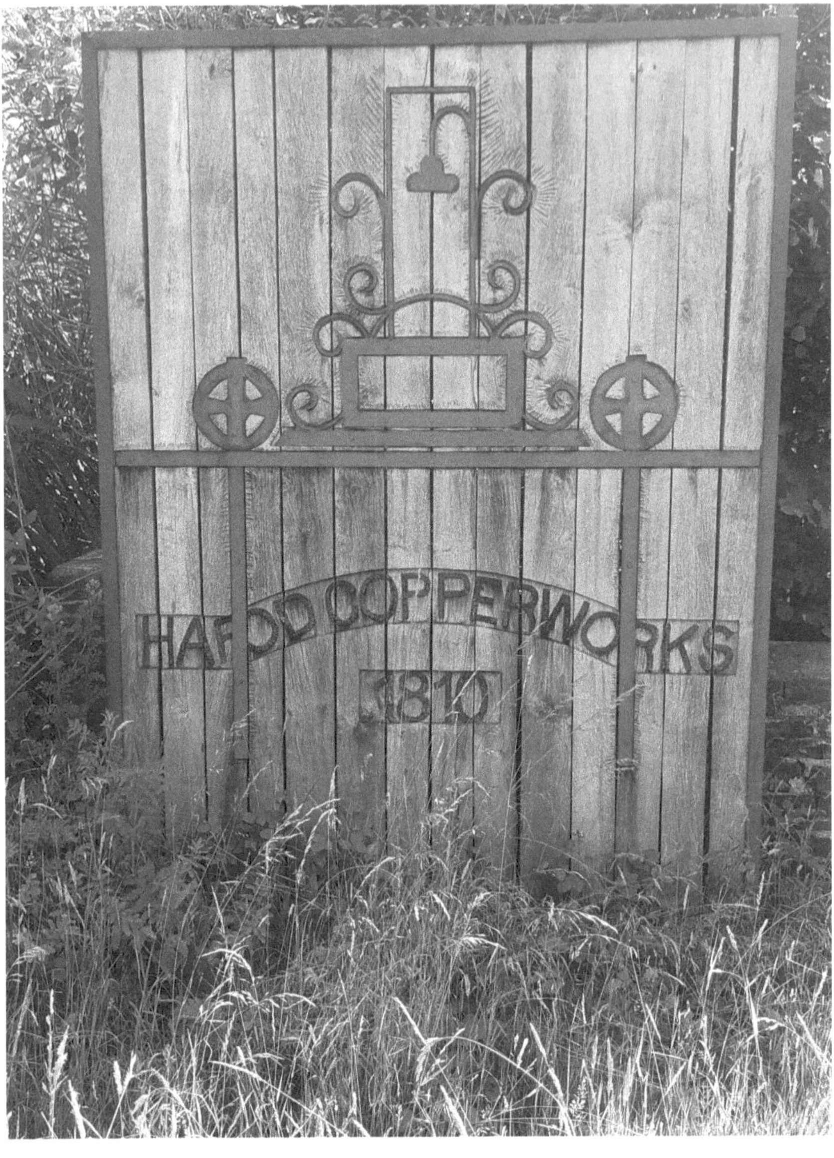

FIGURE 5.3 Heritage interpretation valorising the 19th-century heyday of the copper industry, fading now.

Tracing the post-industrial development of the valley contextualises the industrial past and its complex legacies. A contamination GIS commissioned in the 1990s mapped contamination from records of cultural activity but treated those sources as technical rather than cultural documents (Power et al. 1995, Arup 1997). The reclamation and heritage activities are not mapped – despite the source-pathway-receptor model being used. An update to this, a Historic GIS which includes the subsequent interventions would provide an opportunity to understand these issues in greater depth – for scholars and for citizens.

Heritage, disaster, decontamination

It has been 50 years since the last mills shut in the Swansea Valley and over 100 years since the copper industry went into decline. The landscape has not been frozen at the moment of loss, and nor has it simply been 'recovering'. The technical cultural and heritage interventions continue to make use of that landscape for political purposes which change over time. The story of toxicity in Swansea didn't stop when the metal processing industries collapsed. Subsequent political decisions and their expression through heritage policy and projects are part of how toxicity is managed and experienced.

The initial phase was dominated by the LSVP, led largely by bioscientists at Swansea University, this project aimed to 'deal with dereliction' It was focussed on the decontamination of heavy metals, clearance, stabilisation and the establishment of plant cover on abandoned industrial land. In this phase, the industrial past was framed as a problem to be dealt with. Taking down ruined buildings was a heroic act. The next phase was the Free Enterprise Zone, which decontaminated the economic legacy of the industrial past by linking the area to Thatcherite economic strategy. In this phase, the industrial past was sealed (denied, forgotten?); the focus was on building a future not dependent on that past. Following this, there came heritage-led regeneration, a strategy to revitalise the links between the city and the former industrial area by framing it as an asset, an anchor for the community. In this phase, the industrial past became an asset, the cornerstone of placemaking. Most recently heritage-led regeneration has entered a new phase more closely linked to private enterprise through a partnership with Penderyn distillery which is building a visitor centre in some of the remaining standing buildings. In this phase, the industrial past is a metaphor, a stage and an aesthetic asset.

This description of four phases could be taken to argue that heritage has played an important role in recovering from de-industrialisation and indeed that Swansea has succeeded in overcoming the disaster which it faced in the 1950s. Pasquinelli has recently explored the role of heritage in disaster recovery. Discussing recovery from earthquakes in Italy, she proposes that through heritage practice temporality is rearranged so that the past is defined in relation to the disaster and the 'natural order' is toward reconstruction. The disaster represents a temporal rupture and she suggests three stages to repair it, Dark Heritage, Reconstruction Heritage and Restored Heritage. 'Restored heritage' in which the pre-disaster past is stabilised and valorised is at the end of a linear 'heritage management lifecycle' – while Dark Heritage and Reconstruction heritage – which focus on the disaster precede this stable state (Pasquinelli 2016, 3). The development of heritage narratives in the Lower Swansea Valley follows this pattern. The LSVP and even the Enterprise

Zone focussed on the disaster of industrial collapse and so represent 'Dark Heritage' and 'Reconstruction Heritage'. The two phases of heritage-led regeneration valorise the stable past of industrial prosperity and represent 'Restored Heritage'.

I suggest that Pasquinelli's stages are not naturalised positions related to the passage of time, but political positions, reflecting choices and negotiation. Through the framing of Dark Heritage, the LSVP used the ruined buildings and unmanaged tips to identify deindustrialisation as a disaster to be overcome. The firm focus on the present of the Enterprise Zone moved the disaster into the recent past through the frame of Reconstruction Heritage. Heritage-led regeneration uses Restored Heritage to draw attention away from the disaster linking the industrial past to the post-industrial present without reference to the process of deindustrialisation.

But if the disaster is contamination, not deindustrialisation, toxicity not dereliction, the politics of heritage practice are different. If the disaster is ongoing, can a stable state be reached? Toxic heritage as a frame suggests that the disaster is the contamination – and while that persists the rupture is ongoing. These positions have political consequences and engage different political actors. Heritage has allied itself to economic politics but narratives would be different if it were allied to climate or public health politics. Heritage practice can support the political processes required to face these challenges as effectively as it has engaged economic change; but only if it establishes partnerships with people working on those issues.

If the memory of the copper industry is sanitised, the legacy of contamination can be forgotten. However, the past has not been removed. Like tips used for landscaping, it has been 'sealed' by a cover which allows growth, but the toxic heritage persists.

Bibliography

Arup. 1997. *Swansea-Llanelli earth science information for planning and development.* DoE/Welsh Office research report.

Baeten, J. 2020. "Contamination as artifact: Waste and the presence of absence at the Trout Lake concentrator, Coleraine, Minnesota." In M. A. Rhodes II, William R. Price & A. Walker (Eds.), *Geographies of Post-Industrial Place, Memory, and Heritage*, 85–102. Abingdon: Routledge.

Beatty, C., S. Fothergill & R. Powell. 2007. "Twenty years on: has the economy of the UK coalfields recovered?" *Environment and Planning A* 39(7), 1654–1675.

Betsworth, L., H. Bowen & S. Robinson. 2014. "Performative technologies for heritage site regeneration." *Personal and Ubiquitous Computing* 18, 1631–1650. 10.1007/s00779-014-0766-3

Bridges, E. M. & H. Morgan. 1999. "Dereliction and pollution." In R. A. Griffiths (Ed.), *The city of Swansea: Challenges and change*, 270–291. Swansea: University College of Swansea.

Bromley, Rosemary D. F. & G. Humphrys (eds.) 1979. *Dealing with Dereliction: The Redevelopment of the Lower Swansea Valley*. Swansea: University College of Swansea.

Bromley, Rosemary D. F. & Colin J. Thomas. 1988. "Retail Parks: Spatial and Functional Integration of Retail Units in the Swansea Enterprise Zone." *Transactions of the Institute of British Geographers* 13(1), 4–18. 10.2307/622771

Cadw. 2011. Ysgogiad/driving forces. Wales – The first industrial nation. Pan-Wales heritage interpretation plan. https://cadw.gov.wales/sites/default/files/2019-04/First_Industrial_Nation.pdf

CHART n.d. "Public Heritage and Community Regeneration." https://chart.swansea.ac.uk/projects/public-heritage-and-community-regeneration-project/ [accessed 20/11/2022]

Davies. R. E. 1966. *Report on the preliminary investigation to determined the feasibility of creating an artificial lake in the Lower Swansea Valley Project area.* Swansea: Lower Swansea Valley Project.

Dicks, B. 2019. "Industrial Heritage as Place Making." In S. Berger (Ed.), *Constructing Industrial Pasts: Industrial Heritage Making in Britain, the West and Post-Socialist Countries*, 68–80. Oxford: Berghahn Books.

Evans, R. W. 1981. "The Lower Swansea Valley Enterprise Zone." *Built Environment* 7(1), 20–30.

Evans, C. & L. Miskell. 2020. *Swansea Copper: A Global History.* Baltimore: Johns Hopkins University Press.

Flynn, K. E. 2016. "Welsh Heritage and Cultural Tourism: Engendering Community Regeneration and Environmental Sustainability in the Lower Swansea Valley." In: Hooper, G. (Ed.), *Heritage and Tourism in Britain and Ireland.* London: Palgrave Macmillan. 10.1057/978-1-137-52083-8_7

Goskar, T. 2011. "Pioneers and profits." *Materials World* 19(10), 15–17.

Holt, G. 1966. *Tips and tip working in the lower Swansea valley.* Swansea: Lower Swansea Valley Project.

Jones, C. & M. Munday. 2001. "Blaenavon and United Nations World Heritage Site Status: Is Conservation of Industrial Heritage a Road to Local Economic Development?" *Regional Studies* 35(6), 585–590. 10.1080/00343400120065741

Morrish, Sussie C. & Jonathan H. Deacon. 2011. "A Tale of Two Spirits: Entrepreneurial Marketing at 42Below Vodka and Penderyn Whisky." *Journal of Small Business & Entrepreneurship* 24(1), 113–124. 10.1080/08276331.2011.10593529

Pasquinelli, C. 2016. "Dark, Reconstruction and Restored Heritage: Models of Cultural Heritage and Tourism Management in Post-Disaster Contexts." *ICOT2016 Conference Proceedings.*

Power, S., M. Scott, G. Robinson & I. Statham. 1995. "Database design and data management on the Swansea-Llanelli earth science mapping project." *Geological Society, London, Special Publications* 97(1), 145–155.

Price, William R. 2021. "Mining for Tourists: An Analysis of Industrial Tourism and Government Policy in Wales." *Tourism Planning & Development* 18(4), 435–456. 10.1080/21568316.2020.1829695

Salvo, G. 1989. *Disposal of municipal and toxic wastes in British and Italian law.* [Washington, D.C.: European Law Division, Law Library of Congress] Pdf. https://www.loc.gov/item/2019668871/.

Schilling, K., A. Basu, A. Kaplan & W. Perkins. 2021. "Metal distribution, bioavailability and isotope variations in polluted soils from Lower Swansea Valley, UK." *Environ Geochem Health* 43, 2899–2912. 10.1007/s10653-020-00794-x

Stacey, K. 1962. *The human ecology of the lower Swansea valley.* Swansea: Lower Swansea Valley Project.

Wales Tourist Board (WTB). 2000.Achieving our potential: A tourism strategy for Wales. https://apps.caerphilly.gov.uk/LDP/Examination/PDF/W4-Achieving-Our-Potential-A-Tourism-Strategy-for-Wales.pdf

Walkerdine, V. & L. Jimenez. 2012. *Gender, work and community after de-industrialisation: A psychosocial approach to affect.* London: Palgrave Macmillan.

Waters, C. N., S. J. Price, J. Davies, A. M. Tye, S. E. Brown & D. I. Schofield. 2005. "Urban geology of Swansea: Neath: Port Talbot." In Michael G. Bassett, Valerie K. Deisler & D. Nicol (Eds.), *Urban geology in Wales: 2*, 7–22. Cardiff: National Museum of Wales.

Wetherell, S. 2016. "Duncan Tanner Essay Prize 2015: Freedom Planned: Enterprise Zones and Urban Non-planning in Post-war Britain." *Twentieth Century British History* 27(2), 266–289.

Welsh Copper. n.d. http://www.welshcopper.org.uk/en/about_regeneration.htm [accessed 15/09/22]

Williams, N. 2022. "The big changes at the Hafod-Morfa Copperworks site that's due to become a new visitor attraction for Penderyn Whisky." *Wales Online* January 26 2022 https://www.walesonline.co.uk/news/wales-news/big-changes-hafod-morfa-copperworks-22880088 [accessed 13/10/22]

Wollentz, G., S. May, C. Holtorf & A. Högberg. 2020. "Toxic heritage: Uncertain and unsafe." In R. Harrison, C. DeSilvey, C. Holtorf, S. Macdonald, N. Bartolini, E. Breithoff, H. Fredheim et. al. (Eds.), *Heritage Futures*, 294–312. London: UCL Press.

CASE STUDY 1

GHOST WRECKS OF THE ANTHROPOCENE: AN ENDURING TOXIC LEGACY OF THE PACIFIC WAR

Matthew Carter, Ashley Meredith, Augustine C. Kohler, Ranger Walter, Bill Jeffery, and Paul Heersink

During the Second World War, over 3,800 ships containing as many as 1.5 billion gallons of petrochemicals, and hundreds of thousands of tons of explosive ordnance were sunk throughout the Asia-Pacific Region (Figure 1) (Michel et al., 2005). Over the last 77+ years, these potentially polluting wrecks (PPW) have deteriorated with the threat of oil spills increasing as their corrosion leads inexorably towards structural collapse (Macleod, Selman, & Selman, 2017). These wrecks and their cargoes can be seen as markers of the 'Anthropocene', the concept that through human-induced changes (including in the chemical composition of the atmosphere, oceans and soil), Earth has entered a new epoch in its geological history (Zalasiewicz et al., 2021). While the Anthropocene is a useful term for conceiving the magnitude of the impacts humans have had on the planet, today the toxic legacy of these wrecks is a very real threat to marine ecosystems, cultures and livelihoods across the Blue Pacific (SPREP, 2019).

Chuuk Lagoon of the Federated States of Micronesia (FSM), is the largest lagoon in Micronesia, and has the highest concentration of 'high-environmental risk' PPW in the Pacific Region with at least 50 WWII shipwrecks and numerous aircraft located there (Carter et al., 2021; Jeffery, 2004). Significantly, these potentially polluting wrecks form a tremendous wealth of underwater cultural heritage for Chuuk and also serve the Chuukese people who subsist on the marine resources from the lagoon and revere its species for their clan connections. The underwater cultural heritage and reef biodiversity of the lagoon attract many local and international visitors on an annual basis, including those from Japan who venerate the wrecks as the war graves of their countrymen (Figure 2). Significantly, however, both the cultural and natural resources of Chuuk Lagoon are under threat for a variety of reasons, including climate change, deterioration of the material culture, development and the ever-increasing risk of oil spills from the wrecks.

DOI: 10.4324/9781003365259-10

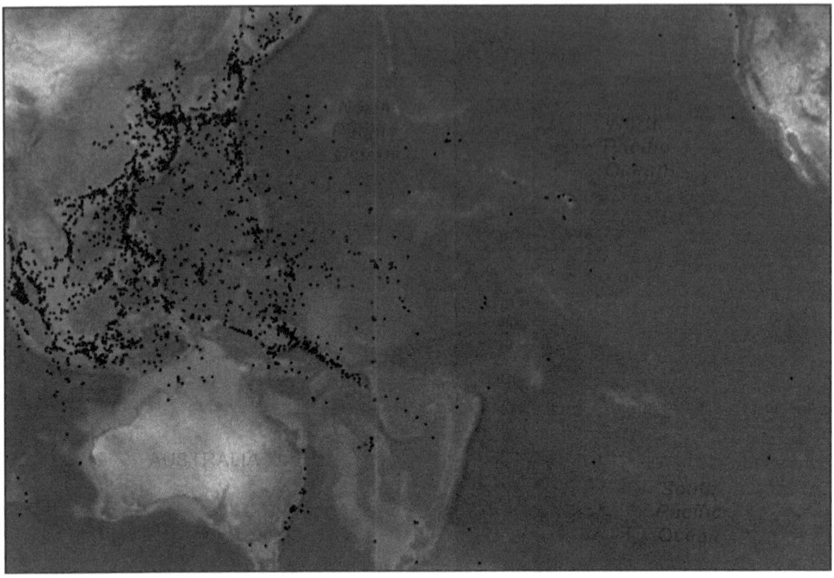

FIGURE 1 Map showing ships lost during WWII across the Asia Pacific Region. Attribution: © Paul Heersink, 2022.

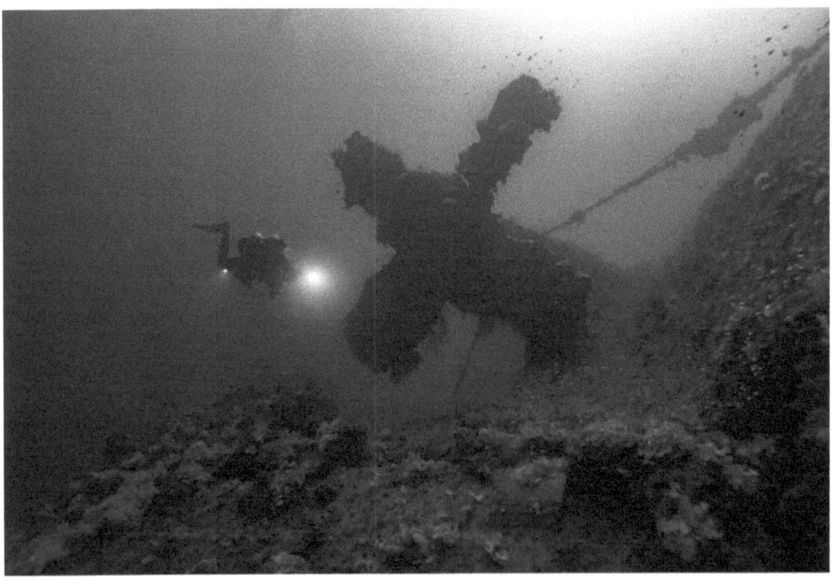

FIGURE 2 The wrecks in Chuuk play an important role as cultural tourism assets, but also pose a threat to the marine environment central to Chuukese life.

War has left another legacy in Chuuk that is as toxic as the oil pollution from the shipwrecks. Many Pacific regions where WWII battles were fought are home to local people who were innocent bystanders in someone else's war. In Chuuk, they were bombed by US aircraft for nearly 18 months, during which time about 1,000 Chuukese were killed. Chuukese helped build the bunkers for Japanese to shelter from the bombing, but were not allowed to use them, they had to dig additional small caves for their own shelter (Aisek cited in Lindemann, 1982, p. 172). With the 18-month blockade of supplies into Chuuk, starvation, malnutrition, and disease spread throughout the 10,000 Chuukese and 40,000 Japanese. The bombing not only sank many ships, but it also destroyed hundreds of the Japanese land-based facilities, as well as many of Chuuk's traditional cultural heritage sites and made much of the usable land for farming unproductive. This legacy has had a lasting impact on Chuukese. They regard land-based sites as a reminder of the death and suffering of their parents/grandparents, in contrast to the shipwrecks, which are a legacy of when the war came to Chuuk, important in Chuukese history, but not seen as heritage sites with the same "sense of place" (Jeffery, 2007, p. 229).

Today the management of Chuuk Lagoon is primarily under the jurisdiction of Chuuk State and is of FSM National concern due to its large marine area and connection to the larger body of water that makes up the FSM. The interconnected and sometimes competing values of the shipwrecks in the Lagoon present authorities with management challenges that cross boundaries both between Departments, and the natural and cultural worlds.

As artificial reefs, the wrecks are home to corals and other marine organisms that provide habitat for organisms and fish taxa that are essential for the Chuukese indigenous way of life. It is estimated that there are more than 300 types of corals in the waters of the FSM and an unknown diversity of octocorals (soft corals and gorgonians) (Rowley et al., 2019). Research to identify and quantify coral species throughout this vulnerable marine environment, their health, and how they contribute to the biodiversity of the Chuuk Lagoon remains to be completed.

The wrecks in Chuuk Lagoon are also important economic resources for the burgeoning dive tourist industry. Prior to the Covid-19 pandemic, Chuuk received 2,768 visitors in 2016 (FSM Statistics Office). These visitors predominantly come from the US, Australia and Europe and have an interest in wreck diving, valuing these sites more for their associations with WWII, than for their nature. In contrast, the wrecks hold particular sensitivity for the smaller number of Japanese tourists who come to Chuuk as a pilgrimage to visit the wrecks as the war graves of their countrymen.

The FSM and Chuuk State Governments recognize that the wrecks in Chuuk Lagoon are of local, national, and international significance for both their natural and cultural values. Significantly, while a number of Chuuk State agencies as well as foreign governments and non-profits contribute to removing oil from the wrecks, the removal of these pollutants occurs slowly due to available human labor and limited resourcing despite recognition of the threat they pose to Chuukese cultural heritage, subsistence way of life and the natural environment.

The wrecks of Chuuk Lagoon present a range of conflicting values; from tourism assets and war graves, to legacies of colonialism and urgent environmental threats. These values present challenging dimensions to the management of these sites for international, national and state authorities. As markers of the Anthropocene these wrecks provide an ongoing case study of how the toxic legacies of the Second World War continue to impact the Blue Pacific.

Bibliography

Carter, M., Goodsir, F., Cundall, P., Devlin, M., Fuller, S., Jeffery, B., & Talouli, A. (2021). Ticking Ecological Time Bombs: Risk Reduction and Remediation of Oil Polluting World War II shipwrecks in Chuuk Lagoon. *Marine Pollution Bulletin*, *164*(Article 112087), 1–9.

Higuchi, W. (n.d.). *War Reparations in Micronesia and Japan's Responsibility*. Micronesian Area Research Center, University of Guam.

Jeffery, B. (2004). World War II Underwater Cultural Heritage Sites in Truk Lagoon: Considering a Case for World Heritage Listing. *International Journal of Nautical Archaeology*, *33*(1), 106–121. 10.1111/j.1095-9270.2004.00009.x

Jeffery, B. (2007). *War graves, munition dumps and pleasure grounds: A post-colonial perspective of Chuuk Lagoon's submerged WW II sites*. Doctoral dissertation, Townsville: James Cook University.

Lindemann, K. P. (1982). *Hailstorm over Truk Lagoon*. Maruzen Asia.

Macleod, I. D., Selman, A., & Selman, C. (2017). Assessing the Impact of Typhoons on Historic Iron Shipwrecks in Chuuk Lagoon Through Changes in the Corrosion Microenvironment. *Conservation and Management of Archaeological Sites*, *19*(4), 269–287. 10.1080/13505033.2017.1378536

Michel, J., Gilbert, T., Etkin, D. S., Urban, R., Waldron, J., & Blocksidge, C. T. (2005). *Potentially polluting wrecks in marine waters. An issue paper prepared for the 2005 international oil spill conference*. Paper presented at the International oil spill conference.

Rowley, S. J., Roberts, T. E., Coleman, R. R., Spalding, H. L., Joseph, E., & Dorricott, M. K. (2019). Pohnpei, Federated States of Micronesia. In *Mesophotic coral ecosystems* (pp. 301–320). New York: Springer.

SPREP. (2019). *SPREP Environment Ministers and High-Level Representatives Talanoa Communique 2019*. Retrieved from Apia, Samoa:

Zalasiewicz, J., Waters, C. N., Ellis, E. C., Head, M. J., Vidas, D., Steffen, W., & Zinke, J. (2021). The Anthropocene: Comparing Its Meaning in Geology (Chronostratigraphy) with Conceptual Approaches Arising in Other Disciplines. *Earth's Future*, *9*(3). 10.1029/2020EF001896

6

MILITARY LEGACIES AND INDIGENOUS HERITAGE IN CANADA'S NEWEST NATIONAL PARK RESERVE

Lisa K. Rankin, Julia Brenan, David M. Finch, Scott Neilsen, and Anatolijs Venovcevs

Introduction

In 2015, the Canadian government announced the creation of the Akami-Uapishk^{u}-KakKasuaak-Mealy Mountain National Park Reserve (MMNPR), comprising nearly 11,000 km^2 on the central coast of Labrador. The intention was to protect the natural landscape and safeguard a 7,000-year history of Indigenous heritage through co-management with local Innu and Inuit populations whose settlements buffer the park borders (Figure 6.1). Ostensibly, these intentions are cause for celebration, demonstrating Canada's commitment to ecological and cultural preservation while addressing reconciliation with Indigenous communities. However, this fails to acknowledge the region's darker heritage whereby Innu and Inuit were coerced from their traditional territories within park boundaries over the course of the 20th century by colonial military activities and resettlement policies. Given that the park exists as an illusion of uninhabited space, it is impossible to separate regional depopulation and colonial control from the creation of the park. In effect, the establishment of the park is a further act of colonial domination, not only over geographical space but also over the Indigenous history and heritage it contains. Archaeological surveys undertaken in the early 2000s on the Porcupine Strand, a scenic sand dune beach along the eastern edge of the park, speak directly to this process. The work recorded over 100 Indigenous archaeological sites, demonstrating the vibrance of land use and settlement between deglaciation and the 20th century in just one small part of the park. The archaeology also correlates the decline of 20th-century Indigenous settlement with an increase in state-sanctioned military activities as the munitions and equipment of war replace Indigenous material culture. Local residents introduced us to the history of the NORAMEX wargames: amphibious exercises held on the Strand by Canadian and American troops in 1949 and 1952 – the infancy of the Cold War.

DOI: 10.4324/9781003365259-11

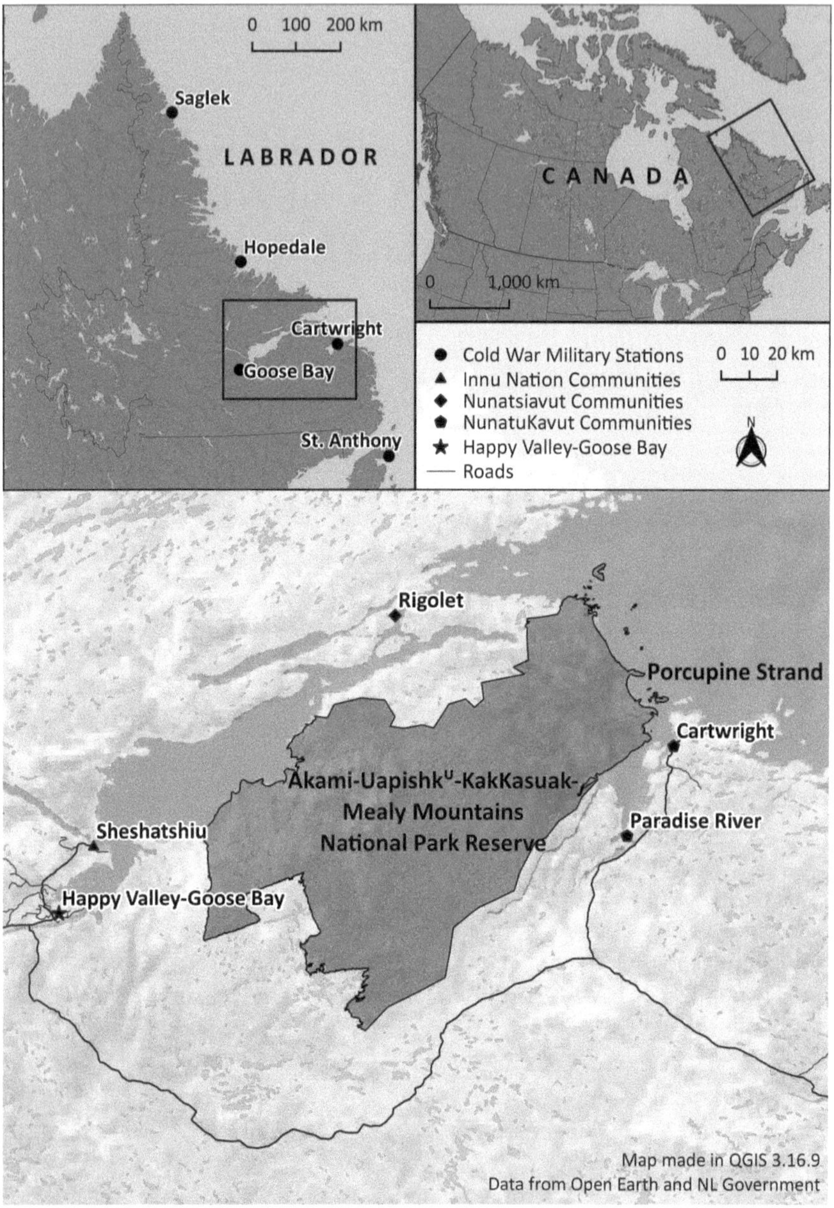

FIGURE 6.1 Mealy Mountains National Park Reserve in local, regional and national context.

The remains of these games encroached on Indigenous archaeological sites and deposited toxic material on the landscape, impacting the ecosystem of the Strand and posing physical risks for both contemporary Indigenous use and visitors.

The MMNPR, therefore, represents much more than a federal exercise in landscape protection and reconciliation. It encapsulates the history of the Polemocene – an age of permanent war readiness that drove the policies of colonial states through the 20th century – as it was played out in the Canadian north. This timescape contains an alternative history of disrupted Indigenous lives, and alienation from their lands, leaving in its wake a potentially hazardous heritage. This draws into question the motivations for park development and whether parks are created to hide (or ignore) certain parts of the past or to reconcile with it. Archaeology, history, and memory are used below to situate the toxic material inside the park in a long-term context and interpret the entanglement of colonialism, military activity and park establishment as well as the impacts on contemporary Innu and Inuit communities. Although the park might be seen as yet another form of Indigenous colonisation, we suggest that in creating the park, Canada may have inadvertently created an opportunity for real reconciliation with Innu and Inuit in Labrador and that there is real value in confronting the history of colonialism and its legacy of cultural and ecological harm.

Background

Parks Canada, the federal agency tasked with protecting natural and cultural heritage, first indicated an interest in creating the MMNPR in the 1970s during an expansion of the national park system (Parks Canada [National Parks Service] 1971). Labradorians initially resisted fearing it was just another exercise in colonial disenfranchisement. The transition of the Mealy Mountain region into a federal park did nothing to address unsettled Indigenous land claims and would infringe on the limited subsistence harvesting Innu and Inuit still relied on. Indigenous distrust of external decision-makers was well-entrenched (Bill 1982; Parks Canada 1977). By 1980, plans for the park were shelved, but in 2001, with land claims advancing, Parks Canada and the province of Newfoundland and Labrador tried again. Following years of negotiation, the MMNPR was established in 2015 (Parks Canada 2018). Today, the MMNPR is co-managed by Parks Canada and three Indigenous administrations with ancestral connections to the region. The Innu Nation, Nunatsiavut Government and NunatuKavut Community Council engage with Parks Canada via cooperative management boards. This model preserves Indigenous access to the park and provides pathways to participation in park management and tourism development. At the same time, the underlying framework (and the park's very existence) are imposed by provincial and federal policymakers with their own agendas founded in colonial-era assumptions of control of both the region and its inhabitants.

Independent of park establishment, between 2001 and 2004 Rankin conducted an archaeological survey of the Porcupine Strand, an area that falls within the traditional territories of all three groups, and recorded 106 Indigenous archaeological sites (Rankin 2002a, 2003). Concentrations of First Nations sites were recorded in the southern and northern reaches of the Strand (19 sites and 42 sites, respectively); one Inuit site was recorded at Cape Porcupine, and 20 more on adjacent islands. Most of the First Nations sites date to the period between 3500–2800 years ago, with some as old as 7000 years. The Innu consider the people who occupied these sites as ancestors and draw on hundreds of Innu toponyms and travel routes associated with hunting, fishing and trading within the MMNPR area as evidence of their ongoing connections to the territory (Pepamuteiati nitassinat 2008). Inuit are relative newcomers to the Porcupine Strand, arriving by the 17th century (Rankin 2015). Inuit presence is manifested today in the lives of Nunatsiavut and NunatuKavut Inuit who reside in communities bordering the MMNPR and continue to use Park lands for subsistence and recreation.

Although Innu and Inuit had travelled south to meet with European fishers and whalers since the 16th century (Rankin et al. 2015), direct colonial influence in the region began with the arrival of British traders in the late 18th century. They and the settlers who followed in the 19th and 20th centuries affected local Indigenous lifeways. By the 19th century, most Inuit were engaged in the fishing and trapping industries, and many lived in small fishing villages along the Strand during the summer (seven of these villages were recorded), shifting to more solitary winter homes elsewhere on the Strand to pursue trapping. Innu remained nomadic until the mid-20th century, travelling to the Strand to fish and trade. Early colonial activities impacted the rhythm of Innu and Inuit lifeways but allowed considerable autonomy that did not sever their connections to the land (Rankin et al. 2012).

This changed in the 20th century as successive colonial governments, entangled in wars abroad, tested and rehearsed their military capacity in the Canadian north. Not coincidentally, this also allowed colonial governments to firmly establish their political sovereignty in Labrador under the guise of protection - particularly after Newfoundland and Labrador, a dominion of the United Kingdom, joined Canada in 1949. This militarism contributed directly and indirectly to the depopulation of the Porcupine Strand between World War I and the early Cold War and led to the deposition of toxic military materials. Like other regions of northern North America, colonial policies of militarism had significant consequences for Indigenous communities that had no voice in these decisions (Hird 2016; Pegues 2021, 143).

Colonialism and the Military

Over the course of the 20th century, the perpetual state of global militarisation defined as the Polemocene (Antonacci 2021; Reno 2019), influenced the decisions of colonial governments located thousands of kilometres from the Porcupine Strand. Consequences of their policies incrementally affected Innu and Inuit both indirectly and directly, causing deaths, depopulating traditional lands and altering cultural practices. Ultimately these communities, already remote from the seats of power, became so marginalised that governments came to see this region as empty and isolated. This paved the way both to Cold War wargames and to the creation of the MMNPR.

The pre-Cold War effects of military colonialism in central Labrador were indirect, as soldiers returning from World War I brought the Spanish flu. The flu arrived on the Porcupine Strand in November 1918. Deaths began within days, but the colonial government in St. John's (1,500 km away) refused to send assistance so late in the shipping season. Following desperate appeals, one minister said: 'Let 'em die: the government will be saved the trouble of feeding them' (Budgell 2018:157). Indigenous communities were left to cope with the pandemic on their own. In Sandwich Bay, on the southeastern edge of the MMNPR, slightly more than 20% of the 320 inhabitants died (Buckle 2003, 111). On the Porcupine Strand itself, the 2001–2004 fieldwork recorded cabins abandoned following flu deaths (Kelvin 2011). As the flu subsided, an assimilationist residential school was opened for orphans. This removed the children from their homes along the Strand, separating them from extended kin and opportunities to learn traditional life skills (Procter 2020). Faced with poverty, many surviving families moved to Cartwright.

World War II further impacted Indigenous settlements on the Strand as Canadian and American militaries constructed infrastructure throughout Labrador to aid Great Britain by moving supplies and personnel by air (Kennedy 2015, 252). In 1941, an airbase was constructed in Goose Bay, while to the south of the Porcupine Strand, Cartwright became part of a network of weather stations and fuel depots supporting coastal reconnaissance. Both installations attracted Innu and Inuit men eager for a regular income (Pace 2008, 42). The wage economy increased economic differences between Indigenous families as some were unable to send men away for work. Ultimately, communities on the Strand and in the MMNPR area became further marginalised, impoverished and isolated (Kennedy 2015, 256).

The peripheralisation of the Strand because of wars half a world away created an illusion of an empty wilderness, and therefore an ideal space for Cold War exercises. In 1948 this 'isolated northern beach' (Rawley 1949) was chosen for a joint Canadian/American amphibious landing exercise intended to advance naval combat readiness for cold weather operations.

Operation NORAMEX took place in October 1949. It began with a reconnaissance of Cape Porcupine by Navy Underwater Demolition Team swimmers (Rawley 1949) and beach strikes by planes launched from an aircraft carrier (White 1949). This was followed by minesweeping, a simulated shore bombardment (White 1949), and an amphibious beach landing by 2,800 marines (Rawley 1949; White 1949). The landing secured a simulated 'enemy' held weather station, airstrip and other outlying installations (White 1949).

NORAMEX II occurred at the north end of the Porcupine Strand, at Sandy Cove, in December 1952 with the same goals (Liles 2010). It included both airborne attacks and assaults by more traditional landing barges (U.S. Department of the Navy, Bureau of Naval Personnel Information 1953; U.S. Department of the Navy, Bureau of Aeronautics 1953).

Considerations were made for safety but were focussed on the exercise itself. Certain areas were designated for bombs and gunfire and safety regulations required the removal of dud ammunition (Canadian Section of the Military Cooperation Committee 1953). Still, 50 years later archaeological surveys recorded unexploded ordnance (UXO) and broken equipment. Remediation has never been fully addressed.

After the NORAMEX operations, military activities shifted from the Porcupine Strand to Cartwright. An American-operated General Surveillance Radar Station was present from 1953 to 1968 as part of the Pine Tree Agreement for Air Defence. During this period, policies of resettlement were formalised as the federal government offered cash incentives to entice the remaining Inuit families from the Strand and into larger centres citing services such as education, medical care and economic opportunities (Kennedy 2015, 265–267). For similar reasons, nomadic Innu were also pressed by both government and church officials to settle into permanent villages. In the 1980s, on the western border of the MMNPR, 5 Wing Goose Bay became a NATO base and interior Labrador was used for low-level flying exercises. This brought new threats to Innu who inhabited bombing ranges, and to the caribou herds central to their subsistence (Chow 1987; Strowbridge 1989; Wadden 2001). The Cartwright Long Range Radar Site replaced the former radar station in 1998. Currently, the Canadian Armed Forces are the only military body in Labrador with the exception of occasional exercises by allied countries. Innu and Inuit use of the Strand again declined following the cod and salmon fishing moratoria of the 1990s, and the primary reason for visits to the Strand became recreation.

The intrusion of American and Canadian militaries onto the Porcupine Strand had both immediate and long-lasting effects on local Indigenous people, their heritage, and the environments in which they lived. While inhabitants complained of the noise, far more disturbing was the failure of the government to notify them that wargames were planned (Davies pers. comm.). Operation planning documents reveal that the Americans were aware of a few Indigenous settlements, mostly far from the Strand, and that warnings would be issued to civilians by Canadian personnel if there was a potential danger (Canadian Section of the Military Cooperation Committee 1953; Lackenbauer and Heidt 2019). However, residents do not remember receiving such warnings. Given that the NORAMEX operations took place in autumn it is possible that people were difficult to locate, as Inuit inhabitants were in the process of 'shifting' from summer to winter settlements along the Strand. It seems more likely that the American and Canadian militaries believed the area to be uninhabited following decades of regional population decline and its 'isolated' setting. Either way, the failure of the military to understand local settlement strategies resulted in the destruction of property and heritage resources, and at least one death. During the 1952 landings, Inuit homes at nearby Plant's Bight were damaged by artillery shelling (Mugford pers. comm.). Flat Waters resident Israel Williams died, supposedly of a heart attack 'brought on by fright' during the operation. His body, curled under the kitchen table, was later recovered by his family (Way pers. comm.).

Between 2001 and 2004 the Porcupine Strand Archaeology Project recorded remnants of the NORAMEX operations at 11 site locations - all located near the site of the 1952 wargames. We recorded machine gun shells, vials of ear plugs, communication cables, razor wire, and blast holes impacting much older Indigenous archaeological sites (Rankin 2002a and b) (Figure 6.2a). In the years since, Inuit visiting the Strand have informed us about many more military artifacts they have recovered or photographed. Jeff Martin, a NunatuKavut beneficiary from Cartwright, has photographed abandoned landing craft, machine parts, bomb casings and campsites (Figure 6.2b and c). In 2008, Martin collected 5 kg of 'unusual rock' that was later identified as dunnite (ammonium picrate), a 20th-century military explosive (CBC News 2008). The Royal Canadian Mounted Police (RCMP) and Department of National Defense (DND) soon became involved, searching the Strand and detonating several stores of the explosive near Cape Porcupine, including one adjacent to an early Inuit winter house. In 2009, DND sent a representative to meet with Rankin's archaeological crew to discuss protocols for further encounters with the explosive. It is well known locally that unexploded ordnance still exists on the Porcupine Strand. Martin recently told us that there is more material visible where he initially collected the dunnite and believes that it regularly washes ashore.[1]

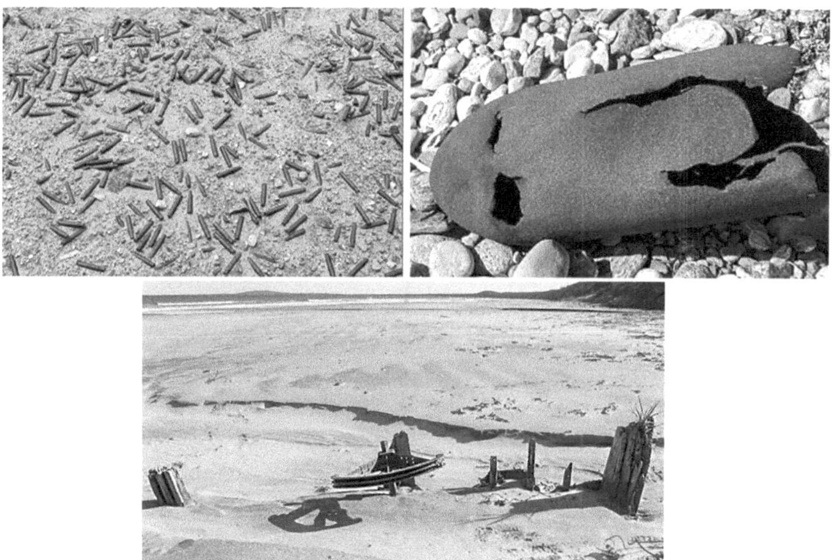

FIGURE 6.2 Remains of Operation Noramex on the Porcupine Stand: a) Machine gun shells atop a Maritime Archaic site; b) Bomb casing, c) Landing vessel.

Clean-up of military infrastructure is an ongoing and costly process for the Canadian government. In 1997 the United States reached a deal with Canada giving Ottawa a $100 million credit to buy arms from the United States over a period of 10 years as compensation for the environmental destruction left behind at former military sites. Yet, the clean-up of the military base in Goose Bay alone is projected to cost $300 million (CAD) (Pugliese 2009). Many of the current and former military sites are considered contaminated or contain UXO and are under assessment or in remediation, including Cape Porcupine. There are no details concerning future remediation or potential health issues that may arise from this kind of contamination.

Local Innu and Inuit were suspicious of Parks Canada's original interest in developing national parks in Labrador in the 1970s, due in part to their prior colonial experiences. Park establishment proposals were set aside until provincial and federal governments began to address the rights and land claims of communities whose history and cultural heritage were entwined in the land. The MMNPR was therefore not established until 2015. At the time of writing, Parks Canada is preparing to celebrate the area's natural and cultural splendours with the world and the MMNPR is co-managed by three Indigenous descendant populations. Yet this is clearly at odds with its history of conflict legacies that emptied the region of permanent inhabitants a mere few decades before the creation of the park reserve. It would seem that the government's agenda is once again based on colonial assumptions of *terra nullius,* of empty and unused land. Little thought seems to have been given to historical injustices or the toxic material abandoned in the reserve. When we asked if they had a strategy to deal with this material, Parks Canada did not have an official response, only that '... [they were] aware that military exercises occurred on the Strand however, further investigation and a deeper understanding of these activities by PC have not yet occurred' (Parks Canada, pers comm.).

Discussion

In this paper, we have framed the dark history and toxic heritage as part of the Polemocene, a contemporary 'age of war' characterised by military profusion through space, material, and time (Antonacci 2021; Reno 2019, 195–201; see also Bonneuil and Fressoz 2016), which is articulated on the Porcupine Strand by human absences and problematic material presences. As outlined above, the deposition of toxic materials by the NORAMEX wargames on the Strand took place in an infinitesimally short moment in relation to the 7,000 years of Indigenous heritage and yet it has a long trajectory of growing entanglement with global military conflict situated within a colonial policy towards the region and its inhabitants.

The recent move to create the MMNPR in this area can thus be seen as part of military environmentalism, where conflict sites can facilitate so-called collateral values – secondary, unexpected benefits for nature-cultural heritage conservation within the scars left by the devastation of war (Lookingbill and Smallwood 2019). This position has been extensively critiqued for greenwashing military spaces through secondary and coincidental benefits while hiding toxic legacies and erasing histories of dispossession and harm to the people most impacted by military developments (Havlick 2007; 2019; Coates et al. 2011). The conflict heritage and its aftermaths on the Porcupine Strand underscore this critique since the creation of the national park has at its heart Polemocene militarism and colonialism.

By entering co-management of the MMNPR with Innu and Inuit communities, Parks Canada has indicated a willingness to address reconciliation for the harm colonialism wrought. But the establishment of a National Park is by its very nature a further stage of colonialism. A 'park' is not a traditional land use. And while the park has been established in consultation with Innu and Inuit, it was not initiated by them. The very idea of the park has been imposed from above, with consultation an afterthought. Fundamentally, the selection of this area as a national park suggests a preference of the federal government to consider this as an empty landscape (for similar scenarios see Reno 2019, 174–178; Valadares 2018). On the contrary, it has very clearly been modified by human agency, both ancient and recent. This alienation is minimally overcome by allowing Innu and Inuit access to the land for traditional activities which may eventually be discouraged through new tourism-based economies. Tourism also brings its own concerns as both ancient and recent archaeological remains are in a dynamic sand dune environment that continually shifts thereby covering and uncovering archaeological objects, including toxic ones, intermittently exposing them to threats by and to visitors.

Finally, the park has done little to address the toxic history of the region or mitigate the dangerous and toxic material found along the Porcupine Strand. In this manner, the codification of the park and the toxic heritage extend the Polemocene impacts on the Porcupine Strand into the future by reinforcing processes of alienation which keep descendent communities from occupying the land in traditional ways in favour of tourism or through the possibility of environmentally complex clean-up efforts of harmful material.

Conclusion

The NORAMEX operations and the initial proposal for the creation of MMNPR in the 1970s both stem from the perception of the area as 'isolated' and largely devoid of people. In contrast, archaeological surveys have highlighted a deep history of Indigenous occupation, but in so doing have exposed the legacies of this disenfranchisement. The scattered, toxic artifacts found on the shores of a new park in northeastern Canada demonstrate that the park's creation is inherently intermingled with militarism and Indigenous erasure under the guise of protection and preparedness. Those same toxic materials could now aid reconciliation with descendant communities by making what was once invisible visible to all, and by remembering and addressing the legacies of dispossession and contamination.

One possible route forward is to constructively employ both the dark history and the military material as a heritage to bring the various stakeholders of the MMNPR together. By using recent heritage within the park boundaries as a discursive object about the area's past, better working relationships may be forged to manage all the area's natural and cultural heritage and to reintroduce Innu and Inuit engagements with the land. This too may be difficult. While technically part of the archaeological Cold War heritage of the park (Hanson 2016), Innu and Inuit do not necessarily view modern military activities within the MMNPR as significant. The Innu Nation, for example, views these resources as contamination that should be cleaned up to alleviate further damage to the cultural and natural resources. Yet for this to happen it may be essential to confront the toxic legacy of the park directly through consultation and remediation. Acknowledging and interpreting Indigenous disenfranchisement in the park may lead to a wider understanding of the effects of militarism and colonialism for both Innu and Inuit communities and the general public, and allow for greater reconciliation. In turn, this could lead to an emphasis on remediation of the toxic materials and encourage further Indigenous engagement with and in the park.

The history of this small corner of Labrador has some implications in considering toxic heritage on a broader scale. It highlights the fact that toxicity takes many forms, is dependent on individual perspectives and takes on different degrees of importance in different political contexts. For a government concerned with the welfare of the area's inhabitants and users, for example, the priority would have to be finding culturally appropriate ways to address the history, disarm the war materials and make the area safe. In contrast, for a government concerned with putting Canada's exceptionalism on display, the toxicity of the war games detritus has no real relevance, since the underlying premise is that the area currently has no inhabitants or users – making it the perfect setting for a park. And yet, there remains a path to real reconciliation to be found within the colonial enterprise. To achieve this the park must make room for alternate histories that cross the boundaries of culture and power and give voice to the dark history as part of its creation narrative.

Acknowledgements

The authors would like to thank their Innu and Inuit partners for kindly sharing their memories and knowledge with us over the years. Thanks also to Jeff Martin and Peter Ramsden for assistance with images. We acknowledge the ongoing support of the Social Sciences and Humanities Council of Canada whose grants made the initial research and this publication possible.

Note

1 Many sites associated with the military in Labrador are now considered contaminated by the Canadian government and have been reported to the Federal Contaminated Sites Inventory (https://www.tbs-sct.gc.ca/fcsi-rscf/home-accueil-eng.aspx). The DND UXO and Legacy Sites program currently lists Cape Porcupine as 'In Assessment' but further details are lacking (https://www.canada.ca/en/department-national-defence/services/uxo/uxo-locations.html).

Works Cited

Antonacci, John Peter. 2021. "Periodizing the Capitalocene as Polemocene." *Journal of World-Systems Research* 27 (2): 439–467. 10.5195/jwsr.2021.1045.

Bill, Richard E. 1982. *Attempts to Establish National Parks in Canada: A Case History in Labrador from 1969 to 1979*. Master's thesis, Carleton University.

Bonneuil, Christophe, and Jean-Baptiste Fressoz. 2016. *The Shock fo the Anthropocene: The Earth, History, and Us*. Translated by David Fernbach. London and Brooklyn: Verso.

Buckle, David V. 2003. *Labrador Diary: 1915–1925, The Gordon Journals*. Cartwright: Anglican Parish of Cartwright.

Budgell, Anne. 2018. *We All Expected to Die: Spanish Influenza in Labrador, 1918–1919*. St. John's: ISER Books.

Canadian Section of the Military Cooperation Committee. 1953. "US Naval Operating Requirement: NORAMEX -54– Cold Weather Amphibious Training Exercise on the Coast of Labrador." Ottawa: Library and Archives Canada.

CBC News. 2008. "Beachcoming Labrador Family Carries Home Wartime Explosive." Last modified September 10, 2008. https://www.cbc.ca/news/canada/newfoundland-labrador/beachcombing-labrador-family-carries-home-wartime-explosive-1.757030

Chow, Thom. 1987. "Low-level Flying Harming Traditional Game of Innu." *Sunday Express* [St. John's]. 22 March, 6.

Coates, Peter, Tim Cole, Marianna Dudley, and Chris Pearson. 2011. "Defending Nation, Defending Nature? Militarized Landscapes and Military Environmentalism in Britain, France, and the United States." *Environmental History* 16 (3): 456–491. 10.1093/envhis/emr038.

Hanson, Todd A. 2016. *The Archaeology of the Cold War*. Gainesville: University Press of Florida.

Havlick, David. 2007. "Logics of Change for Military-to-Wildlife Conversions in the United States." *GeoJournal* 69 (3): 151–164. 10.1007/s10708-007-9086-8

Havlick, David. 2019. "Legitimizing Militarization or Legitimate Conservation? Collateral Value and Landscapes of the Iron Curtain Borderlands." In *Collateral Values: The Natural Capital Created by Landscapes of War*, edited by Todd R. Lookingbill, and Peter D. Smallwood, 189–212. Cham: Springer.

Hird, Myra J. 2016. "The DEW Line and Canada's Arctic Waste: Legacy and Futurity." *The Northern Review* 42: 23–45.

"Joint United States-Canadian Operation 'Noramex', 1949." 1951. *Polar Record* 6 (41): 102. 10.1017/S0032247400041085

Kelvin, Laura. 2011. *The Inuit-Metis of Sandwich Bay: Oral Histories and Archaeology*. MA Thesis, Memorial University of Newfoundland and Labrador.

Kennedy, John C. 2015. *Encounters: An Anthropological History of Southeastern Labrador*. Montreal: McGill-Queen's University Press.

Lackenbauer, P Whitney, and Daniel Heidt. 2019. *The Advisory Committee on Northern Development: Context and Meeting Minutes, 1948–71*. Calgary: Documents on Canadian Arctic Sovereignty and Security (DCASS), Centre for Military and Strategic Studies.

Liles, Christian. 2010. *The Development of Amphibious/Expeditionary Warfare in the United States and the United Kingdom, 1945–1968: A Study in Comparison, Contrast and Compromise*. PhD thesis, King's College, London.

Lookingbill, Todd R., and Peter D. Smallwood (eds.) 2019. *Collateral Values: The Natural Capital Created by Landscapes of War*. Cham: Springer.

MacKenzie, D. 1986. *Inside the Atlantic Triangle: Canada and the Entrance of Newfoundland into Confederation 1939–1949*. Toronto: University of Toronto Press.

Pace, Jessica. 2008. This is Where I Live, but it's Not My Home: Archaeology and Identity in Sandwich Bay, Labrador. MA Thesis, Department of Anthropology and Archaeology. Memorial Univeristy of Newfoundland and Labrador, St. John's.

Parks Canada. 1976. *Preliminary National Park Considerations for Labrador: A Basis for Discussion.* Ottawa: Indian and Northern Affairs, Parks Canada.

Parks Canada [National Parks Service]. 1971. *National Parks System Planning Manual.* Ottawa: National Parks Service, National and Historic Parks Branch.

Parks Canada. 1977. *A Summary Report of the Proceedings of Open Meetings Held in Twelve Labrador Communities with Local Residents and Officials of Parks Canada and the Provincial Department of Tourism to Discuss the Possibility of National Parks in Labrador.* Ottawa: Indian and Northern Affairs, Parks Canada.

Parks Canada. 2018. "Akami-Uapishk^{U}-KakKasuak-Mealy Mountains National Park Reserve - Park management." Last modified March 28, 2018. https://www.pc.gc.ca/en/pn-np/nl/mealy/info/index

Pegues, Julianna Hu. 2021. *Space-Time Colonialism: Alaska's Indigenous and Asian Entanglements.* Chapel Hill: University of North Carolina Press.

"Pepamuteiati nitassinat: As we walk across our land - About Place Names." Innu Nation and Sheshatshiu Innu First Nation, 2008. Accessed October 20, 2022. https://www.innuplaces.ca/aboutPlaceNames.php?lang=en

Procter, Andrea. 2020. *A Long Journey: Residential Schools in Labrador and Newfoundand.* St. John's: ISER Books.

Pugliese, David. 2009. "Defence Department Getting Rready to Clear out of Goose Bay?" *Ottawa Citizen.* July 13, 2009. https://ottawacitizen.com/news/national/defence-watch/defence-department-getting-ready-to-clear-out-of-goose-bay

Rankin, Lisa, Marianne Stopp, and Amanda Crompton (eds). 2015. *Études/Inuit Studies* 39 (1).

Rankin, Lisa, Matthew Beaudoin, and Natalie Brewster. 2012. "Southern Exposure: The Inuit of Sandwich Bay, Labrador." In *Settlement, Subsistence and Change Among the Labrador Inuit: the Nunatsiavummuit Experience,* edited by David C. Natcher, Larry Felt, and Andrea Proctor, 61–84. Winnipeg: University of Manitoba Press.

Rankin, Lisa. 2015. "Identity Markers: Interpreting Sod-house Occupation in Sandwich Bay, Labrador." *Études/Inuit Studies* 39 (1): 91–116.

Rankin, Lisa. 2003. *Interim Report for Archaeological Permit 02.04.* Report on file at the Provincial Archaeology Office, Department of Tourism, NL Report on File NL Department of Tourism, St. John's.

Rankin, Lisa. 2002a. *Field Notes Porcupine Strand Archaeology Survey.* On file at the Department of Archaeology, Memorial University of Newfoundland and Labrador, St. John's.

Rankin, Lisa. 2002b. *Interim Report, Permit No. 01.04. Porcupine Strand Archaeology Survey, Summer 2001.* Report on file at the NL Department of Tourism, St. John's.

Rawley, Diana. 1949. "Operation Noramex." *The Arctic Circular* 2 (8): 107.

Reno, Joshua O. 2019. *Military Waste: The Unexpected Consequences of Permanent War Readiness.* Oakland: University of California Press.

Strowbridge, Linda. 1989. "Military Sanctions Supersonic Dogfights in Skies Over Labrador." *The Sunday Express,* February 5, 1989, 1: 12.

U.S. Department of the Navy, Bureau of Naval Personnel. 1953. "Today's Navy. 'Airphibious' Assault." *All Hands: The Bureau of Naval Personnel Information Bulletin* 431: 35.

U.S. Department of the Navy, Bureau of Aeronautics. 1953. "Marines Use Helicopters in Arctic." *Naval Aviation News* (April 1953): 18–19.

Valadares, Desiree. 2018. "Disposseing the Wilderness: Contesting Canada's National Park Narrative." In *Cultural Contestation: Heritage, Identity and the Role of Government*, edited by Jeroen Rodenberg, and Pieter Wagenaar, 139–153. Palgrave Macmillan.

Wadden, Marie. 2001. *Nitassinan: The Innu Struggle to Recalim their Homeland*. Vancouver: Douglas and McIntyre.

White, Cpl. M.J. 1949. "'NORAMEX' Completed: 9th Marines (Reinf.) Return to Lejeune." *The Globe* [Camp Lejeune, North Carolina], November 17, 1949, 1: 6.

CASE STUDY 2

TRASH FIRES AS TOXIC HERITAGE IN PALESTINE

Sophia Stamatopoulou-Robbins

How can the term "toxic heritage" help us understand trash fires on occupied territory?

Trash fires' health and environmental effects are ways in which fires leave indexical signs (e.g. Crossland: 2014, 44) after they stop burning. Yet heritage is more than an index of something having occurred. Heritage connotes ownership. It requires human affiliation. As a form of waste (Ross and Angel: 2020; Sosna and Brunclíková: 2016), trash has a similar defining characteristic: its existence implicates the existence of those who produced it or who relate to it after production (Stamatopoulou-Robbins: 2019, 134). Both heritage and trash thus raise questions about who continues to relate to them. Trash fires as toxic heritage invite us to consider who or what constitutes fires' possessor after extinguishment, and what politics possessorship enacts.

Amahl Bishara describes how Palestinians in the West Bank prefer to destroy Israel's apartheid Wall – for example by burning it – rather than to adorn it, for example through murals (Bishara: 2013, 233-259; Bishara: 2020, 94). Acts of destruction have specific materialities. In Palestine, fire fuel is often trash. An urban context like 'Aida refugee camp adjacent to the West Bank city of Bethlehem is strewn with household trash composed of cardboard, plastics, glass, and food scraps. Shoved up against the Wall's base, they become fuel for fire as a "hermeneutic and technology" (Chance: 2015) of struggle between a community and its occupiers.

The term "toxic heritage" invites us to think about trash fires' temporalities. It is easy to think of fires as fleeting. Fires appear spectacular because they are understood as events. They mark protest, ritual, or disaster. But trash fires last after extinguishment. When Palestinians burn the Wall, trash's material components combine with oxygen and emit bright light and heat. Smoke blackens concrete. It carries dioxins, a group of related compounds that exhibit health effects including

DOI: 10.4324/9781003365259-12

cancer, reproductive and developmental problems, immune system damage, and interference with hormones. Dioxins remain in the environment for a long time. They travel long distances, depositing onto land, surface water, and vegetation (see also MacLeish and Wool: 2022; Rubaii: 2018).

The term "toxic heritage" forces the question of possessorship to butt up against the damage caused by that which is possessed. Trash fires are the heritage of the bodies that carry the dioxins they emit: the residents of 'Aida camp, especially children who play there, young people who confront soldiers there, parents who carry groceries, and elders enjoying time outdoors. The occasional tourist may return home with dioxins as a secret souvenir. Israeli soldiers operating the Wall's tower may inhale the smoke, though many sit inside glass-encased rooms to avoid the physical hazards of their job.

Trash fires are also heritage for those who set the fires. Fire making can be understood as a heritage practice that yields marks on the Wall as residues of that practice. For as long as young 'Aida camp residents are alive to bear witness to the blackened Wall, the residues are among the elements of their heritage. While images protestors post on social media circulate, like Bishara's ethnographic rendering, protestors' descendants can claim the marks as their heritage as well (see also Abu Hatoum: 2021).

How can we think about the fact that the same communities who could claim this heritage are physically impacted by it? For the young people in 'Aida camp, trash fires may be as life-affirming as they are toxic. Or perhaps they are life-affirming *because* they are toxic. They may resemble the trash fires I learned about while conducting fieldwork in the West Bank. Some communities harnessed fires to create sensory irritations for those who governed or oppressed them. Faqu'a residents used to set fire to household trash at Faqu'a's edge, allowing smoke to waft toward nearby settlements. Burning dumpsters and tires often feature in protests. Trash fires articulated a *toxic* life politics (cf. Stamatopoulou-Robbins: 2021) that harnessed the potential for ecological damage to articulate and materialize political claims.

Yet trash fires are also the toxic heritage of settler colonialism (see also Stamatopoulou-Robbins: 2019: 202). Fires marking nonsovereign conditions are a heritage for those perpetuating nonsovereignty. Those people are disbursed through the Israeli state's apparatuses and its supporters beyond its borders. They are in international agencies orienting the Palestinian Authority toward some issues (e.g. compliance with Israel) and neglecting others (e.g. support to communities like 'Aida). They are in Palestinian Authority offices that decide to saturate Palestinian markets with disposable goods, which end up as trash, and to censor protesters, which results, among other things, in the Wall-burning that becomes an unevenly accounted-for and toxic heritage.

References

Abu Hatoum, Nayrouz, "For 'A No-state Yet to Come': Palestinian Urban Place-making in Kufr Aqab, Jerusalem," Environment and Planning E: Nature and Space 4, no. 1 (2021): 85–108.

Bishara, Amahl, Back Stories: US News Production and Palestinian Politics. Palo Alto: Stanford University Press, 2013.

Bishara, Amahl, "An Ongoing Violence, a Sustained Resistance: Israel's Racist Separation Wall at Aida Refugee Camp," in Walling in and Walling Out: Why are We Building New Barriers to Divide Us? edited by Laura McAckney and Randall H. McGuire. Albuquerque: University of New Mexico Press, 2020.

Chance, Kerry Ryan, "'Where there is Fire, there is Politics': Ungovernability and Material Life in Urban South Africa," Cultural Anthropology 30, no. 3 (2015): 394–423. 10.14506/ca30.3.03.

Crossland, Zoë, Ancestral Encounters in Highland Madagascar: Material Signs and Traces of the Dead. Cambridge: Cambridge University Press, 2014.

MacLeish, Kenneth, and Zoë Wool, "Burn Pits: US Military Waste as War Violence," Theorizing the Contemporary, Fieldsights, January 25, 2022. https://culanth.org/fieldsights/burn-pits-us-military-waste-as-war-violence

Ross, Susan, and Victoria Angel, "Heritage and Waste: Introduction," Journal of Cultural Heritage Management and Sustainable Development, (2020).

Rubaii, Kali, Counterinsurgency and the Ethical Life of Material Things in Iraq's Anbar Province. Santa Cruz: University of California, 2018.

Sosna, Daniel, and Lenka Brunclíková, Archaeologies of Waste: Encounters with the Unwanted. Oxford: Oxbow Books, 2016.

Stamatopoulou-Robbins, Sophia, Waste Siege: The Life of Infrastructure in Palestine. Palo Alto: Stanford University Press, 2019.

Stamatopoulou-Robbins, Sophia, "Excrement, Scale, and the Politics of Neglect: How life Politics became Progressive Critique," Confluences Mediterranee 2 (2021): 89–100.

7

POLITICS OF MINING: TOXIC HERITAGE IN THE ATACAMA DESERT

Marina Weinberg and Valentina Figueroa

Introduction

Most planetary technologies to reduce emissions and tackle climate change would be impossible without materials such as copper and lithium. Although the development of new technologies seems to justify the increasing demand for these metals, global dependence on these resources is actively and profoundly reshaping societies and ecologies in the places from which they are extracted. Copper is essential for the development of massive infrastructure projects, maritime, terrestrial and air transportation as well as for the manufacture of countless products used in contemporary life. Lithium has also stood out in the last decade as a key element due to the growing demands of the electromobility innovation, which seeks to replace fossil fuels and thus reduce carbon emissions. Both play a fundamental role in the development of different types of renewable energy and clean storage technologies; thus they are central to energy transition projects that push towards a so-called green future.

Copper and lithium mining in the Atacama Desert has produced high levels of toxicity that threaten local ecologies. Although the copper extraction process has intensified and is currently removing unprecedented volumes of material, and the extraction rate of brine to process lithium into lithium carbonate and hydroxide is also increasing steadily, extractive activities in the Atacama Desert are not new. Chile's Norte Grande, in particular the desert region, has been the scene of industrial mining-related processes for more than a century. In the context of the socio-environmental conflicts originating in the Atacama Desert, it is essential to think about these conceptualizations within the frame of politics of toxic heritage. This proposal considers the importance of power relations, tensions produced through access, and unequal use of natural resources and lands (Bebbington, 2009;

DOI: 10.4324/9781003365259-13

Perreault, 2013, Prieto, 2015), the uneven distribution of the impacts, and therefore the different capacities of stakeholders to operate within the extraction system.

The impact of the saltpetre–copper–lithium mining cycle has had different material correlates over the last hundred years, but it has crystallised into its contemporary form of exploitation by dispossession since the installation of the neoliberal model during the dictatorship of Augusto Pinochet (1973–1990). Deforestation, desiccation, hydrocide, accumulation and dispossession, proletarianisation, pollution and environmental violence have become widespread in most parts of Chile (Prieto, 2016a, 2017; Bolados, 2016; Aránguiz y Prieto, 2021). In these genealogies of mining cycle technologies, some pride and collective memories have been reclaimed, and discourses about the future have been also shaped. Thus, the progression of mining operations is constantly transforming the present, reifying the past and building upon the illusion of a 'better' future. In the most recent mining cycle, which has witnessed unprecedented levels of extraction and production over the last decade, the incessant growth of infrastructure that supports the constant development of these extractive activities, continues to produce mining heritages under promises of futuristic desires pushed forward by the global energy transition paradigm.[1] Under this context, we consider mining heritages a diverse compound that entails manifestations both tangible and intangible.

Throughout the essay, we identify diverse mining traditions in the region that materialise through objects, practices, narratives, rituals, affections and claims, based on the need to reinforce a mining worker, regional and/or national identity (which is also passed on generationally) and creating and embracing a mining heritage alongside the complexities created by the same activity (Baird, 2022). The patrimonialisation of mining is heterogeneous since it refers to different phenomena: the nationalist narrative of Chilean copper, the representation of the mining worker, the tradition of mining cults, the authorised mining heritage centred on architecture, the musealisation of contemporary work sites, and toxic heritage, are some of the cases we observe.

Establishing a dialogue between archaeology and anthropology, this chapter reflects on past and present toxic impacts of mining activities in the Atacama Desert by following their tangible and intangible traces and remains since the establishment of the Chilean neoliberal dictatorship in 1973. While these marks, or collateral effects, are sometimes perceived and observable, at other times they act so intimately in everyday practices and even at a cellular level that they become part of a local biopolitical heritage (Galaz-Mandakovic, 2013; Weinberg, 2021). However, following Murphy's (2017) observation in the context of related studies in North America, it is not our intention to pathologise already dispossessed communities. As she points out, 'it is hard to perceive the infrastructures of chemical violence in the world at the same time that research attends to molecular manifestations in bodies and communities already living in hostile conditions' (Murphy, 2017: 496).

In the Atacama Desert, as we will illustrate by exploring the cases of copper and lithium extraction, many of the marks produced by the development of these mining activities are enacted, appropriated, and contested as situated heritage by

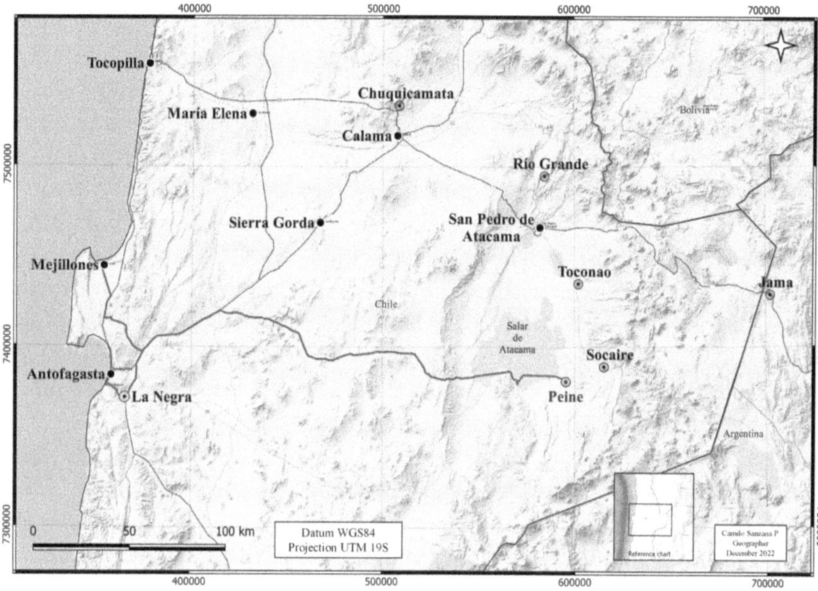

FIGURE 7.1 Map of the studied area.

local communities, who live in contested socio-environmental places (Figure 7.1). Moreover, this heritage-making process can be observed as complex, relational and in constant transformation. In particular, the prevalent technical-engineering narrative tends to support the idea of a successful technology and is less condescending than other ways of perceiving mining in the Atacama Desert. In this region, the long-term entanglement between the extractive industry activity, toxicities and climatic and geochemical conditions has created and sustained particular habitats and species, while others have been lost, adaptively transforming the socio-ecological heritage of the desert landscape.

Through engaging with decolonial feminism and eco-feminism (Bidaseca & Vázquez, 2011; Segato, 2011; Svampa, 2015; Ulloa, 2016) we develop our concept of toxic heritage, highlighting care practices (Ureta & Flores, 2018; see also Mol, 2002, 2008; Bellacasa, 2017; Biskupovic et al., 2023). In its attention to care, this reflection allows us to analyse copper and lithium extraction not only through the traditional frameworks of environmental decay and political action, but also by considering ways of life intrinsic to the domestic spaces most affected by extractivism. Given the power of massive industrial intervention, these domestic spaces often go unnoticed in academic research (Salinas et al., 2012; Segovia & Salinas 2020; Segovia et al., 2023). In its exploration of practices typically ignored by hegemonic discourses, our framework is inspired by feminist research that deconstructs the masculinised, technocentric and geoengineered armours (Haraway, 1995; Haraway, 2016) that are built on discourses of 'innovation' and 'development' (Biskupovic et al., 2023).

Politics of Toxic Heritage

We live in a world in which it is no longer possible to find a human or non-human community or even an individual exempt from toxic traces. This condition, described as a 'new age of toxicity' (Walker, 2011: xi), is unevenly shared across the length and breadth of the planet (Liboiron et al., 2018: 333). The Atacama Desert, with just over a century of sustained mining activities that have intensified in recent decades, can be identified as a large toxic zone. Regardless of its meanings, this conceptualisation allows for the framing, identification, and classification of a place for the purpose of questioning extractivist productive activities perceived as destructive (Bolados, 2016; Holifield & Day, 2017; Hormazabal-Poblete et al., 2019; Galaz-Mandakovic, 2021). In this sense, it is important to identify a politics of toxic heritage that has been sustained and is constantly updated in order to maintain a dominant structure that requires sustaining the modern capitalist system (Figure 7.2).

Markets, mining operations, tailings, smelters and human and non-human communities are structurally intertwined. The relationship we seek to delve into in this work is that between the alleged concern about the levels of toxicity within the mining environment on the one hand and the constant promotion of the mining industry on the other. The large data bank of techno-engineering advances, which overlaps with many forms of biocitizenship, is far removed from inhabitants' ability to respond.[2] In this context, Murphy's (2004, 2008, 2017) work is enlightening for

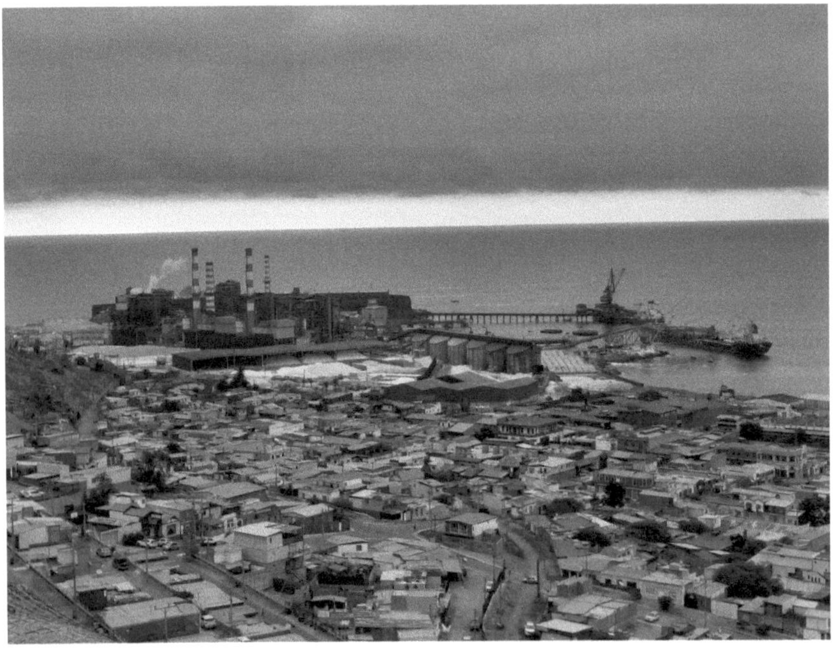

FIGURE 7.2 Tocopilla, by the authors.

considering the political dimensions of toxicity and the privilege of imperceptibility. Society is set up to protect its privileged sectors from toxic events. In fact, if we map chemical violence in the Atacama Desert, the relationship between chemical violence and dispossession is located precisely in areas where other forms of dispossession have long existed, such as migrant neighbourhoods, slums and indigenous communities. It is crucial to keep in mind the political dimension of toxic mining waste. Its location in itself represents a political cartography (Hecht, 2018). One of the best examples involves the waste from the Swedish company Boliden[3], which was transported by ship during the Chilean dictatorship and deposited in a northern Chilean coastal city's central sector, on which houses were built and residents were affected by various diseases due to exposure to the toxic waste.[4]

Although in recent years, important waste management measures have been taken in Chile to address toxicity and toxic waste in mining–metallurgical contexts, focusing on the daily practices of the affected subjects, a blind eye is still turned to the eco-feminist/indigenous critique which perceives that those affected are still victimised. In the same sense, in a more recent work, Murphy (2017) notes that despite the good intentions of studies on harm in affected communities, they often 'resurrect racist, misogynistic, and homophobic portrayals of poor, black, indigenous, female, and queer lives and communities as damaged and doomed, as inhabitants of irreparable states' (2017: 496). In this sense, in the present chapter, we are interested in problematising the conceptualisation of these areas as toxic in a polysemic manner that not only considers these zones undeniable toxicity but also brings attention to the politics that have portrayed them as toxic without paying attention to local complexities, and especially to the people living in those areas.

Toxic Mining Heritages: Between Effects and Affection

In this analysis, we want to discuss two regimes. The first has to do with mining heritage, and the second relates to perceptions and management of toxic mining heritage. Although these are deeply linked, we can recognise their different trajectories. In the first regime, a historical relationship exists between heritage, progress and national identity in Chile, which materialise through the patrimonialisation of mining operations. These processes have reified the hegemony of the monumental heritage of Chile '*como país minero*' (as a mining country), creating a narrative that portrays mining technology as part of the national ethos. This is evident in the abandoned mining sites that are part of visiting circuits, mining museums, the musealisation of contemporary mines and company towns and technological mining devices, a real aesthetic that materialises in male workers and technologies (Weinberg, 2021). Although few examples of critical museography on mining heritage exist, we can cite some museum narratives that show the poor working conditions of the proletarianised workers in the saltpetre mines (e.g. the Augusto Capdeville Museum in Taltal, the Regional Museum of Antofagasta), or of several saltpetre mining enclaves that were used as torture centres during the dictatorship, such as Chacabuco

(Vilches, 2011). These examples offer a critical understanding of the complexities of mining heritage at a local level; this heritage distances itself from the dominant idea of mining patrimonialisation, which has been related to techno-progressive, hetero-normative and patriarchal discourses (Barrientos et al., 2009; Silva, 2008; Pavez y Hernández, 2014; Silva y Salinas, 2016). In short, while on the one hand, we can easily observe that Chilean mining aesthetics are based on the progress of mining cycles, on the other hand, there is a mining space that is unknown and outside the authorised heritage discourse and that relates to the intangible heritage of mining, which we found in the duality consisting of the 'negative' effects of mining and the affection for mining.

To address the second regime, it is necessary to situate the discussion of toxic mining heritages in broader critical heritage debates. Regarding the 'toxic' dimension of mining and its link with the notion of heritage, we are interested in its contemporary political dimension. Our interest focused on the recent past, the present and the near future. Wollentz et al. (2020) point out that the toxicity of heritage is not so much related to its level of life or death, but rather to its management and the narratives in which it is used. Returning to the case of the Atacama Desert, no discussion has been held around managing toxic mining heritages (similar to the debate on nuclear waste in Europe, for example). What has prevailed is the management of toxicity through the prism of risk technologies and corporate social responsibility.

In mining, there are different levels, categories, scales and natures of toxicities. We have focused on the understanding of toxic mining heritages as a form of material/immaterial environmental violence that became naturalised in the Atacama Desert with the military dictatorship and that has become a *habitus* of large-scale mining since then. One of the most representative examples of the material culture of mining extractivism is tailings. The geo-engineering management of tailings categorises them into active, non-active and abandoned sites (Sernageomin[5]). This distinction between the different natures of the 'lives' of toxic wastes is a provocation that makes the complexity of these wastes invisible (Hetch, 2014). This situation evokes not only the relationship between these mining landscapes and local communities, but also with their global character, considering that in this radicalised modernity, technologies bring together communities separated by time and space (Arboleda, 2021).

While there is a socio-political dimension that has traces in the miningscapes (Méndez et al., 2020), the accumulation resulting from this dispossession is currently nourishing univocal green narratives that feed the energy transition paradigm. As we will present in the next two sections (Copper and Lithium), in the Atacama Desert, we see a crucial tension: while local and regional identities mobilise mining heritage as a form of recognition and belonging, the mining industry is simultaneously creating dramatic traces and evidence of destruction and new dystopian socioscapes. We can observe friction between the local need to create and embrace mining heritage as a way of embodying identity politics and (regional and national) belongings, and the traces of toxicity and destruction that are produced by the same activities in the spaces of extraction that allow for these notions of belonging to exist.

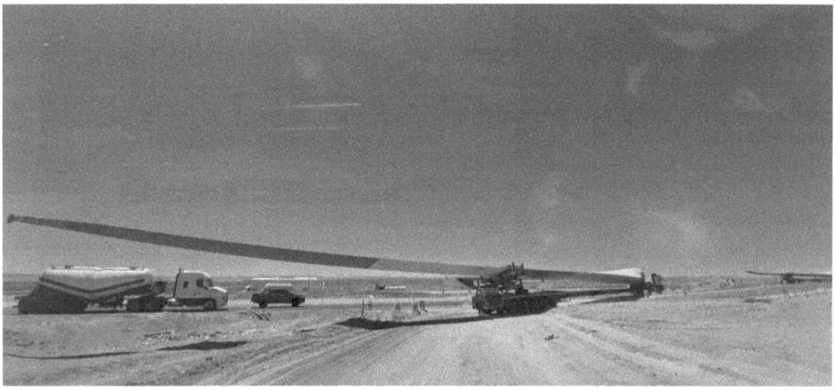

FIGURE 7.3 Transporting a windmill blade of a wind turbine generator, by the authors.

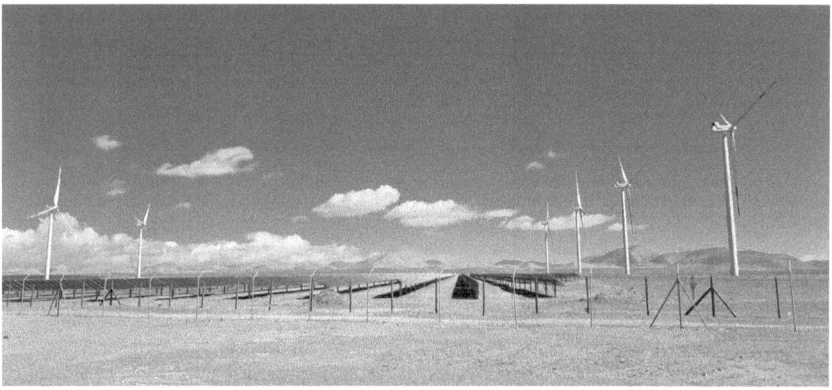

FIGURE 7.4 Solar panels of a wind turbine generator, by the authors.

In addition, in recent years, these scenarios have been affected not only by mining ventures but also by a massive energy infrastructure that has been fundamental to the maintenance and development of extractive industries in northern Chile, and that has once again affected heritage-construction processes (Figures 7.3 and 7.4).

Copper

Despite the relevance of copper to the Chilean national economy and identity, no significant funding exists for the field of mining heritage in its most classical sense. In fact, the most important projects have been local initiatives started by former workers, their families, unions and local associations as a way of maintaining the sites (as in the cases of Chuquicamata and Sierra Gorda, for example). However, it is not the case that no funds are earmarked for heritage, but simply that they are elsewhere. Although it may seem paradoxical, large-scale mining is a major producer of heritage (Baird, 2022). Although it may seem paradoxical, large-scale mining is a major producer of heritage. In the context of mining, a huge market exists for contract archaeology and heritage management associated with environmental impact assessments. This market is even bigger if we also include the infrastructure that supports large-scale mining, such as roads and renewable energies projects. In this setting, the most common development in recent years, apart from the expansion of mining districts, involves foreign investment in Chile in 'clean' energies such as photovoltaic panels and wind farms. We do not yet know the exact percentage, but the Chilean heritage management model must be similar to the one described by Zorzin (2015) in his analysis of the Canadian model, where private archaeology represents 95% of the occupation.

The Antofagasta Region is a quintessential leader in Chilean copper extraction, yielding 2.88 million tons of fine copper, representing 52% of national production (Cochilco, 2018). This copper is extracted from one of the driest places on Earth: the Atacama Desert (Jordan et al., 2014), home to the largest porphyry copper deposits in the world (Sillitoe, 2010). Over the past fifty years, the Chilean state has exploited these deposits through Codelco (the National Copper Corporation of Chile), a public copper company considered fundamental to the Chilean state. In the current context of climate crises and political transformations, the strategic role of Codelco is increasingly relevant to Chilean social and economic organisation and has major implications for global energy supply chains. Under the prevailing logic that more mining is required to guarantee global energy transitions, Chuquicamata – one of the oldest and most important divisions of Codelco – transitioned its open-pit production to underground mining in 2019. Chuquicamata has played a fundamental role in organising social, political, and economic life in northern Chile.

Chuquicamata is central to contemporary Chilean politics and development, and to global energy supply chains. It has also played a central role in the historic development of Chile as a 'mining country'. More than one hundred years after the start of the industrial extraction process, the technology used continues to be modernised to increase the company's performance, even to the detriment of the environment and the health of the miners and the local population (there are 180,000 inhabitants in Calama according to the 2017 census).

During the first decades of its operation (1915–1923) in the hands of the North American businessman Guggenheim representing the firm Chile Exploration Company (Chilex), and then until the end of the 1960s under the firm Anaconda Company (Carrasco, 2015, 2020), the political and economic logic from which the hegemonic administration was shaped, turned Chuquicamata into a bubble of privilege in the context of northern Chile, although it also fostered a space in which freedoms were compressed and restricted (Figueroa, 1928; Finn, 1998; Galaz-Mandakovic, 2013, 2017; Latcham, 1926). Some research has examined the internal coercive practices and property regulation systems that not only guaranteed the appropriation of natural resources but also legalised the use of force against miners. Consequently, these 'artificially stable conditions' allowed the open-pit copper mining exploitation model to produce unequal territories in the region (Méndez et. al., 2020, 11).

In 1992, after nearly eighty years of existence, the Chuquicamata town was declared 'a zone saturated by sulfur dioxide and breathable particles' which residents had to evacuate. The entire population was moved to the city of Calama (Santolaya et al., 1995). The original strategy of setting up the camp in the extraction space – which promoted continuous relations between workers, families, and copper – was replaced by socioeconomic incentives when the town was moved to Calama, such as housing loans, scholarships and health insurance (Pérez-Bustamante & Wolf, 2014). For over a century, Chuquicamata has fostered an intriguing sense of belonging and pride among its workers, as well as in the larger Chilean imagination. To this day, the camp opens once a year to celebrate the anniversary of its inauguration with workers, former workers and their families (Weinberg, 2021). It is in this sense, we precisely pointed out the tensions. Whereas the mining camp was shut down given toxic reasons that made it unliveable, until this day ex-workers and their families, which used to live there and probably are sick or have lost some member of their families for this reason, have enormous affection and sense of belonging, and really treasure remembering and sharing with younger generations their past in Chuquicamata.

In Chuquicamata, we observe, working conditions and note the significant presence of diseases and deaths that have occurred in various periods in the history of the mine (Vergara, 2005; Galaz-Mandakovic, 2017, 2021). Biopower and necropolitics work together through control mechanisms that have been generated by the extractive companies. And over the decades these have also been internalised by the workers themselves. With the stark recognition that the miners and their families exchange 'health for money', the legitimacy of the biopower that can be seen in Chuquicamata, lies not only in its capacity to optimise life opportunities, but in how death is regulated (Foucault, 1976, 2008; Mbembe 2003; Galaz, 2013, 2021). In Chuquicamata, we observe control and co-optation of the population through their bodies, at a capillary, almost imperceptible level, using biopolitical surveillance mechanisms that have been internalised in an effective and complex manner over the decades (sensu Foucault, 1976). The corps are valued economically for their work by receiving good salaries, but they also get sick and, in some cases, die (Weinberg, 2021; Galaz-Mandakovic, 2021).

Lithium

The univocal assumption that the production of lithium and its by-products will yield a transition towards 'clean–green energy' erases the real environmental impacts of lithium mining in the regions of the planet from which this soft metal is extracted, including the Atacama Desert. Several fundamental studies have detailed how this water-intensive extractive activity relies on fragile local ecologies (Aránguiz et al., 2020; Babidge, 2021; Bustos & Prieto, 2019; Göbel & Gundermann, 2018; Flores-Fernández, 2020; Garcés & Álvarez, 2020; Bonelli & Dorador, 2021; Prieto, 2016, 2017; Weinberg & Bonelli, 2021).

The widespread imaginary of brine extraction areas romanticises extensive salt flats and striking blue-green pools. However, extraction processes have also generated massive infrastructure, material systems, intangible damage, effects on human and non-human beings and the establishment of socio-technical imaginaries, which have been normalised over time. Undeniably, the further one moves away from the world-famous brine pools, the 'dirtier' the landscape becomes.

So far, no certified correlation has been established between exposure to lithium carbonate and human diseases, as has been done in the case of direct contact or inhalation of metallic particulate matter in the case of concentrated copper (Galaz y Tapia, 2022; Vergara, 2005, 2008; Bernales et al., 2008; Weinberg, 2021). However, 'the European Chemicals Agency (ECHA)'s Risk Assessment Committee (RAC) at the end of 2021 published its opinion that it agreed with French proposals to classify three lithium salts as Category 1A reproductive toxicants. This publication determined that lithium carbonate, lithium hydroxide and lithium chloride should be classified under the Classification, Labelling & Packaging (CLP) Regulation as substances that may damage fertility and unborn children. It also agreed that the substances may harm breastfed children'.[6] Following this new potential harmful status for lithium salts, opens the possibility of undermining the European Union's attempt to create and support a local supply chain for battery materials. However, this same declaration does not seem to have had any impact on the policies in place in the north of Chile, once again revealing the 'unevenly toxic world', where 'toxic harm also maintains systems, including those that produce inequity and sacrifice' (Liboiron et al., 2018: 332).

When we look at the Atacama Desert, it becomes evident that to guarantee the transition to 'clean energy', not only will local ecologies within the salt flats continue to be affected, but the infrastructure throughout the desert will continue to expand. These material arrangements have a much wider geographical coverage than the extraction area itself, including workers' camps, company offices, processing plants, ports, electric power lines, land, rail, air and water routes, and so on. These historically disputed extractivist territories, marked by socio-environmental conflicts, are now being reshaped by the installation of energy enterprises. Due to the rising prices of hydrocarbons in neighbouring countries and to the 'natural' advantages of northern Chile (with its high direct solar radiation and atmospheric attenuation), the country is presented as a laboratory, as one of the region's areas with the greatest promise of successfully transforming its energy matrix. But for now, what we see is a renewable energy generation model whose only aim is to power mining industries; thus, we are witnessing the production of a variety of clean energy materials that maintain an extractive mining model, which continues to be sustained mainly by fossil fuels.

In the global south, what energy transitions really do is produce new energy expenditures that continue to ensure the expansion of capital produced by resource extraction. Green moralities imposed as a counterpart to the paradigm of energy transitions seem to develop in those planetary spaces and socioeconomic sectors that require mining development to ensure prevailing inequality. However, in local communities, the only choices available are to work in a toxic environment or to not work at all.

Conclusions: After Mining?

The proliferation of hegemonic discourses describing green utopias makes invisible the catastrophe and urgency of the present impacts on the territories of extraction. The north of Chile appears in this scenario as a space that struggles between life and destruction. It is a vital, necessary and sacred space for many, and a natural and pristine place for others, but it has become mostly a subaltern place, subalternised at the service of global capital. The micro-politics of mining are concerned with production costs, in which the so-called 'liabilities' and 'assets', both social and environmental, are completely monetised. But this cost of production certainly does not include infinite concerns such as death.

Lithium and copper mining have developed vis-à-vis explanations of inevitability, but mainly as a way to achieve a future for the few. While uninterrupted economic growth in the global north promises a chimaera of green horizons, we see that in the region where lithium, copper and so many other elements are obtained, massive extraction exacerbates inequality, dispossession and the destruction of ecosystems and ways of life. The mining boom that has accelerated in the last decade is based on potentialities: paradigms of not *yet* green-transitions pronounce a future that *will arrive*. Not only has this future not yet arrived, but it is also likely to be a future that is inaccessible to many. Conceptualizing the future has become a commonplace of diverse and heterogeneous narratives (public, private, political, associative, individual) that support green morality as an ideology that corrects, remedies and repairs socio-environmental problems. However, access to material existence in different types of future would require the security of capital in the present, which would ensure some kind of link with the future.

In the quotidian experience of living in the Atacama Desert, we observe that there is a distance bordering perversion between assessments and concerns over futures. On the one hand, the overvaluation of a futuristic metaphysics mobilised by companies, state bodies, and expert knowledge, concerned with scenarios and communities of the future in unknown territories. On the other, actors whose future is less aesthetic, their future is an immediate one, worrying about household subsistence contingencies that must find strategies to access clean air, water and food to get through the day. Moreover, the governance of the future only makes invisible the communities affected by toxic air poisoning and lack of clean water in the present, and in this sense, we believe that issues related to energy transition and future sustainability should be carefully invoked from the social sciences, as they often contribute to abandoning the contradictions of the present.

The politics of toxic heritage conceptualization we have proposed, invite us to think about the processes of this work from a perspective that considers power relations and the uneven distribution of the impacts of mining activities at the forefront. The toxicity of mining can be thought of as a part of capitalism's legacy, but not necessarily analysed as binary from a moral perspective. As we have shown by paying attention to the multiplicity of experiences in mining territories, toxic heritage might be approached as a multivocal concept. While sustaining the modern capitalist system, it is also a way of life (and sometimes the only option) for most of the population in the Atacama Desert. It turns out that extractivism in northern Chile is much more than a mode of production; it is also a way of life, a heritage for many, a dependence for others and one of the few livelihoods that has allowed people far from the centre of the country to feel some kind of national belonging. This paradox is difficult to unravel, as the people have affection for their territories even though these have been labelled as toxic. It could be thought of as a kind of 'conflicted subjectivity' (*sensu* Navaro-Yashin, 2009), or it could be concluded that the Cornelian dilemma of producing a toxic environment and living in it, is unavoidable in a mining space where mining *is* life.

This chapter is part of the ERC Project 'Worlds of Lithium. A transnational study of people and materials transitioning toward post fossil fuel societies' that has received funding from the European Research Council (ERC) under the European Union's Horizon 2020 research and innovation programme (grant agreement No. [853133])" and ANID Fondecyt Project 'Paisajes mineros prehispánicos en el desierto de Atacama: hacia un estudio del uso de los minerales de cobre y su vinculación con la producción de tecnologías rituales (N° 1201603).

Notes

1 With the establishment of the Intergovernmental Panel on Climate Change (IPCC), created by the World Meteorological Organization and the United Nations Environment Programme in 1998, and the organisation of the United Nations Conference on Environment and Development, or the Earth Summit, in Rio de Janeiro in 1992, a new era of climate change awareness began. Since then, environmental deterioration has been unstoppable as the result of large-scale industrial development vis-a-vis the constant implementation of policies and protocols that have been unsuccessful at tackling the crisis.

2 In spite of our pessimistic hypothesis on the supremacy of techno-engineering expertise, we can point out, following Murphy (2008), that 'since there is a regime of imperceptibility that has been purposively assembled around synthetic molecular relations, efforts to render visible such relations – by scientists, by bureaucrats, by community groups, or by NGOs – are political acts'.

3 We thank Anders Berglund for sharing the "Sveriges ansvar för giftigt avfall i Chile" Interpellation (https://www.riksdagen.se/sv/dokument-lagar/dokument/interpellation/sveriges-ansvar-for-giftigt-avfall-i-chile_H810772)
4 'Boliden decided to ship 20,000 tons of toxic mining waste to Chile in the mid-1980s. In this way, the company could circumvent Swedish environmental requirements and save money. Boliden sent the waste to the Chilean company Promel in Arica' (Delgado Varas, 2020).
5 https://www.sernageomin.cl/datos-publicos-deposito-de-relaves/
6 realclearenergy.org/2022/06/22/energy_transition_goals_at_risk_eu_label_lithium_as_toxic_838018.html

References

Aránguiz-Acuña, Adriana, José A. Luque, Héctor Pizarro, Mauricio Cerda, Inger Heine-Fuster, Jorge Valdés, Emma Fernández-Galego yVölker Wennrich. 2020. "Aquatic community structure as sentinel of recent environmental changes unraveled from lake sedimentary records from the Atacama Desert, Chile." *PLoS ONE* 15(2): e0229453.

Aránguiz-Acuña, Adriana yManuel Prieto. 2021. "Justicia ambiental en Chile: ¿La última oportunidad?" *Cuadernos Médico-Sociales* (Chile) 61(3): 53–57.

Arboleda, Sebastián. 2021. *Planetary Mine: Territories of Extraction Under Late Capitalism*. Verso.

Babidge, Sally. 2021. "Seeing water slow resistance and the material enigma of extractive effects on society and ecology." *HAU: Journal of Ethnographic Theory* 11 (2): 395–411. 10.1086/715788

Baird, Melissa. 2022. "Waste Sits in Places. Post-Extractive Landscapes as Heritage." In Giacomo Pettenati (Ed.), *Landscape as Heritage. International Critical Perspectives*. London: Routledge, 204–205.

Barrientos Delgado, Jaime, Paulina Meruane, Pablo Rojas Varas, and Patricio Meza Opazo. 2009. "Minería, género y cultura. Una aproximación etnográfica a espacios de esparcimiento y diversión masculina en el norte de Chile." *Revista Antropológica Iberoamericana* 4 (3): 385–408.

Bebbington, Anthony. 2009. "The New Extraction: Rewriting the Political Ecology of the Andes?." *NACLA Report on the Americas* 42 (5): 12–20.

Bellacasa, María Puig de la . 2017. Matters of Care: Speculative Ethics in More Than Human Worlds. Minneapolis and London: University of Minnesota Press.

Bernales, Bélgica, Juan Alcaíno yRodrigo Solís. 2008. "Situación de exposición laboral a sílice en Chile." *Ciencia y Trabajo* 10 (27): 1–6.

Bidaseca, K., and Vazquez Laba, V. (comps). 2011. *Feminismos y poscolonialidad : descolonizando el feminismo desde y en América Latina*. Buenos Aires: Ediciones Godot.

Biskupovic, Consuelo, Marina Weinberg, Adriana Aranguiz, Jimena Cruz, Valentina Figueroa, Fernanda Kalazich, and Paulina Salinas. 2023. "Territorios de cuidado. Propuestas desde un enfoque feminista ante el extractivismo en el desierto de Atacama, Chile." *Latin American Perspectives*.

Bolados, Paola. 2016. "Conflictos socio-ambientales/territoriales y el surgimiento de identidades post neoliberales (Valparaíso-Chile)." *Izquierdas* 31: 102–129.

Bonelli, Cristóbal yCristina Dorador. 2021. "Endangered Salares: micro-disasters in Northern Chile." *Tapuya: Latin American Science, Technology and Society* 4(1): 1–21.

Bustos, Beatriz yManuel Prieto. 2019. "Ecología Política en (desde y por) Chile: Posibilidades, desafíos y contribuciones." In A. Núñez, E. Aliste yR. Molina (Eds.), (Las) *Otras Geografías en Chile. Perspectivas sociales y enfoques críticos*. Santiago de Chile: LOM, 85–104.

Carrasco, Anita. 2015. "Jobs and Kindness: W. E. Rudolph's Role in the Shaping of Perceptions of Mining Company-Indigenous Community Relations in the Atacama Desert, Chile." *The Extractive Industries and Society* 2: 352–359.

Carrasco, Anita. 2020. *Embracing the Anaconda: A Chronicle of Atacameño Life and Mining the Andes*. Lanham,Maryland: Lexington Books.

Cochilco. 2018. *Balance de gestión integral 2018*. Cochilco: Gobierno de Chile.

Figueroa, Marcial. 1928. Chuquicamata. *La tumba del chileno*. Antofagasta: Imprenta Castellana.

Finn, Janet L. (1998). *Tracing the Veins of Copper, Culture and Community from Butte to Chuquicamata*. Berkeley: University of California Press.

Flores-Fernández, Cristián. 2020. "The Chilean energy '"transition"': between successful policy and the assimilation of a post-political energy condition." *Innovation: The European Journal of Social Science Research* 33(2): 173–193.

Foucault, Michel. 1976. *Historia de la sexualidad*. México: Siglo XXI.

Foucault, M. 2008. *The Birth of Biopolitics*. Nueva York: Picador-Palgrave Macmillan.

Galaz-Mandakovic, Damir. 2013. *Migración y Biopolítica. Dos escenas del siglo XX tocopillano*. Tocopilla: Retruécanos Ediciones.

Galaz-Mandakovic, Damir. 2017. "Turbinas y electricidad para lamina, lámparas a parafina para la población. Crónica de una asimetría del capitalismo minero en Tocopilla (1914–1942)." *Estudios Atacameños* 54: 179–200.

Galaz-Mandacovic, Damir. 2021. "La necroeconomía de la generación de electricidad para la minería. El impacto comunitario del uso del petcoke en Tocopilla (Chile, 2000-2015)." CUHSO. 10.7770/cuhso-v31n2-art2155

Galaz-Mandakovic, Damir yVíctor Tapia Araya. 2022. "El panteón de la montaña de cobre. Trabajo, ambiente y causas de muerte en la mina de Chuquicamata durante la etapa Guggenheim. (Chile, 1915–1923)." *História Unisinos* 26(2): 312–329.

Garces, Ingrid and Gabriel Alvarez. 2020. "Water mining and extractivism in the Salar the Atacama." *WIT Transactions on Ecology and the Environment* 245: 189–199. 10.2495/EID200181

Gundermann, Hans yBarbara Göbel. 2018. "Comunidades indígenas, empresas del litio y sus relaciones en el Salar de Atacama." *Chungará*. 10.4067/S0717-73562018005001602

Haraway, D. 2016. *Staying with the Trouble: Making Kin in the Chthulucene*. Durham, NC: Duke University Press.

Haraway, Donna. 1995. *Ciencia, cyborgs y mujeres: la reinvención de la naturaleza*. Madrid: Cátedra.

Hecht, Gabrielle. 2018. "Interscalar vehicles for an African Anthropocene: On waste, temporality, and violence." *Cultural Anthropology* 33(1): 109–141.

Hecht, Gabrielle. 2014. "Invisible Production and the Production of Invisibility: Cleaning, Maintenance, and Mining in the Nuclear Sector." In Daniel Kleinman and Kelly Moore (Eds.), *Handbook on Science, Technology and Society*. Routledge, 346–361.

Holifield, Ryan and Michael Day. 2017. "A framework for a critical physical geography of 'sacrifice zones': Physical landscapes and discursive spaces of frac sand mining in western Wisconsin." *Geoforum* 85: 269–279.

Hormazabal-Poblete, Nina, Sandro Maino, Magdalena Herrera, and Matias Vergara. 2019. "Habitar en una zona de sacrificio: Análisis multiescalar de la comuna de Puchuncaví." *Revista Hábitat Sustentable* 9: 6–15. 10.22320/07190700.2019.09.02.01

Jordan, Teresa, Kirk-Lawlor, Naomi, Blanco, Nicolás, Rech, Jason , and Nicolás Cosentino . 2014. "Landscape modification in response to repeated onset of hyperarid paleoclimate states since 14 Ma, Atacama Desert, Chile." Geological Society of America Bulletin 126, 1016–1046. 10.1130/B30978.1

Latcham, Ricardo A. 1926. *Chuquicamata. Estado Yankee*. Santiago: Editorial Nascimiento.
Liboiron, Max, Manuel Tironi, and Nerea Calvillo. 2018. "Toxic Politics: Acting in a Permanently Polluted World." *Social Studies of Science* 48(3): 331–349.
Mbembe, Achille. 2003. "Necropolitics". *Public Culture* 15 (1): 11–40.
Méndez, M., Prieto, M., and D. Galaz-Mandakovic. 2020. "Tele-production of Miningscapes in the open-pit era: The case of low-grade copper, Bingham Canyon,US and Chuquicamata, Chile (1903–1923)." *The Extractive Industries and Society*. https://doi.org/10.1016/j.exis.2020.10.013
Mol, A. 2002. The Body Multiple: Ontology in Medical Practice. Durham, NC and London: Duke University Press.
Mol, Annemarie. 2008. *The Logic of Care: Health and the Problem of Patient Choice*. London: Routledge.
Murphy, Michelle. 2004. Uncertain Exposures and the Privilege of Imperception: Activist Scientists and Race at the U.S. Environmental Protection Agency. Osiris, 2nd Series, Vol. 19, Landscapes of Exposure: Knowledge and Illness in Modern Environments, pp. 266–282.
Murphy, Michelle. 2008. "Chemical Regimes ofLiving." *Environmental History* 13 (4): 697–703.
Murphy, Michelle. 2017. "Alterlife and Decolonial Chemical Relations." *Cultural Anthropology* 32(4): 494–503.
Navaro-Yashin, Y. 2009. "Affective Spaces, Melancholic Objects: Ruination and the Production of Anthropological Knowledge." *Journal of the Royal Anthropological Institute* 15 (1): 1–18.
Pavez, Jorge yGerardo Hernández. 2014. "Regímenes de trabajo, relaciones laborales y masculinidades en la gran minería del cobre (norte de Chile)." *En Trabajos y familias en el neoliberalismo: Hombres y mujeres en faenas de la uva, el salmón y el cobre*, X. In L. Valdés et al., Santiago: LOM, 169–263.
Pérez-Bustamante, Leonel, and Wolf. 2014. "Chuquicamata: crónica de un desalojo / cierre, despedida y últimos años de funcionamiento." In María Isabel López & Leonel Pérez-Bustamante (ed.), *Patrimonio minero y sustentabilidad: propuestas y experiencias de re-utilización* (pp. 34–47). Concepción: Ediciones Universidad del Bío-Bío & CYTED.
Perreault, Tom. 2013. "Dispossession by Accumulation? Mining, Water and the Nature of Enclosure on the Bolivian Altiplano." *Antipode* 45 (5): 1050–1069.
Prieto, Manuel. 2015. "Privatizing Water in the Chilean Andes: The Case of Las Vegas de Chiu-Chiu." *Mountain Research and Development* 35: 220–229. 10.1659/MRD-JOURNAL-D-14-00033.1
Prieto, Manuel. 2016a. "Transando el agua, produciendo territorios e identidades indígenas: El modelo de aguas chileno y los atacameños de Calama." *Revista de Estudios Sociales* (55): 88–103.
Prieto, Manuel. 2016b. "Practicing costumbres and the decommodification of nature: The Chilean water markets and the Atacameño people." *Geoforum*.
Prieto, Manuel. 2017. The irrigation system that market doesn't want to see: the history of the water's dispossession in the communities of Lasana and Chiu- Chiu (Atacama Desert, Chile)
Salinas Meruane, Paulina, Barrientos Delgado, Jaime, and Rojas Varas, Pablo. 2012. "Discursos sobre la discriminación de género en los trabajadores mineros del norte de Chile." *Atenea* 505: 139–158.
Santolaya Biondi, R., L. Salazar Corco, R. Santolaya Cohen, M. Sandoval Martínez, and R. Alfaro Torrico. 1995. "Arsénico. Impacto sobre el hombre y su entorno. II Región de Chile (Antofagasta). Chuquicamata: Programa Ambiente 02 (1992–1993)." Centro de Investigaciones Ecobiológicas y Médicas de Altura C.I.E.M.A.

Segato, R. L. 2011. Género y colonialidad: en busca de claves de lectura y de un vocabulario estratégico descolonial. In K. Bidasecaand V. Vazquez Laba. (comps). *Feminismos y poscolonialidad : descolonizando el feminismo desde y en América Latina.* (pp. 11–40). Ediciones Godot: Buenos Aires.
Segovia, Jimena Luz, and Paulina, Salinas. 2020. ""Me canso de ser hombre" Paternidad periférica en el trabajo minero chileno." *Sociologia, Problemas e Práticas* 93: 31. 10.7458/SPP20209312014
Segovia, Jimena Luz, Paulina, Salinas, and Estefany Castillo Ravanal. 2023. "Habitus of Masculinity in Chilean Miners: Efficiency, Control, and Consumption of the Bodies." *Social Sciences* 12: 119. 10.3390/socsci12030119
Sillitoe, R. H. 2010. "Porphyry Copper Systems." *Economic Geology* 105 (1): 3. https://doi.org/10.2113/gsecongeo.105.1.3
Silva, Jimena *et al.* 2008. "Tacones cercanos. Diagnóstico y situación de la mujer que ejerce comercio sexual, región de Antofagasta, Chile." Servicio Nacional de la Mujer, Chile.
Silva, Jimena yPaulina Salinas. 2016. "With the Mine in the Veins: Emotional Adjustments in Female Partners of Chilean MiningWorkers." *Gender, Place & Culture* 23 (12): 1677–1688.
Svampa, M. 2015. "Feminismos del Sur y ecofeminismo." *Nueva Sociedad* 256: 129, 130.
Ulloa, Astrid. 2016. "Feminismos territoriales en américa latina: defensas de la vida frente a los extractivismos." *Nómadas* 45: 123–139.
Ureta, S., and P. Flores. 2018. "Don't wake up the dragon! Monstrous geontologies in a mining waste impoundment." *Environment and Planning D: Society & Space* 36 (6): 1063–1080.
Vergara Marshall, Angela. 2005. "The Recognition of Silicosis: Labor Unions and Physicians in the Chilean Copper Industry, 1930s–1960s." *Bulletin of the History of Medicine* 79 (4): 723–748.
Vergara Marshall, Angela. 2008. *CopperWorkers, International Business, and Domestic Politics in ColdWar Chile.* University Park, PA: The Pennsylvania State University Press.
Vilches, F. 2011. "From nitrate town to internment camp: The cultural biography of Chacabuco, northern Chile." *Journal of Material Culture* 16 (3): 241–263.
Walker, Brett. 2011. *Toxic Archipelago. A History of Industrial Disease in Japan.* Washington: University of Washington Press.
Weinberg, Marina. 2021. "Cuerpos de cobre: Extractivismo en Chuquicamata, Chile." *Journal of Latin American and Caribbean Anthropology* 26(2): 200–218.
Weinberg, Marina and Cristóbal Bonelli. 2021. "Lithium: Towards a Theory of Bipolar Transitions (Litio: hacia una teoría de las transiciones bipolares)." In Francisco Díaz, Anastasia Kubrak, and Marina Otero Verzier (Eds), *Lithium: states of exhaustion (Lithium: Estados de agotamiento).* Rotterdam: het Nieuwe Instituut; Santiago: Ediciones ARQ, 52–55.
Wollentz, Gustav, May, Sarah, Holtorf, Cornelius, and Anders Högberg. 2020. *Toxic heritage: Uncertain and unsafe.* 10.2307/j.ctv13xps9m.25
Zorzin, Nicolas. 2015. "Dystopian archaeologies: the implementation of the logic of capital in heritage management." *International Journal of Historical Archaeology* 19: 791–809.

CASE STUDY 3

STICKY, STINKY, SQUALID: THE TOXIC LEACHATE OF HOUSEHOLDS' WASTE IN AN AREA OF URBAN DECAY IN TEHRAN (IRAN)

Leila Papoli-Yazdi

Areas of Urban Decay in a Crowded City: Tehran

Tehran is a megapolis with over 8.5 million population. It is estimated that nearly 20% of the city population lives in subordinated areas. The city itself is divided into 22 districts (Figure 1). On the northern side of the city, where districts 1, 2 and 3 are located, the well-off people inhabit, while the southern side is occupied by low-income statuses.

In 2018, my colleagues and I conducted a garbology project in district 17 (Papoli-Yazdi, 2021). This place is known to be an area where the impoverished and working classes live. The western zones of the district are chronologically the oldest and date back at least 100 years ago. The newer buildings have been built adjacent to the older buildings without any premeditated urban plan. The Qanat (ancient water channels) still existed beneath the newer buildings. These channels cause much trouble for the inhabitants. Mice dig holes under the houses, which make the foundation of the houses so weak that the inhabitants live with the constant fear of their residence areas collapsing.

An overlooked yet severe problem of people living in the area is their challenges with daily garbage. The inhabitants dispose of their everyday garbage bags in the containers set by the municipality on every avenue. The garbage containers are made of metal and have four wheels set to stand on metal or tile-made frames. In the area of urban decay, each avenue has two narrow water channels on both sides on which the garbage container is usually set. These containers would be full before noontime usually and are stinky, worsening the hygiene of the district.

The interviews with the employees of the municipality revealed that the main problem is a long-term conflict between different sections of the government, which are responsible for the renovation of the historical urban decay. This conflict has led to disastrous waste management in places labelled "heritage."

DOI: 10.4324/9781003365259-14

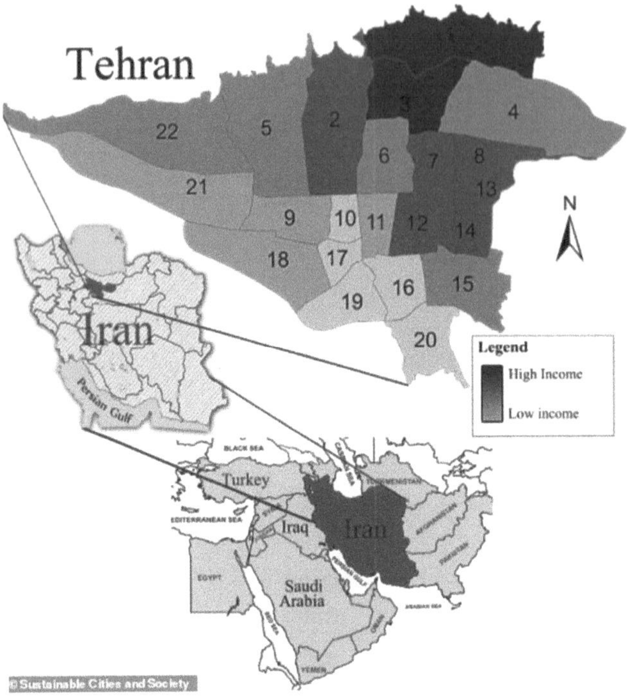

FIGURE 1 Tehran and Iran and income levels in districts of Tehran (Daily Mail: shorturl.at/kuIJL).

Waste Management in Historical Areas of Urban Decay

Some alleys in district 17 are so narrow that two people cannot walk in the alley simultaneously (Figure 2). This has made the waste management service very complicated. At the same time, the municipality ignores the impoverished areas and prefers to focus on more crowded streets, shopping malls, and streets that wealthier people occupy.

Some businesses, like auto shops, have been transferred to modest areas since the well-off people do not desire to have them in their neighbourhoods. Waste such as pieces of car engines and gasoline are discarded in the same containers where people dispose of their daily garbage. The liquid leaking from the wet waste, rain, and sometimes engine oil combines with the accumulation of daily garbage and forms layers of toxic leachate. It sticks to the wheels of the containers and runs in the water channels. When it is warm, the stench overpowers. It is of note that some impoverished inhabitants search in the containers to collect garbage and sell them (Figure 3). Digging into the containers, they tear up the garbage bags, which causes their bodies to be contaminated and the container to get more polluted.

Our studies determined that low-income inhabitants tend to segregate their discarded items more than the municipality or well-off citizens assumed. The reason

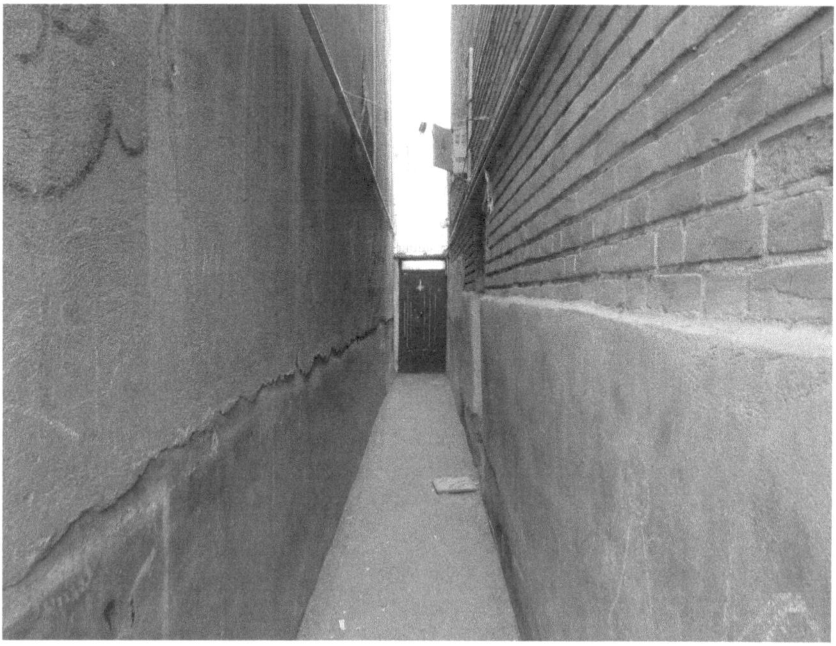

FIGURE 2 Narrow alleys in district 17 (Photo by Garbology project of Tehran).

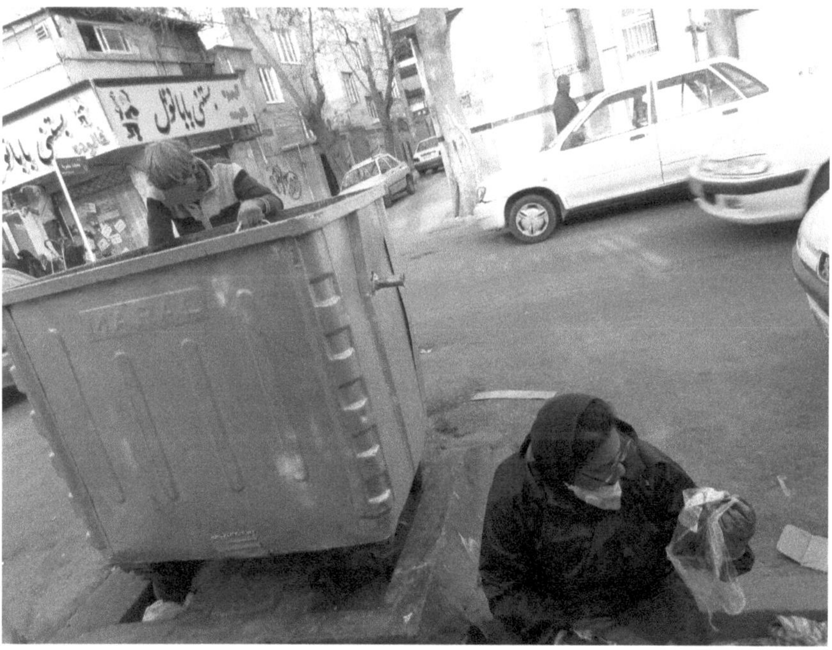

FIGURE 3 A citizen searching a garbage container for still-usable objects (Photo by Garbology project of Tehran).

is that they usually sell plastic, glass, and metal objects to garbage collectors. On the other hand, low-income households produce, on average less than 1:3 of the garbage made by well-off households. It is why I believe that the root of the problem is in how the waste management system organizes its services in modest areas of urban decay. The difference between the "historical districts" located on the northern side of the city and the areas of urban decay on the southern side is that the northern ones obtain annual and occasional renovation and waste management services. One reason for this distinction might be that the city's southern side is no tourist attraction. All of these indicate the dual policy of MCHHT about heritage which changes based on the poverty or richness of the owners of the historic buildings or districts.

Sticky, Stinky, and ... Heritage

Two major sections of the government are responsible for reconstructing and preserving urban decay in Iran. The first is the "Ministry of Cultural Heritage, Handicrafts and Tourism (MCHHT)," and the second is the municipalities' "Reconstruction and development of areas of urban decay" sections. The conflict arises when urban decay is labelled "heritage" by MCHHT while people still live in that area. Being "heritage" means that the neighbourhood's buildings would be located under the Antiquities law (see Falahfar, 2015). Based on the law, it is illegal to reconstruct a "heritage" monument or area of urban decay, and only minor repairs have been anticipated to be performed in such areas.

The case of district 17 shows that historical areas need a particular plan for waste management. Otherwise, the accumulation of daily garbage can potentially create toxicity in the area. Nevertheless, the organized discrimination between the "northern heritages" and "southern heritage" indicates that the government focuses on particular areas and ignores other parts of the city even if they are also heritage. In other words, the stinky squalid areas of urban decay in district 17 result from a policy in Iran: all heritages are equal, but some heritages are more equal!

References

Falahfar, Saeed. 2015. *1309: Ghanoon-e hefz-e athar-e melli va atigqat [1931: The laws of Antiquities and preservation]*. Tehran: Avineh (in Persian).

Papoli-Yazdi, Leila. 2021. The archaeology of a marginal neighborhood in Tehran, Iran: garbage, class, and identity. *World Archaeology*, 53(3): 547–562.

8

TOXIC LANDMARKING AND TECHNOPRECARIOUS HERITAGE IN GHANA

Peter Carskadon Little and Grace Abena Akese

Introduction

Toxic heritage is a way of understanding social and environmental transformation. In this chapter, we engage electronic waste (e-waste) toxicity and urban transformation in Ghana to critically unpack how heritage thinking and rethinking might help us reckon with a postcolonial site that is always in flux. In particular, we explore dimensions of toxic heritage in Agbogbloshie, Ghana, a site known for e-waste ruination, scrap business, and extreme toxicity. Our guiding questions include: How does e-waste toxicity get reproduced alongside and in response to urban transformation? What remains in a site of large-scale urban decongestion and socio-environmental conversion? How do erasure, resettlement, and advocacy art projects figure in the toxic heritage imaginaries at the site? Finally, what can the shifting toxic e-waste landscape in Ghana teach us about what we shall call *technoprecarious heritage* more generally?

The global electronics industry involves contentious residues of toxicity and discards (Smith, Sonnenfeld, and Pellow 2006; Lepawsky 2018). Following the insight of the techno-feminist Precarity Lab, these toxic traces of tech remind us of the broader need to "turn to how toxicity—the spread of environmental harm and vulnerability in the depletion economy—is the condition for digital production today" (Precarity Lab 2020, 24). Here we want to explore how this tech toxicity is meshed with histories of urban transformation and precarious "development" in the Global South. For us, precarity can be understood as "a state of being and lived experience of insecurity, loss of control, and unpredictability of one's world" (ibid 2020, 100). The chapter draws on ethnographic research to critically examine and reinterpret traces of urban land management and transformation in an area of Accra, Ghana, known as Agbogbloshie. This vibrant urban market became a legacy e-waste hot zone in the

DOI: 10.4324/9781003365259-15

early 2000s. More recently, the market has been relocated, calling for a renewed understanding of toxic e-waste heritage that also includes complex dynamics of erasure, resettlement, and landmarking.

Our primary contribution to "toxic heritage" theory is to more fully engage shifting urban geographies of e-waste and what we shall call *technoprecarious heritage*. We shall argue that the concept fits the "toxic heritage" theory for a few reasons. First, technoprecarious heritage is an approach to heritage that strongly considers the interlocking materiality and social precarity that emerges in a fully globalised tech industry, with a planetary discard situation where toxicity is the condition of production. It pulls analysis and critique of the tech industry into broader discussions and theories of social and environmental precarity. Second, the explosion of concern with digital discards and their toxic legacies is tied to deeper dynamics of precarity that involve infrastructural inequalities and basic income struggles. Additionally, we contend that urban transformations and resettlement unfolding in Agbogbloshie call for a perspective of toxicity and contamination that also deals with how "residual materialism" (Boudia et al. 2022) exposes deeper heritages of land management, heritages of pollution response, heritages of urban neoliberalisation, and heritages conditioned by certain geographies of forced eviction and displacement (Morrison and Davies 2017).

When we engage toxic heritage in a transformative place and space like Agbogbloshie, a multidimensional sense of plunder, transformation, and remainder becomes critical. This is largely the case because these sites and urban zones of technoprecarious heritage reinforce the power of precarity in shaping what we know about toxic e-waste legacies and the critical and ongoing transformations emerging in Ghana's urban scrap metal economy. In this way, our approach to technoprecariuos heritage seeks to push the conceptual boundaries of "residual materialism" (Boudia et al. 2022). This approach to toxic heritage is also sensitive to related theories of toxic legacy. Toxic heritage, like toxic legacy, is "more than material (after)life; legacy is lived through the cultural meanings of place and community and the routinisation of the social systems—regulations, zoning, and evaluations of health among others. Legacy is inherited but continuously reinforced" (Boudia et al. 2022, 23). In Agbogbloshie, many of the struggles that come from these processes of inheritance and reinforcement are what play up the centrality of precarity in matters of toxic heritage.

As members of the Precarity Lab remind us, "to examine precarity in our contemporary moment is to be attuned to the damages, the stubborn remainders (or reminders) of modes of power invested in the differential management of human and nonhuman lifeworlds" (Precarity Lab 2020, 28). Toxic heritage, in this sense, involves precarity at its core. It involves the movement and settlement of damage, a landscape of shifting patterns of "wastelanding" (Voyles 2015), of "e-wastelanding" (Little 2022), and a place with ongoing and overlapping disasters (Akese and Little 2018). Agbogbloshie has been described as a location of "twin disasters"; disasters of fire and flood, and disasters emerging from neoliberal urbanisation (Oteng-Ababio and Grant 2019). This ongoing neoliberalisation is deeply fused with elusive urban regeneration, sceptical slum upgrading, and contentious decongestion politics that reinforce, rather than mitigate, power relations.

The rest of the chapter is organised as follows. We begin with a brief overview of Agbogbloshie, discussing the urban reordering and environmental transformations that have shaped this area of Accra. Next, we draw on ethnographic fieldwork to report on local worker experiences with market demolition and relocation efforts which started in the summer of 2021. We engage the lived experiences of those displaced from the former e-waste processing site and showcase how "toxic heritage" touches workers and their bodies, even if it is emplaced and located or dislocated due to urban planning and other land management processes. Additionally, we attend to how toxic legacy persists amid large-scale urban reordering and even creative neoliberal slum interventions inspired by, among other things, international art and finance.

Brief History of the Agbogbloshie Area

Agbogbloshie is in constant flux and conversion. At the time of writing, the infamous "the world's largest e-waste dump" is no more. On July 1st, as part of the Greater Accra Regional Ministry's urban transformation campaign of #letsmakeAccrawork, the scrapyard was torn down, and market activities there completely stopped. What appears to be a complete demolition of the scrapyard had precedents in evictions and similar violent demolitions in 1993, 2002, 2004, and 2015 (Lepawsky and Akese 2015; Little 2022). The Agbogbloshie scrapyard was located on a strategic and highly contested land in Accra (Grant 2006; Akese and Little 2018). The scrapyard sits on the banks of the Korle-Odaw Basin, a water receptacle that drains about 60% of the Accra metropolis. (Karikari, Asante, and Biney 2006). As an important cultural and socioeconomic asset of the Indigenous Ga people of Accra, the lagoon supported local fishing, leisure activities, and religious practices in the past. During the colonial era, the British feared that the lagoon and surrounding environment were a "native reservoir" that unleashed dangerous tropical fevers and malaria parasites (Roberts 2010). They planned to transform the lagoon into a deep-water harbour to facilitate cocoa export for the colony. This vision of urban health and colonial wealth clashed with the spiritual practices and beliefs of the local Ga population, for whom the lagoon was sacred (Roberts 2010). The harbour project never materialised, and upon independence, the Ghanaian government designated the Korle and its environs (including where Agbogbloshie sits now) as a green wedge – green space on the city's fringes. Indigenous Ga's still revered the Korle Lagoon as a deity; as such, the area features prominently in their cultural heritage (Roberts 2010).

Developments in post-independence Accra steadily changed the lush, green environment of the lagoon. Struggling economically and with a robust socialist agenda, the government allocated parts of the area as a light industrial zone in support of import substitution industrialisation. Also, as rapid urbanisation post-independence resulted in housing shortages and growing informal commerce

(Yankson and Bertrand 2012), migrant communities built informal housing and livelihoods outside the formal regulatory and planning systems on the fringes of the lagoon (Arku 2009). This marked the beginning of the Agbogbloshie area as a vibrant informal zone where commercial, industrial, and residential activities overlap (Figure 8.1).

The broader Agbogbloshie area comprises four main activity zones: (1) the Agbogbloshie scrapyard where e-waste processing happens; (2) the Agbogbloshie food market; (3) Old Fadama, an informal settlement; and (4) a light industrial area. The Agbogbloshie scrapyard and Old Fadama function as an extended settlement: about 90% of the 4500–6500 estimated workers at the scrapyard make Old Fadama their home (Prakash and Manhart 2010). Residents do not have statutory rights to the land, although some of them have lived there for over four decades and thus have "a collective property claim" to this urban space due to "sustained use and appropriation" (Grant 2006; Gillespie 2016, 69).

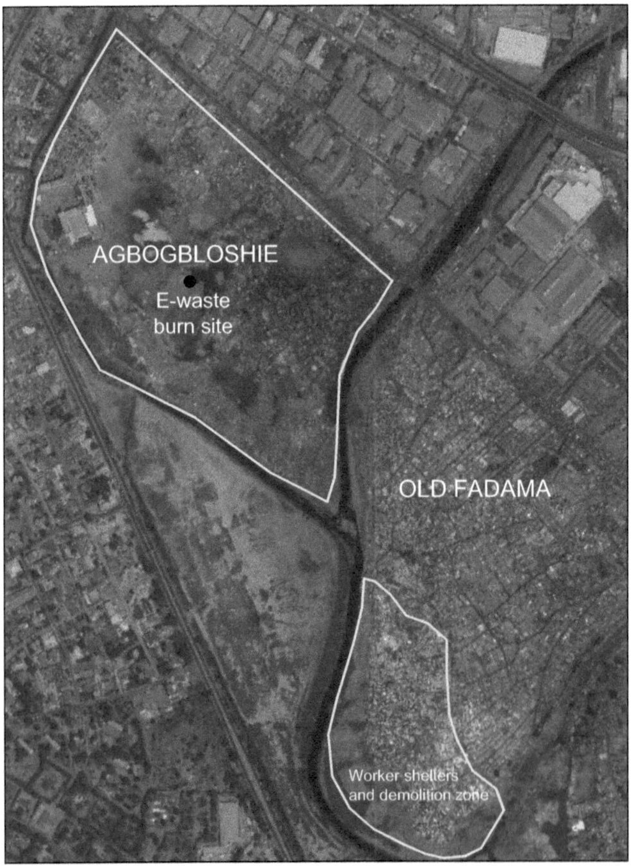

FIGURE 8.1 Area map of Agbogbloshie.

Although colonial entanglements and multiple socio-ecological struggles remain salient within the broader Agbogbloshie area (see Akese and Little 2018; Little 2022), what has become contentious in recent times is the sites' toxicity. The chemical residues from e-waste processing are well documented. Toxicological studies have identified pathways of multiple exposures: water, air, and food. The Korle Lagoon and Odaw River have higher than normal amounts of lead, cadmium, mercury, and other organic compounds (Caravanos et al. 2011; Hosoda et al. 2014; Wittsiepe et al. 2015). Without any protective gears and with this rudimentary burning process, workers' bodies biologically accumulate toxicants as they visibly demonstrate tearing and breakages (Caravanos et al. 2013; Wittsiepe et al. 2015; Issah et al. 2021). Food produced and sold in the area is also contaminated by e-waste processing (Petrlik et al. 2019; Takyi et al. 2020). Beyond toxic residues from e-waste activities, the Korle Odaw Basin is also a sink for the city's other toxic waste, including untreated fecal and mortuary waste (Boadi and Kuitunen 2002; Karikari et al. 2006). Agbogbloshie cannot escape a toxic landscaping imagination even as attempts to clean up the site and reduce the sources of pollution, as articulated in projects such as the Korle Lagoon Ecological Restoration Project (KLERP), exist. Toxic accumulation persists and endures amidst lagoon restoration efforts and other creative neoliberal e-waste risks mitigation projects, such as the German development organisation GIZ's Technical Training Centre and Health Post built in 2019, as well as Pure Earth and Green Advocacy Ghana's model e-waste recycling facility (Little and Akese 2019; Little 2016, 2022).

Landmarking Agbogbloshie: Remains of Toxic Resettlement

When the Agbogbloshie market was bulldozed and relocated in the summer of 2021, the space converted quickly from a place of "waste" or "refuse" dumping and acceptance to a space of explicit waste refusal (See Figure 8.2). While workers are busy constructing a wall around the old Agbogbsloshie scrap market, the path that crosses this area of land is still used as a footpath for people crossing from Ring Road (the major roadway that encircles Accra). What is interesting, for our purposes here, is not just how quickly this market was erased and relocated – this process of relocation has been ongoing for more than a decade – but rather how workers remained, re-organised, and remade the market, and how amidst removal and relocation, urban technoprecarity is still widespread, multi-sited, and toxicity remains.

Making sense of toxic heritage in e-waste sites like Agbogbloshie highlights the need to consider the complex ways urban land conversion happens amidst toxic residue and enduring scrap metal salvaging (See Figure 8.3). For Mohammed, a scrap worker from Tamale in northern Ghana, the technical training centre and e-waste recycling facility are close enough to see and touch. Still, their goals of "formalising" don't change his situation. Salvaging scrap metal in the earthen layers of long-term e-waste and scrap metal economic activity is a way of labouring in a toxic heritage site, even in a space targeted for widespread erasure, relocation, and socio-economic

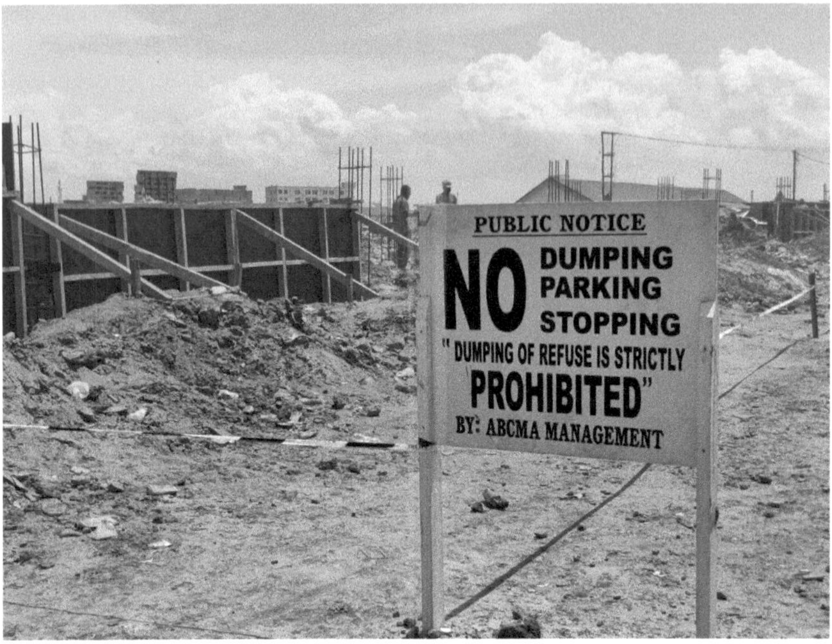

FIGURE 8.2 New Rules, New Walls. Photo by Peter C. Little.

rearrangement. The continued reuse of erased land tells us something important about e-waste heritage and the ongoing lure of scrap metal in this area of Accra. This enduring scrap metal labour, then, is situated and precarious labour that reminds us of the complex and ongoing heritages of "e-wastelanding" (Little 2022) processes in Ghana.

Among workers in Agbogbloshie, there is confusion about what the bulldozed land will become. Some of this confusion stems from unclear local media coverage. For example, some early reports on the ground in August 2021 hint that the redevelopment of the land is being folded into a government-supported project called Agenda 111, which prioritises health infrastructure projects in Ghana. Importantly, Agenda 111 involves a regional hospital consulting team led by HIG (Hospital Infrastructure Group Ltd), an Accra-based project management, consulting, and financing agency that "seeks to provide excellent healthcare infrastructure as a turnkey solution for both government and private clients … Our delivery solution is technology-driven from design, financing, project management, and construction, offering our clients a seamless turnkey approach" (https://agenda111gh.com/services/). A year after the demolition, there is still mystery over what will become of the Agbogbloshie land. Even the recently erected billboard at the front gate of the newly constructed fence wall doesn't give a clue about future plans (See Figure 8.4). As noted on the board, the only project commissioned is "the construction of the fence wall for the 'reclaimed' Agbogbloshie land."

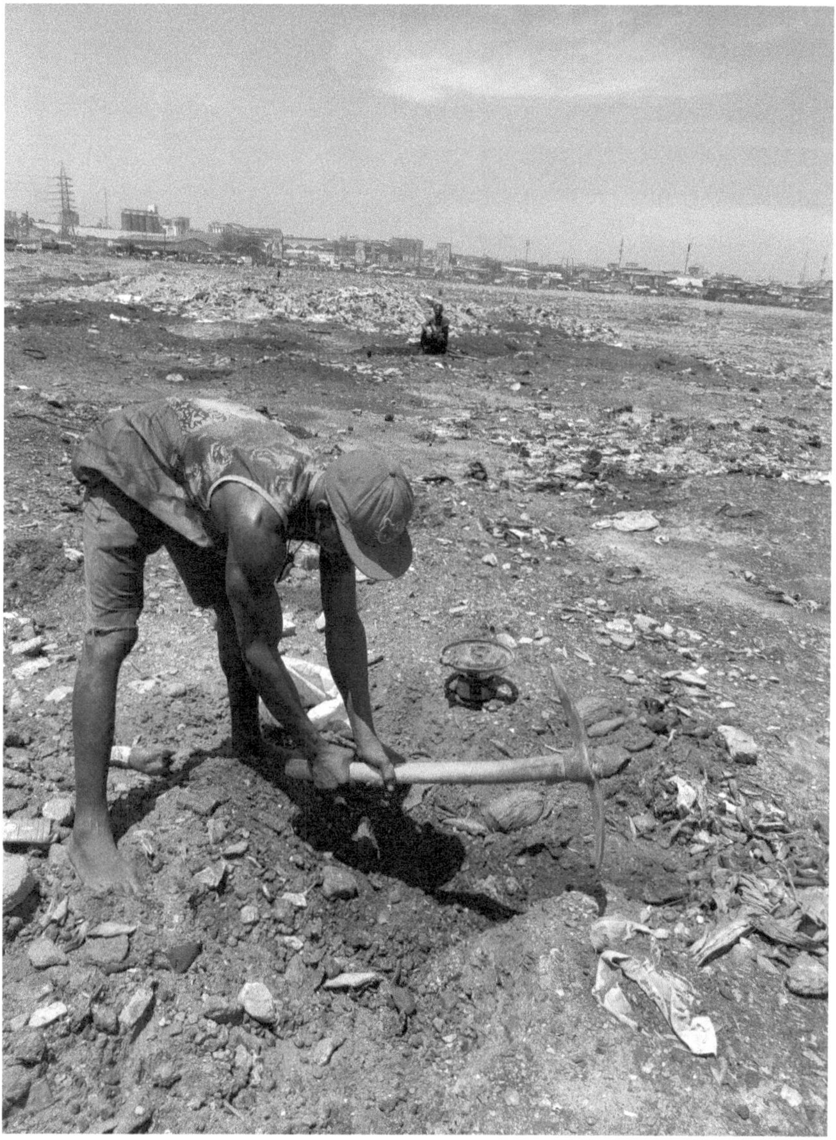

FIGURE 8.3 Mohammed digging for iron scraps in bulldozed Agbogbloshie. Photo by Peter C. Little.

But there is no confusion about what remains at this e-waste heritage site. The remaining buildings and projects include a Technical Training Centre and Health Post established in 2019 (See Figure 8.5) and a public-private partnership focused on recycling e-wastes such as aluminium-containing cables, mixed low-grade cables, and iron-containing cables. A project launched in June 2020 and overseen

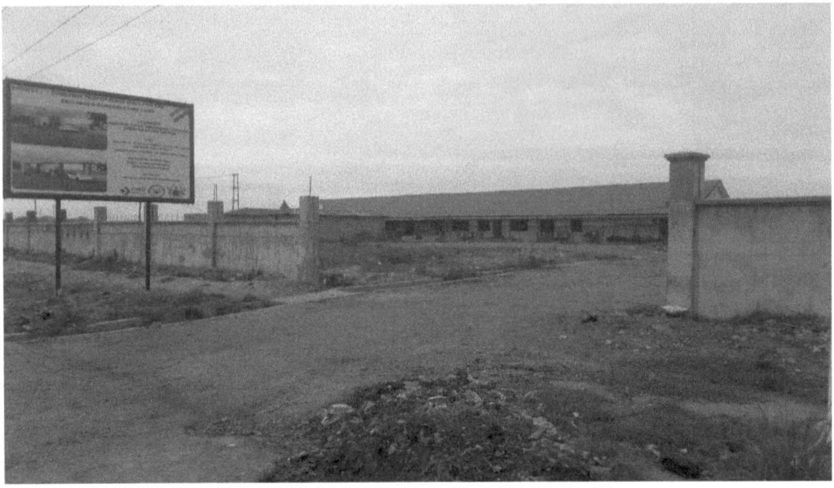

FIGURE 8.4 Completed fence wall around the Agbogbloshie land. Photo by Grace Akese.

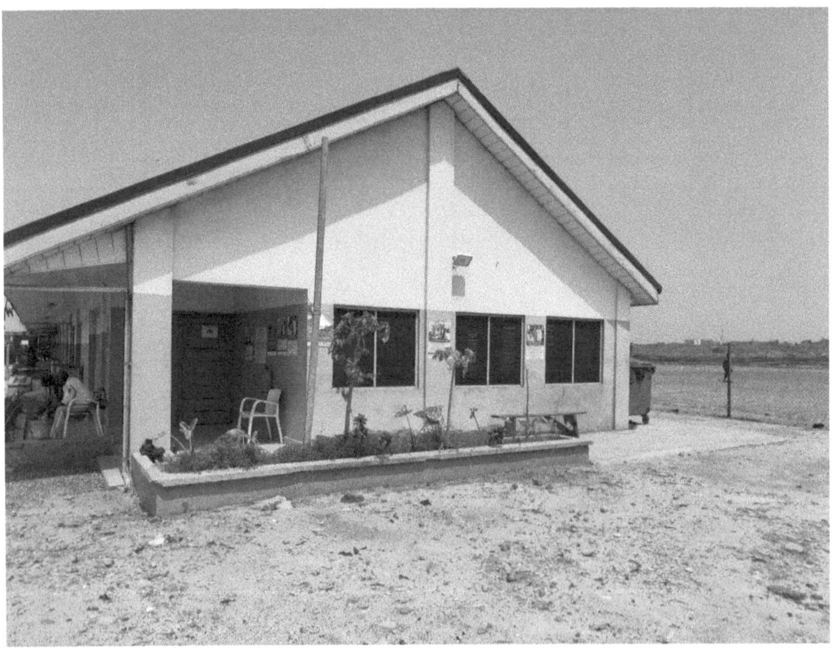

FIGURE 8.5 The Remaining Technical Training Centre and Health Post. Photo by Peter C. Little.

by Ghana's Ministry of Environment, Science, Technology, and Innovation, the facilities and projects work in concert with each other to achieve several shared goals. Most importantly, there is an effort to "minimise negative environmental and human health impacts from improper management of electronic wastes" and "transition the most polluting e-waste recovery activities from the informal sector to the formal sector without removing jobs from the lowest income group." While drawing attention to the environmental-health-economic relationship more explicitly in Agbogbloshie is an effort that needs increased attention and support, several points of friction and precarity remain.

First, what is concerning is that the actual removal of people from the land has removed the target population in focus in this ongoing project. E-waste workers mostly fall within this lowest income group, and nowhere are they to be seen at these enduring centres and posts. And second, the minimisation of toxic exposures and environmental health risks has not occurred at all. Those removed are now more concentrated, experiencing an even more dense work environment that is even more at risk of flooding in this area of the Korle Lagoon. Based on ethnographic research in May and August 2022, congestion in the new market space is a reported source of problems. As one worker noted, "There be too many workers here now. It is no good. Everyone be together now. Trucks be stuck all the time now." The road connecting the new scrap metal market and the nearest major road (Abosei-Okei Road) is packed with scrap worker stations, and trucks moving in and out have little room to manoeuver. As one worker reported, everyone in the new scrap site near the main road is expecting to be moved again soon, as the congestion has gotten unmanageable for many of the scrap metal workers.

Additionally, the materiality of "settlement" itself is important to our techno-precarious heritage approach. For example, workers living in Old Fadama and near Bola Top and the relocated scrap metal market mostly live in makeshift wooden structures, unlike the cement buildings remaining at the Agbogbloshie site. As one worker put it, "wood be permanent, cement forever." Infrastructure tells a story of power and risk in this critical zone of Ghana's urban margins, highlighting what landmarks remain and those that have been discarded and erased. Of those discarded include the Agbogbloshie Makerspace Platform (AMP), a collaborative initiative to create an innovation hub at Agbogbloshie by bringing together researchers, workers, and students in the field of science and technology (Osseo-Asare and Abbas 2015). AMP operated through a 'spacecraft' structure and concept, which they describe as a "small-scale, mobile, incremental, low-cost and open-source platform … … offering a set of tools and equipment [for workers and collaborators] to craft space and terraform their environment" (qamp.net, 2022).

What shall we make of a remaining technical training centre and health post in a toxic landscape that for two decades served as a robust urban market dominated by scrap metal, goats, and onions? These infrastructures have become a certain landmark. They are buildings with plaques that confirm their founding and the founders involved. They are the only buildings that remain in what is, otherwise, a bulldozed landscape. Of course, the politics of landmarking is not new to critical heritage studies, but the politics of landmarking in zones of toxic heritage in Accra is. That new terrain of heritage-making is what we are trying to engage and highlight as a central node in e-waste toxic heritage studies more broadly. For many years Agbogbloshie was an active symbol of toxic heritage in Ghana, and e-waste was a centrepiece of this dominant narrative (Little 2022). But now, it is an urban space emptied of people but remains a toxic heritage site of idle, slow, and precarious development and intervention. It contains empty facilities which, ironically, were built to serve workers pushed elsewhere and further marginalised from "formalised" socio-economic change and local-level development projects aiming to assist with actual labour "transition."

Scrap metal and e-waste market resettlement, in this case, *is* toxic resettlement and relocation. Workers who were removed from Agbogbloshie in the summer of 2021 have created a new e-waste burn site, a new toxic legacy site in the making just across the river from the old Agbogbloshie market. The copper wire burners reorganised to form one primary group, a mixture of Dogomba and Frafra-speaking workers from Bolgatanga. The new work site, "Bola Top"– *bola* has a Yoruba origin, meaning "wealth" – sits below a massive mound of discard. Bola Top is located on the southwest side of a massive mound of refuse. One worker reports that the mound has been forming for about six years. In Bola Top, as in the old Agbogbloshie market, we are dealing with a flexible and precarious toxic heritage, a moveable and flexible social economy. Workers haven't just become the subjects of some outside "toxic heritage," but instead exist within and in response to ongoing technoprecarious conditions that endlessly remind us of life in the work of legacy. In other words, we ought to see toxic heritage and legacy in Agbogbloshie *as* the ongoing situatedness of living bodies doing not just the hard and toxic work of metal scrap recovery and recycling but also the difficult labour of precarious market resettlement.

A central theme of Agbogbloshis's toxic e-waste legacy is the broader political legacy of erasure, removal, and land control. Following Lepawsky and Akese (2015), we can't even imagine e-waste, or e-waste relocation politics for that matter, in Ghana without careful attention to the "struggles over land rights in Accra that have nothing to do with where the 'West's' e-waste goes to die." Few studies on e-waste at Agbogbloshie directly engage with the site's broader local political economy as relevant to understanding e-waste legacy. Andersson underscores this neglect when he notes that "representation of e-waste is characterized by portrayals of e-waste dumping grounds as strange, alien spaces … with no clear and comprehensive histories" (2017, 2). At Agbogbloshie, histories of erasure and land struggles are central in the play out of toxic legacies (Akese and little 2018).

Landmarking Agbogbloshie: Creative Neoliberal Installments

Based on ethnographic fieldwork carried out in May 2022, follow-up interviews with workers removed from the old Agbogbloshie scrap metal market report continued copper wire burning. To some, this may be a surprise because one primary goal of the removal project was to mitigate the risks of wire burning. But toxic burning continues in full force at Bola Top, the newest toxic legacy site in the making. Toxic heritage has also taken a more creative neoliberal turn in recent years. For example, in recent years, Agbogbloshie has captured the attention of global artists interested in slum advocacy art projects. For example, a partnership spearheaded by the successful Japanese artist Nagasaka Mago (or MAGO) attempts to bridge local e-waste workers and art. According to the artist's website, "Agbogbloshie is an impoverished area in Ghana known as the world's largest graveyard of electronic devices. Under the slogan 'Sustainable Capitalism,' artist Nagasaka Mago (MAGO) has set out to change the problems of poverty and the environment in this area with the power of art." The *Full Moon* piece (See Figure 8.6), made entirely of local discard, including e-waste, was installed in the new scrap metal market in Old Fadama in 2018 and remained in place. A resident of Old Fadama and a scrap market worker report that it is sometimes illuminated at night, and someone watches over the art installation to ensure it is not vandalised.

MAGO held a solo exhibition in Los Angeles in 2019, an exhibit that was described as follows: "You will be able to see many truths in this work: capitalism, the truth of environmental destruction, true love beyond race, and why art is necessary to the world." MAGO is also the director of a new film *Still A Black Star*. Crowdfunding for the movie surpassed its target, raising $300,200 from 395 supporters. As a result of the film, he has sold 150 artworks. The film is also under reproduction and is to be made into a Hollywood movie. Aiming to be a stimulator for local jobs and getting wire burners to stop burning, MAGO Creations established a facility that hires workers to sort plastics from e-waste. Next to this facility is also an exhibit space featuring the art of Old Fadama's youth. Apparently, the revenue generated from art sales goes directly to those living in Old Fadama. In fact, according to his website, his aim is to raise, by 2030, 10 billion yen – roughly $74 million dollars – and "give it to the people in Old Fadama." This current situation in Agbogbloshie, then, brings the whole concern with toxic heritage and resettlement into the realm of creative slum development and the deeper-rooted environment of scepticism this new wave of intervention brings with it. For us, this new situation hints at the unexpected forms of precarity emerging in a transformative urban environment that remains a significant anchor point for broader interrogations and critiques of the e-waste toxic legacy in Ghana (Little 2022; Akese and Little 2018).

Conclusion

Addressing "toxic heritage" in Agbogbloshie involves acknowledging multiple project traces and heritages. Local heritages of scrap metal toxicity, removal and erasure, resettlement and creative instalment, and rearrange and break down the boundaries between heritages. Heritage, like contamination itself, "carries the sense

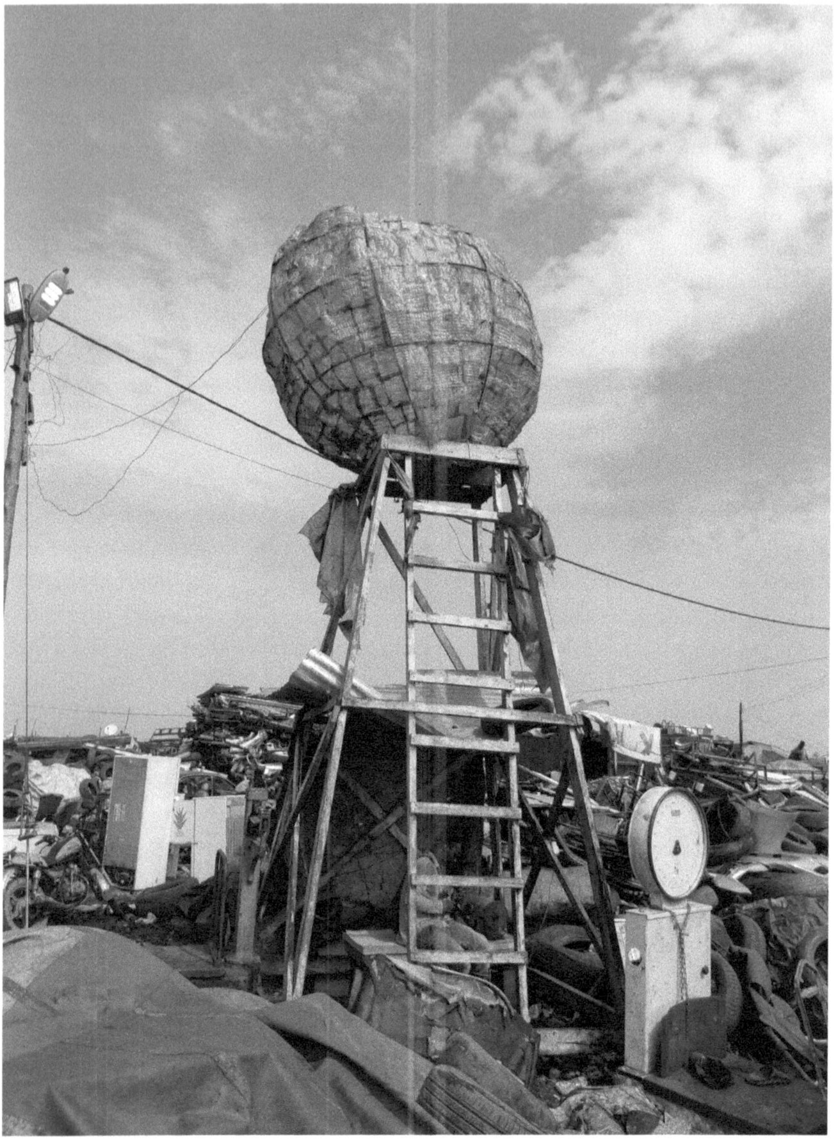

FIGURE 8.6 Art installment in the new scrap metal market in Old Fadama. Photo by Peter C. Little.

that boundaries may no longer hold, that the future is as open-ended as our lives have become, and that purity is no longer an option" (Bond 2021, 1). Toxic heritage has a location yet is everywhere at the same time. For this reason, toxicity doesn't just remind us that a spill happened, a toxic export or import happened, or that heavy industrial activity and ruination took place somewhere for some period of time. Toxicity is now an always and forever thing and challenge not just for some communities but worldwide (Nading 2020). That toxic struggle is dynamic and always a lived experience that involves various degrees and moments of oppression and resistance, development, and change.

Despite all the urban regeneration plans, decongestion campaigns, and intervention projects – from formal waste management to international art investments – precarity in Agbogbloshie persists. Under these conditions and practices of neoliberal urbanisation, "Agbogbloshie, therefore, remains intact as a poor, spontaneous sprawling settlement" (Oteng-Ababio 2019, 7). Toxic heritage, in a sense, is simultaneously precarity heritage in this case. Toxic tech and precarity are equally intact, despite urban transformations. This mix of processes, for us, has made Abgogbloshie a complex space with a living *technoprecarious heritage*. While toxicity is undoubtedly in the ground, air, water, and food, and ever-present in this ever-changing urban environment, the overemphasis on toxic heritage critically downplays certain dimensions of cultural heritage and legacies of land management in Accra. As we have explored here, knowing the toxic heritage in Agbogbloshie involves critically accounting for, remembering, futuring, and connecting toxicity and precarity to shifting land-power relations and dynamics. Refocusing on Agbogbloshie as a place of shifting technoprecarious heritage, rather than simply toxic e-waste legacy, reminds us of the need to rethink, recentre, and continuously engage critical heritage perspectives in zones of e-waste toxicity.

References

Akese, Grace A., and Peter C. Little. 2018. "Electronic Waste and the Environmental Justice Challenge in Agbogbloshie." *Environmental Justice* 11 (2): 77–83. 10.1089/env.2017.0039.

Arku, Godwin. 2009. "Housing Policy Changes in Ghana in the 1990s." *Housing Studies* 24 (2): 261–272. 10.1080/02673030902719763

Boadi, Kwasi Owusu, and Markku Kuitunen. 2002. "Urban Waste Pollution in the Korle Lagoon, Accra, Ghana." *The Environmentalist* 22: 301–309.

Bond, David. 2021. "Contamination in theory and protest." *American Ethnologist* 48(4): 386–403. 10.1111/amet.13035

Boudia, Soraya et al. 2022. *Residues: Thinking through Chemical Environments*. New Brunswick, NJ: Rutgers University Press.

Caravanos, Jack, Edith Clark, Richard Fuller, and Calah Lambertson. 2011. "Assessing Worker and Environmental Chemical Exposure Risks at an E-Waste Recycling and Disposal Site in Accra, Ghana." *Journal of Health and Pollution* 1 (1): 16–25.

Caravanos, Jack, Edith E. Clarke, Carl S. Osei, and Yaw Amoyaw-Osei. 2013. "Exploratory Health Assessment of Chemical Exposures at E-Waste Recycling and Scrapyard Facility in Ghana." *Journal of Health and Pollution* 3 (4): 11–22.

Gillespie, Tom. 2016 "Accumulation by Urban Dispossession: Struggles over Urban Space in Accra, Ghana." *Transactions of the Institute of British Geographers* 41 (1): 66–77. 10.1111/tran.12105.

Grant, Richard. 2006. "Out of Place? Global Citizens in Local Spaces: A Study of the Informal Settlements in the Korle Lagoon Environs in Accra, Ghana." *Urban Forum* 17: 124. 10.1007/BF02681256

Hosoda, Junki, John Ofosu-Anim, Edward Benjamin Sabi, Lailah Gifty Akita, Siaw Onwona-Agyeman, Rei Yamashita, and Hideshige Takada. 2014. "Monitoring of Organic Micropollutants in Ghana by Combination of Pellet Watch with Sediment Analysis: E-Waste as a Source of PCBs." *Marine Pollution Bulletin* 86 (1): 575–581. 10.1016/j.marpolbul.2014.06.008

Issah, Ibrahim, John Arko-Mensah, Laura S. Rozek, Katie R. Zarins, Thomas P. Agyekum, Duah Dwomoh, Niladri Basu, Stuart Batterman, Thomas G. Robins , and Julius N. Fobil. 2021. "Global DNA (LINE-1) methylation is associated with lead exposure and certain job tasks performed by electronic waste workers." *International Archives of Occupational and Environmental Health* 94: 1931–1944.

Karikari, Anthony, Kwadwo A. Asante, and Charles Augustus Biney. 2006. "Water Quality Charateristics at the Estuary of Korle Lagoon in Ghana." *West African Journal of Applied Ecology* 10(1). 10.4314/wajae.v10i1.45700

Lepawsky, Josh. 2018. *Reassembling Rubbish: Worlding Electronic Waste*. Cambridge, MA: MIT Press.

Lepawsky, Josh, and Grace Akese. 2015. "Sweeping Away Agbogbloshie Again." *Discard Studies* (blog). Accessed November 11, 2022. https://discardstudies.com/2015/06/23/sweeping-away-agbogbloshie-again/

Little, Peter C. 2022. *Burning Matters: Life, Labor, and E-Waste Pyropolitics in Ghana*. New York: Oxford University Press.

Little, Peter C., and Grace Abena Akese. 2019. "Centering the Korle Lagoon: Exploring Blue Political Ecologies of E-Waste in Ghana." *Journal of Political Ecology* 26(1):448–465. 10.2458/v26i1.22988

Morrison, N., and M. Davies. 2017. "Struggling for the right to be recognized: The informal settlement at Old Fadama, Accra, Ghana." In *Geographies of Forced Evictions*. K. Brickell, A. Fernandez, and M. Vasudevan, eds., pp. 25–45. London: Palgrave Macmillan.

Nading, Alex. 2020. "Living in a Toxic World." *Annual Review of Anthropology* 49: 209–224. 10.1146/annurev-anthro-010220-074557

Osseo-Asare, D. K. , and Abbas Yasmine. 2015. Investigating 3E-materials at Agbogbloshie in Accra, Ghana. In *2015 Conference on Raising Awareness for the Societal and Environmental Role of Engineering and (Re)Training Engineers for Participatory Design (Engineering4Society), Leuven, Belgium*. pp. 41–50.

Oteng-Ababio, Martin, and Richard Grant. 2019. "Ideological traces in Ghana's urban slums: How do traces get worked out in Agbogbloshie, Accra?" *Habitat International* 83: 1–10. 10.1016/j.habitatint.2018.10.007

Prakash, Siddharth, and Andreas Manhart. 2010. "In Cooperation with Yaw Amoyaw-Osei (Green Advocacy Ghana) Obed Opoku Agyekum (Green Advocacy Ghana)." Socioeconomic Assessment and Feasibility Study on Sustainable E-waste Management in Ghana. Accessed November 11, 2022. https://www.sustainable-recycling.org/wpcontent/uploads/2015/07/Ghana_Oeko_2010.pdf

Precarity Lab. 2020. *Technoprecarious*. London: Goldsmiths Press.

Petrlik, Jindrich, Jim Puckett, Lee Bell, and Joe DiGangi. 2019. "Weak Controls: European E-Waste Poisons Africa's Food Chain." Accessed November 11, 2022. https://ipen.org/sites/default/files/documents/final_ghana-egg-report-v1_6-web_copy.pdf

Qamp.net – Agbogbloshie Makerspace Platform project website. 2022 Accessed November 11, 2022. https://qamp.net/

Robers, Jonathan. 2010. "Korle and the Mosquito: Histories and Memories of the Anti-Malarie Campaign in Accra, 1942–5." *The Journal of African History* 51(3): 343–365. 10.1017/S0021853710000502

Smith, T., D. A. Sonnenfeld, and D. N. Pellow, eds. 2006. *Challenging the Chip: Labor Rights and Environmental Justice in the Global Electronics Industry*. Philadelphia: Temple University Press.

Takyi, Sylvia, N. Basu, J. Arko-Mensah, D. Dwomoh, A. Nti, L. Kwarteng, A. A. Acquah, T. G. Robins, and J. N. Fobil (2020). "Micronutrient Status of Electronic Waste Recyclers at Agbogbloshie, Ghana." *International Journal of Environmental Research and Public Health* 17(24): 9575. 10.3390/ijerph17249575

Voyles, Traci Brynne. 2015. *Wastelanding: Legacies of Uranium Mining in Navajo Country*. Minneapolis: University of Minnesota Press.

Wittsiepe, Jürgen, Julius N. Fobil, Holger Till, Gerd-Dieter Burchard, Michael Wilhelm, and Torsten Feldt. 2015. "Levels of Polychlorinated Dibenzo-p-Dioxins, Dibenzofurans (PCDD/Fs) and Biphenyls (PCBs) in Blood of Informal e-Waste Recycling Workers from Agbogbloshie, Ghana, and Controls." *Environment International* 79: 65–73. 10.1016/j.envint.2015.03.008

Yankson, Paul W. K., and Monique Bertrand. 2012. "Challenges of Urbanization in Ghana." In *The Mobile City of Accra*. Ed by Elizabeth Ardayfio-Schandorf, P. W. K. Yankson, and Monique Bertrand, 25–44.

SECTION 3

INTRODUCTION: AFFECTED COMMUNITIES, ACTIVISM, AND AGENCY

Elizabeth Kryder-Reid and Sarah May

This section foregrounds the impact of environmental harm on communities and the ways in which people, as biological citizens, community activists, and environmental justice frontline workers, have responded. A central premise of this work is that toxic materials are not only unevenly distributed, but unevenly perceived. For example, industries may use their privilege to sanitise places with dangerous legacies. Communities may resist or become habituated to risks that are unavoidable and unwanted legacies that must simply be endured (Wateau et al., Browning). The community-centred studies highlight the agency of people to advocate for their rights and to be advocates for their communities in the face of systemic and structural forces. They may wield heritage as a tool for self-determination and advocacy (Fiske) and navigate toxic heritage as it becomes a bargaining chip between communities and developers negotiating the future of contaminated places (Lou).

Heritage as a form of cultural practice is inherently concerned with community engagement and participation. An important issue that toxic heritage brings to the fore is the fact that the management of toxic materials often excludes members of the public from sites and landscapes. As Cusack-McVeigh argues in the case of Indigenous objects treated with arsenic in museum collections, it is not merely the object that is poisoned, but the relationships with ancestors and spiritual integrity of the community that are endangered. Evia et al. wrestle with the tensions of reckoning with the damage of past practices when that same history is also the legacy of family, work, lifeways, etc. This section also raises important issues about the conditions and constraints of engaging with toxic materials as heritage and how that differs from other types of heritage. Toxic heritage can be something to be remediated, faced (Evia et al.), avoided (Wateau et al.), or indeed compensated

DOI: 10.4324/9781003365259-16

(Lou) depending on the context. For example, there are ethical concerns to navigate and sensitivities to the experiences of affected communities that may be impacted by gentrification and displacement, even as the sources of contamination are removed (Browning), as well as the dangers of stigmitising contaminated communities (Evia et al.). Fiske explores, for example, how tourism can engage with toxic heritage without valorising or sensationalising it.

These contributions raise questions about how we identify and define affected communities as the harm from toxic materials crosses borders and transgresses temporal boundaries. It also recognizes the importance of identity and perception, demonstrating that not all those harmed feel part of the same community. The multiscalar impacts of toxicity experienced by communities also raise questions about how harms are experienced in the slow violence of long-term exposures – premature death, illness from birth, unexplained chronic illness, and other consequences of bioaccummulation – and the ways those harms are manifested in oral histories, private collections, and other forms of memory practices that circulate outside of mainstream channels.

9

RELUCTANT RETURNS: REPATRIATING A POISONED PAST

Holly Cusack-McVeigh

In 2018 I got a call from an FBI agent in New York City. Hundreds of cultural heritage items had been seized as part of an ongoing investigation. Would I identify the cultural origin of the items in question and help facilitate returns? This was not a new request. Museums and other agencies are often in possession of items whose origin is a mystery. When I received the images, I carefully studied every detail and diagnostic feature that might shed light on where to begin with the consultation and repatriation process. One object commanded my immediate attention. It was a Grizzly bear head decorated with round metal disks on either temple. The metal was tinged with powdery turquoise – a hint that this piece *might* be old.[1]

From these disks hawk feathers freely dangled. Bear's eyes were encircled with deep crimson paint. Atop his head, a series of small metal bells adorned his forehead. I remember thinking, "is this real or is it a fake?" Fakes, I knew, are fabricated to deceive eager collectors. Fakes can be very convincing. Creators go to great lengths to "age" pieces – to mimic the wear of time. Knowing that bears are sacred, powerful beings in many Indigenous cultures, I wondered how he came to be in the hands of a private collector. I wondered where he'd been all these years. Bear's story would be complex. If as old as I suspected, if he had survived time, I knew he'd likely been treated with a poisonous substance sometime in his past. Was this the reason Bear had survived all these years? If contaminated, this return would be even more challenging.

Museums, academic institutions, and government agencies around the world are stepping up their repatriation efforts with the recognition that each return brings them closer to building a more just and equitable world, but this chapter highlights the tensions that arise when these returns have been made toxic. Over three decades have passed since the Native American Grave Protection and Repatriation Act (NAGPRA) became law. Under this federal act, Native American tribes reclaim

DOI: 10.4324/9781003365259-17

their sacred items, funerary objects, and objects of cultural patrimony. It also ensures the return of ancestors who have been torn from their resting places. Repatriation is often framed as an important form of restorative justice in the wake of a colonial past (Meloche 2022), but how can these returns support social justice and equality when these very returns threaten the health and safety of the living descendant communities receiving this toxic heritage? This contamination is a form of violence that disrupts the reciprocal relationship Indigenous communities strive to maintain with the land, the water, and the spirit world. Are these returns really an act of decolonisation or do these toxic returns simply further the legacy of colonisation?

Understanding the Problem: A History of Toxic Treatments

To understand the prevalence of this problem, we must delve more deeply into the how and why of poisoned heritage. For centuries museums have collected scientific specimens and cultural heritage objects central to their mission. These tangible collections serve to tell compelling stories of people and places through time. However, their long-term storage and special environmental needs present a variety of ongoing challenges for those charged with their care and preservation (Deering et al. 2020; Landrey 2009; Simmons and Kiser 2020; Szcepanowska 2013).

While there was widespread public recognition that health hazards existed in a variety of household items including clothing and wallpaper[2] as early as the mid-1800s (Figure 9.1) museums often collected without an understanding of the inherent and acquired dangers collections posed to human health (Hawks 2001). Inherent hazards include a range of threats including dyes, pigments, silica dust, radioactive elements in geological specimens, etc. (Hawks and Makos 2000; Hawks et al. 2010; Pool, Odegaard and Huber 2005).

In the 1800s, collections consisting of organic artifacts, ethnology, and natural history specimens were routinely treated with heavy metal pesticides (arsenic and mercuric salts). By the 1940s, synthetic organic pesticides (DDT)[3] were also widely employed to prevent pest destruction.[4] Especially impacted were objects made of feathers, fur, skin, leather, wood, grass, and other natural fibres (Hawks and Makos 2000; Hawks 2001). Unfortunately, those applying toxins rarely kept records. "Many acquired hazards are unknown to the user, as documentation of preservative treatments is often poor" (Makos 2001). Arsenic and mercury have significant human toxicity and although these toxins leave substantial quantities of residue on treated objects, these hazards are not always observable. In some cases, powder-form arsenic is visible, but more often there is no visual que of contamination. Alarmingly, a wide range of chemicals now known to be carcinogens was routinely applied to collection items by museum staff who were unaware of the serious health risks (Hawks 2001). Today we know that pathways for poisons include ingestion, dermal exposure, and inhalation. Poisoning may be acute, chronic, or cancer/reproductive related (Boyer et al. 2005).

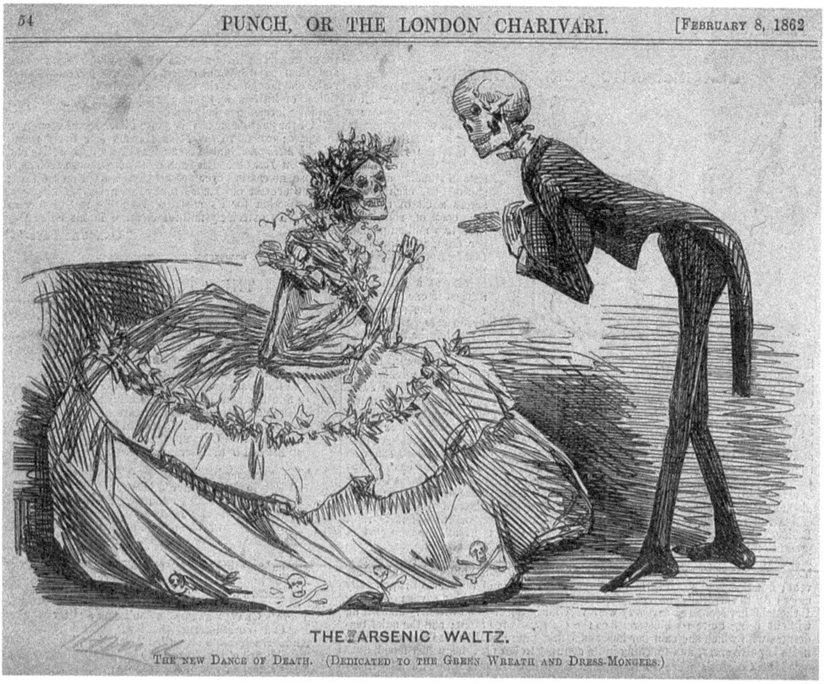

FIGURE 9.1 Punch cartoon warning of arsenical dyes and pigments in clothing and accessories. Wood engraving, 1862. Wellcome Collection. Public Domain Mark https://wellcomecollection.org/works/awbr7whm/items.

"With Dignity and Respect"

In the early years of NAGPRA, tribes encountered immense obstacles as they fought to get their ancestors and their belongings back. The passing of this federal law in 1990, was a watershed event. Although repatriation efforts long pre-dated NAGPRA (Zimmerman 2013) this provided, for the first time, a solid legal mechanism for protecting burial sites and facilitating returns. Prior to NAGPRA, tribes relied on the "good will" of museums to "do the right thing." In most cases, an ethical argument built on the moral ground did not get claimants very far. With this law, Congress acknowledged "that human remains of any ancestry 'must at all times be treated with dignity and respect'" (NAGPRA 1990).

National NAGPRA Program Manager, Melanie O'Brien, reports that 84,677 individual sets of ancestral remains have completed the regulatory process since 1990. To date, over 1.8 million associated funerary objects, 349,000 unassociated funerary objects, and 21,600 cultural items (sacred and patrimony) have been returned (personal communication 25 June 2022). This single law, far from perfect, is rooted in common law understanding that all people, everywhere, have the right to protect their dead along with the safeguarding of ancestral belongings and sacred items needed to perpetuate a community's worldview and way of life. These are basic human rights (Hutt and Riddle 2007), and while much has been accomplished in way of respectful returns, "less than half of the Native American ancestral remains in collections have been repatriated to their traditional caretakers. Over 117,576 Native American individuals are still in museum and Federal agency collections" (Beasley 2022:2, 2).

A respected Elder recently stated, "there is still *a lot* of hard work to be done," on the repatriation front but what happens when these sacred items, and even the ancestral remains, have been made toxic, as was likely to be the case with Bear? Although museums no longer address pest management with toxic treatments,[5] past methods remain an issue for those charged with the care and return of this toxic heritage. This legacy presents even greater challenges for Indigenous communities. After years of fighting to reclaim their ancestors and sacred items, many communities are forced to delay these important returns. Others reluctantly bring them home without adequate storage or options for safe reburial. How can this toxic heritage be properly reclaimed, returned, and reburied? Is it always possible? And, if it cannot be returned to their homeland, how might this disrupt the balance in the natural and spirit world? How do these reluctant returns impact the well-being of contemporary communities?

A Homecoming Deferred

There are hundreds, perhaps thousands, of cases where returns are postponed as communities struggle to resolve this problem. The Hopi Tribe's Cultural Preservation Office was forced to delay the physical return of many cultural items until they can be tested for contaminants. Only after testing reveals the items are "reasonably believed to be safe and clean" will they come home (HCPO 2022). Under NAGPRA, museums and other agencies must inform descendant communities of any *known* treatments. Failure to give proper notification to "the recipients of repatriations of any presently known treatment of the cultural items with pesticides, preservatives, or other substances that represent a potential hazard to the objects or to persons handling the objects," means they face a civil penalty by the Department of the Interior (NAGPRA 2022).

"Damned If You Do"

What the law does not require is routine testing before items are returned to descendant communities. The onus falls on those working to bring their cultural heritage and ancestors home. Most do not have the resources to do the necessary testing and those who do are faced with an even greater challenge: What to do with sacred items and other cultural heritage so hard won when they test positive for contaminants. The sacred items and objects of cultural patrimony are often reclaimed so that they can once again enter ceremonial life. Masks may be returned one day and dancing in a ceremony the next. Ancestral remains and sacred items deemed unsafe for return pose yet another conundrum for communities who have a spiritual obligation to their ancestors and spirit beings. If their toxicity means they must remain on a museum shelf, communities cannot fulfil their cultural obligations to these beings. This too can be dangerous and cause disruption between the living and the spirit worlds.

Expressing his frustration over this conundrum, a community member remarked that, "You are damned if you do and damned if you don't." When ancestors who were torn from their graves are returned to their people, they must be reburied following specific cultural protocols. How can they be properly cared for when the fear is that putting them back into the ground with their personal belongings (funerary items) might harm the living? Tribes express concerns that reburial might contaminate the earth where they grow their food or poison their community's groundwater. How much is too much? One item, one hundred, one thousand? These critical questions routinely come up during the tribal consultation and repatriation process. Imagine the ways that this spirit world is disrupted, then, if the land and water are poisoned. Contamination is a kind of disrespect. It disrupts the reciprocal relationship that so many Indigenous communities maintain with the land, the water, and the spirit world. Just as "objects" are viewed as living, sentient beings, so too are the land and water. In 2017, for example, the country of New Zealand finally granted legal personhood to the *Whanganui* River. For the Iwi Māori people, this river *is* an ancestor (*tupuna*) deserving of the same respect as any other living being.

"Why Did You Give Them Back?"

Katsinam are the *living* embodiment of spirits ... and, as such, they are living, breathing beings. Imagine then the painful legacy of knowing that these beings have been poisoned. In the face of centuries of museums treating their sacred beings and cultural patrimony with toxic substances, Indigenous communities are responding with initiatives to raise concerns and advocate for their communities. On the Hopi Tribe's Cultural Preservation Office website, one such initiative is highlighted under "Hopi Study Hazardous Museum Chemicals." This was one of the most comprehensive attempts to address this critical issue back in the late 1990s when NAGPRA was but a few years old. Taking the lead in this effort was the Hopi Tribe who became increasingly concerned about toxic returns to their own reservation. Lee Wayne Lomayestewa, who is the NAGPRA Repatriation Coordinator for the Hopi Tribe explained:

> Hopi was the first tribe to bring up this issue because when we visited museums they always told us to put on gloves, lab coats, face masks and cover our hair. The elders were upset and didn't know why they had to wear them. They wanted to touch their sacred items that have been gone so long. When we brought them

> home, we gave them back to the villages where they were taken from and only to find out these were preserved in some way with pesticides, mainly arsenic and other preservatives. We had to go back to the villages to retrieve them and the caretakers were very upset at us saying "Why did you give them back to us only to take them back?" The Hopi tribe has worked with many museums that got NAGPRA grants and now test Hopi items before the sacred items are sent back home to re-use in our ceremonies (personal communication, 7 June, 2022).

Working with the University of Arizona, the State Museum, and Poison Center they created the "Hopi Kachina Friends Pesticide Documentary Project" to address ways to "neutralize toxins on objects" (HCPO 2022). This collaboration also resulted in a "Contaminated Cultural Material in Museum Collections" workshop (Pool et al. 2005, xv). Still, the Hopi continue to encounter this problem with each new return knowing that "a significant number of returns involve sacred objects meant to be handled or worn in ceremony" (HCPO 2022).

"Giving Voice to Bear"

Turning back to Bear's story, comparative research led to a similar western Sioux bear headdress featured in "Giving Voice to Bear" (Rockwell 2003). When I finally got eyes on Bear two things struck me. First, this was an adult Grizzly – it was enormous. Secondly, it was just too pristine. No signs of insect damage on the fur, hide, or hawk feathers. Collaborating with colleagues from Purdue University's School of Public Health, X-ray fluorescence (XRF) was employed to detect the presence of heavy metal contaminants. Testing quickly revealed what I feared most - Bear had been heavily treated with arsenic! Knowing that communities are reluctant to reclaim poisoned cultural heritage, I wondered if Bear would ever get home. In this case, "giving voice to Bear" means knowing his treatment history.

"Identify and Isolate"

There is often an inherent tension between Indigenous cultural protocols for care and museum best practices. In the museum world, emerging collections professionals are trained to recognise potential toxins. Where suspect (presence of white mystery powder), we isolate to prevent toxins from becoming airborne in collections or exhibit spaces. Museums around the world house sacred objects following *their* understanding of preservation needs, but, as Indigenous scholars have long argued, museum best practice does not take into account "the diversity of tribal cultural practice" (Sadongei 2004, 17). Nor does it privilege Indigenous systems of care. We are trained to "identify and isolate" toxins the same way we manage a pest infestation, but how can living, breathing beings be sealed in plastic? At a repatriation conference, one tribal Elder likened this collection practice to placing a newborn in an airtight bag. This is horrific and unimaginable, and yet, this is exactly how Indigenous community members feel when they enter a museum space and see their ancestors and sacred items sealed in polyethylene bags. For many, their cultural heritage "items" are alive in the *same* way as creatures who draw breathe. What then do I do with Bear? He is a sentient being who roamed the western Plains before becoming an integral part of ceremonial life. Museum best practice says to "isolate and seal" toxic heritage. My understanding and deep respect for Indigenous communities mean that I must treat this sacred being with reverence.[6]

A Global Problem

Working with partners at Purdue University, I also tested foreign items that, like Bear, awaited return. The authenticated mummy tested negative, as did the spears simply labelled "Oceania," but the Vanuatu "over modelled skull" (labeled "tribal art" by collectors) was "hot!" At some point in its history, between its creation and acquisition, it had been treated. Given its composition of largely clay and decorative paint/pigment, it didn't seem like a good target for pests who are proteinaceous feeders. This serves as a cautionary tale. You cannot spot these hidden hazards, nor can you predict the why and when of these historical treatments. This Indigenous ancestor, *Ni-Vanuatu*, will return home in a shipping crate marked with a skull and crossbones tag – the universal symbol for poison.

Iwi Māori scholars in New Zealand report similar findings with their ancestral returns. Dr. Amber Aranui, a former researcher at Te Papa National Museum's *Karanga Aotearoa* repatriation program reported that the staff faced similar issues with "*Toi moko* (preserved human heads) coming home from overseas institutions" (personal communication, 11 June 2022). "Following the 'discovery' and early study of New Zealand by Europeans after Cook's voyages, scientists took a great interest in Māori" (Aranui 2020, 402), leading to the collection of *Toi moko*. Similarly, from the other side of the world, it is reported that Sámi ethnographic collections were routinely treated with various toxic chemicals in response to webbing clothes moth infestations at the Norsk Folkemuseum in Oslo, Norway (Rinke Bangstad 2022). From Africa's Congo region, repatriated collections also reveal treatment residues from biocides (Arndt 2021).

Colonisation's Painful Legacy

The emotional toll of this dilemma is difficult to quantify. This is a global human rights issue. To fight so hard for the return of what is rightfully yours only to discover that, while you have the ethical and legal right to these returns, your community does not have the resources to safely bring them home and care for them following your cultural protocols. Museum collections from around the world were born out of colonisation and scholars are increasingly acknowledging the power imbalances that persist (Arndt 2021; Turnbull 2020; Wergin 2021).

In the wake of a colonial past, these toxins represent yet another wave of relentless assault and injustice (Simms and McIntyre 2014). Reflecting on this painful legacy,

Te Herekiekie Haerehuka Herewini, Te Papa's Pouwhakahaere Kaupapa Pūtere Kōiwi (Head of Repatriation), stated:

> There are many lessons I've learned over the last 14 years as part of the journey of seeking the return of Māori and Moriori ancestral remains. This includes the importance of appropriate spiritual and physical well-being to achieve meaningful repatriation, restitution and reconciliation. My whānau (family) and kaumātua (elders) taught me the importance of good spiritual care, however, the immediate induction process of engaging with Māori and Moriori ancestors at the Museum of New Zealand Te Papa Tongarewa, was to look for signs of arsenic on ancestral remains. From there on, I realised there was the potential of intense physical and health danger by engaging with my ancestors, not for spiritual reasons, but due to the toxic treatments they received as part of their trade, collection and imprisonment within museums and academic institutions. My advice is to proceed with caution and care when engaging with ancestral remains (personal communication, 26 June, 2022).

Moving Forward: Collaboration is Key

I am certainly not the first to call attention to this old and complex problem. A lot of excellent work has already been done on this front. The work of the Hopi Tribe

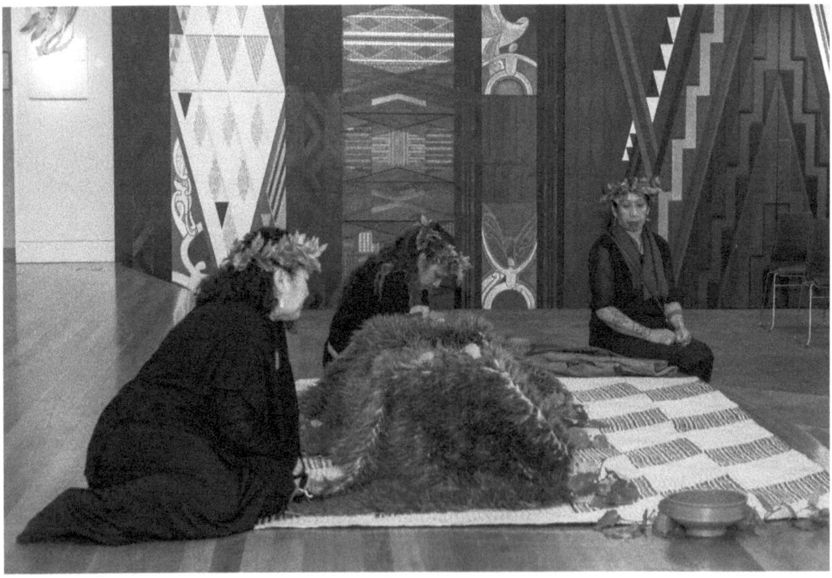

FIGURE 9.2 Te Papa staff comforting tūpuna (Māori ancestral remains) repatriated from Germany in October 2020. From left are Haley Stewart, Moana Parata, and Lisa Osborne. (Photo Courtesy of Te Papa).

and Arizona State Museum (Odegaard and Sadongei 2000) paved the way for what needs to be a worldwide effort to address hazardous collections (Hawks 2001). In 2001 the Society for the Preservation of Natural History Collections held a symposium titled "Contaminated Collections: Preservation, Access, and Use" (SPNHC 2001). This meeting, comprised of conservators, scientists, attorneys, public health/safety officials, Native Americans, and other preservation professionals, "showed that collaboration can work and work well in dealing with these thorny issues" (Hawks personal communication, 19 September 2022).

While these earlier efforts are to be applauded, we must further this dialogue and continue to seek new solutions to this problem. What is needed is continued education on all fronts including international and national conferences, public outreach, and ongoing education for museums. Most importantly we need to provide support and regular non-invasive testing for Indigenous communities when they request it. Better standards for such returns should include mandatory testing, but might this get in the way of those returns? Giving museums and other institutions yet another excuse to delay the repatriation process under the law?

We cannot eliminate or even mitigate these toxins in all situations. Once introduced, they remain and, in most cases, pose serious health risks. The best approach is an interdisciplinary, team-based approach. Continued dialogue and collaboration are key, but what does that look like? In 2022, with the hopes of establishing new collaborations on this front, the American Industrial Hygiene Association's Museum and Cultural Heritage Working Group, presented this problem to the Workplace Health Without Borders, an international organisation that "works to advance the protection of workers and promote best practices to improve worker safety and health." This meeting revealed that most in the industry knew nothing about this problem. One appalled board member remarked that these toxic returns were analogous to distributing "smallpox exposed blankets."

> Tribal community members, as well as the toxicologists and public health officials with whom tribes will be consulting, need to compare source contaminant data in order to make rational statements as to potential risks. Critical to this process is the need for standardized … assessment protocols that take into consideration the restrictions placed on traditional industrial hygiene sampling methods on sacred objects and reflect an appreciation of the cultural issues surrounding the object's intended use
>
> (Makos 2001).

While it is suggested that museums and agencies do their due diligence to test for the presence of toxins, it is not mandated. While NAGPRA states that they must reveal any *known* treatments, there is nothing requiring institutions to test as a routine part of the repatriation process. Simply put, it is easy to take a "don't ask, don't tell" approach avoiding the time and expense of testing. And, what about the thousands of sacred items that have already been returned to descendant communities?

Rebecca Tsosie, a faculty member for the Indigenous Peoples' Law and Policy Program at the University of Arizona and Yaqui descendant, argues that:

> the issue of repatriation of contaminated objects requires a restructuring of existing law and policy directives to achieve a coherent legal solution. However, before such a restructuring is possible, it is necessary to examine the legal and ethical dimensions of the problem through an intercultural lens. The nature of the problem is one that threatens human health and safety, requiring scientific study of the health effects of such contamination given the patterns of use employed by Native people. However, it is also one that requires recognition of cultural harm and the inadequacy of existing tort law to quantify the damages that are being suffered by Native people
>
> (Tsosie 2001, abstract).

In a July 2021 DOI press release, Interior Secretary Deb Haaland[7] stated that "Changes to NAGPRA regulations are long overdue..." She concluded her address with "I'm hopeful this process will eliminate unnecessary burdens to the repatriation process and allow Indigenous peoples greater access to their ancestors' remains and sacred items (U.S. Department of the Interior 2021)" Unfortunately, these much-needed regulatory changes did not address the need for mandatory testing of toxic heritage. Melanie O'Brien, Manager of the National NAGPRA Program did note that:

> In the existing regulations at §10.10(e), museums and Federal agencies must inform Indian Tribes and NHOs about any potentially hazardous substances used to treat any of the objects only after repatriation has occurred. The proposed revisions would require a museum or Federal agency to disclose this information first in the summary or inventory and second in the related notice. This change will also allow for civil penalties to be alleged against a museum that fails to include information regarding the known presence of a potentially hazardous substance used to treat human remains or cultural items (personal communication, 25 June, 2022).

It has been over twenty years since Nancy Odegaard and Alyce Sadongei testified to the NAGPRA Review Committee calling "for a united front to combat this issue" (Odegaard et al. 2005). In their plea they too emphasised the need for a collaborative approach that would bring together "a whole diverse group of professions … a cooperative team should include tribal representatives, medical toxicologists, conservators, chemists, museum collections professionals, and/or industrial hygienists or public health professionals" (Odegaard et al. 2005). Indigenous communities should be at the centre of this continued dialogue and collective effort. In the end, it is their heritage and their right to determine how this problem is addressed case by case.

As this chapter goes to press, Bear sits silently on a collection storage shelf awaiting his return. He was once an integral part of sacred ceremonies. His people know he is there, but they do not have the resources or a safe location to allow for his homecoming. In the wake of colonisation, repatriation is a vitally important form of restitution – a chance for community healing, but these reluctant returns continue to have an impact on living communities, the natural environment, and the spirit world. When sacred objects can't be returned to their ceremonial roles, when places are feared to have been contaminated because of reburial, and subsequent repatriation efforts are hindered or halted altogether the imbalance and disruption caused by their removal are furthered. This is so often misunderstood, minimised, or overlooked altogether by museum professionals and academics involved in the repatriation process. As long as descendant communities continue to shoulder this burden, the deeply painful legacy of colonisation will prevail. As long as sacred items continue to be repatriated without a full understanding of their treatment history and hidden health hazards, the museum world will not be decolonised. Instead, well-intentioned museums will perpetuate the legacy of colonisation.

Notes

1 This occurs when brass, bronze, or copper oxidizes.
2 Treating private collections, including household textiles, with arsenic was common in the 16th-19th centuries and well into the early 20th century.
3 DDT was widely used up until the early 1970s when the EPA issued a cancellation order because of adverse impacts on the environment and human health. For a list of organic pesticides see Hawks et. al (2010, 227-354).
4 Today, museums avoid chemical treatments whenever possible, instead implementing "Integrated Pest Management" preventive care to limit damage … " (AIC 2022).
5 Strang notes that "to effectively contribute to pest control for cultural property one must work to formulate a 'most-good with least harm' action" (Strang 2012, 2). "To reduce health hazards … alternatives such as thermal treatment and controlled atmosphere fumigation have replaced applied residual chemicals … " (Strang abstract, 2012).
6 "In some instances, tribal religious leaders have carefully instructed these museums about the care of objects … it is important for non-tribal museums to know the difference between active practice and passive accommodation" (Sadongei 2004, 18).
7 Haaland, a member of the Laguna Pueblo community, is the first Native American to be appointed as Cabinet secretary.

Bibliography

Aranui, Amber. 2020. "USES AND ABUSES: Indigenous human remains and the development of European sciences: an Aotearoa/New Zealand case study." In *Routledge Companion to Indigenous Repatriation: Return, Reconcile, Renew*, edited by Cressida Fforde, C. Timothy McKeown, and Honor Keeler, 400–412. Springer International Publishing.

Arndt, Lotte. 2021. "The toxic afterlives of colonial collections." [Les survivances toxiques des collections colonials.] *Trouble dans les Collections* (2) Sept 2021. URL: https://troublesdanslescollections.fr/2246-2/

Beasley, Joy. 2022. "The Long Journey Home: Advancing the Native American Graves Protection and Repatriation Act's Promise After 30 years of Practice." Statement of Joy

Beasley, Associate Director, Cultural Resources, Partnerships and Science, National Park Service United States Department of the Interior Before the Senate Committee on Indian Affairs. February 2. URL: https://www.doi.gov/ocl/native-american-graves-protection.

Boyer, Leslie, Steven Seifert, Nancy Odegaard, Marilen Pool, and G. Edward Burroughs. 2005. "Understanding the Hazards: Toxicity and Safety." In *Old Poisons, New Problems: A Museum Resource for Managing Contaminated Cultural Materials*, edited by Nancy Odegaard, and Alyce Sadongei, 73. Walnut Creek, CA: AltaMira Press.

Deering, Katharina, Elise Spiegel, Christiane Quaisserc, Dennis Nowak, Stefan Rakete, Mercè Garíad, and Stephan Bose-O'Reilly. 2020. "Exposure assessment of toxic metals and organochlorine pesticides among employees of a natural history museum." *Environmental Research* Volume 184 May 2020, https://www.sciencedirect.com/science/article/abs/pii/S001393512030164X?via%3Dihub

Hawks, Catharine, and Kathryn Makos. 2000. "Inherent and Acquired Hazards in Museum Objects: Implications for Care and Use of Collections." *Cultural Resource Management* 22 (5): 31–37. URL: http://npshistory.com/newsletters/crm/crm-v23n5.pdf

Hawks, Catharine. 2001. "Historical Survey of the Sources of Contamination of Ethnographic Materials in Museum Collections." *Collection Forum* 16 (1&2): 2–11.

Hawks, C., M. McCann, K. Makos, L. Goldberg, D. Hinkamp, D. Ertel, and P. Silence, eds. 2010. *Health & Safety for Museum Professionals*. New York: Health & Safety Committee of the American Institute for Conservation of Historic & Artistic Works and Society for the Preservation of Natural History Collections.

Hopi Cultural Preservation Office. 2022. "Past Projects." *Hopi Cultural Preservation Office*. Accessed 7 June 2022. https://www8.nau.edu/hcpo-p/pestDocProj.html.

Hutt, S., and J. Riddle. 2007. "The law of human remains and burials." In *Human Remains: Guide for Museums and Academic Institutions*, edited by Vicki Cassman, Nancy Odegaard, and Joseph F. Powell, 223–243. New York: AltaMira Press.

Landrey, Gregory J., and Henry Francis du Pont Winterthur Museum . 2009. In *The Winterthur Guide to Caring for Your Collection Updated*, edited by Winterthur Del Hanover N.H. Henry Francis du Pont Winterthur Museum, University Press of New England distributor.

Makos, Kathryn A. 2001. "Hazard Identification and Exposure Assessment Related to Handling and Use of Contaminated Collection Materials and Sacred Objects." *Collection Forum* 17 (1&2). https://spnhc.org/wp-content/uploads/2018/11/CFAbs_2001_V17N12.pdf

Meloche, Chelsea. 2022. "What happens next? Exploring connections between repatriation, restorative justice, and reconciliation in Canada." Simon Fraser University.

Native American Graves Protection and Repatriation Act, 25 U.S.C. 3001 et seq. [Nov. 16, 1990]. In *National NAGPRA*. National Park Service, U.S. Department of the Interior. Available at www.nps.gov/nagpra/ (accessed 14 May 2022).

Odegaard, Nancy, and Alyce Sadongei. 2000. "Contaminated Cultural Materials in Museum Collections: Reflections and Recommendations for a NAGPRA Issue." *Western Association for Art Conservation (WAAC) Newsletter* 22 (2). https://cool.culturalheritage.org/waac/wn/wn22/wn22-2/wn22-207.html

Pool, Marilen, Nancy Odegaard, and Melissa Huber. 2005. "Identifying the Pesticides: Pesticide Names, Classification, and History of Use." In *Old Poisons, New Problems: A Museum Resource for Managing Contaminated Materials*, edited by Nancy Odegaard, and Alyce Sadongei, 5–31. Walnut Creek, CA: Altamira Press.

Rinke Bangstad, Torgeir. 2022. "Pollution and permanence: museum repair in toxic worlds." In *Museums and Social Issues* 10 (1): 1–15.

Rockwell, David. 2003. *Giving Voice to Bear: North American Indian Myths, Rituals, and Images of the Bear.* Random House Value Publishing.
Sadongei, Alyce. 2004. "What About Sacred Objects?" In *Caring for American Indian Objects: A Practical and Cultural Guide*, edited by Sherelyn Ogden. St. Paul: Minnesota Historical Society Press.
Simmons, John E., and Toni M. Kiser, eds. 2020. *MRM6 Museum Registration Methods.* Lanham, Maryland: Rowman & Littlefield Publishers.
Simms, S. Jordan, and John D. McIntyre. 2014. "Toxic Representations: Museum Collections and the Contamination of Native Culture." *The Canadian Journal of Native Studies* 34(1): 151–169.
SPNHC Society for the Preservation of Natural History Collections. 2001. Contaminated Collections: Preservation Access and Use: Proceedings of a Symposium Held at the National Conservation Training Center (NCTC), Shepherdstown, West Virginia. 6–9 April 2001. J. S. Johnson, ed. Collection Forum, vol. 17.
Strang, Thomas. 2012. *Studies in Pest Control for Cultural Property.* Gothenburg: University of Gothenburg.
Szcepanowska, Hanna. 2013. *Conservation of Cultural Heritage: Key Principles and Approaches.* New York: Routledge.
Tsosie, Rebecca. 2001. "Contaminated Collections: An Overview of the Legal, Ethical and Regulatory Issues." *Collection Forum* 17 (1&2). https://spnhc.org/wp-content/uploads/2018/11/CFAbs_2001_V17N12.pdf.
Turnbull, Paul. 2020. "Collecting and Colonial Violence." In *Routledge Companion to Indigenous Repatriation: Return, Reconcile, Renew*, edited by Cressida Fford, C. Timothy McKeown, and Honor Keeler, 400–412. Springer International Publishing.
U.S. Department of the Interior. 2021. "Interior Department Announces Tribal and Native Hawaiian Consultations to Discuss Updates to Native American Graves Protection and Repatriation Act." *DOI Press Release,* 15 July. Accessed 19 May 2022. https://www.doi.gov/pressreleases/interior-department-announces-tribal-and-native-hawaiian-consultations-discuss-updates.
Wergin, Carsten. 2021. "Healing through Heritage? The Repatriation of Human Remains from European Collections as Potential Sites of Reconciliation." *Anthropological Journal of European Cultures* 30(1): 123–133.
Zimmerman, Larry J. 2013. "Repatriation Acts: Before NAGPRA." In *Encyclopedia of Global Archaeology*, edited by Claire Smith, 6298–6303. New York: Springer.

CASE STUDY 4

PUBLIC MEMORY OF TOXIC DISPLACEMENT: HEAVY METAL CONTAMINATION AND SUPERFUND REMEDIATION IN FEDERALLY ASSISTED HOUSING COMMUNITIES

Elizabeth Grennan Browning

In 2017, the US Environmental Protection Agency and the US Department of Housing and Urban Development reported that 70% of Superfund sites sit within one mile of a public housing or HUD multifamily housing complex. Across the span of multiple generations, low-income individuals residing in federally assisted housing have disproportionately faced environmental inequities born of exposures to environmental contamination (HUD 2021). This case study considers the slow violence that has unraveled within two public housing communities located in Superfund sites, inflicting not only physical traumas of contamination but also collective disturbances stemming from the erasure of social networks (Nixon 2011). What meanings does toxic heritage evoke for public housing communities displaced by remediation, and how might it contribute to a dynamic framing of public memory for Superfund sites, where cleanup processes have largely advanced narratives of disremembering?

In March 1994, NAACP leaders convened at the 160-unit Washington Park Public Housing Project (WPPHP) in Portsmouth, Virginia, to launch a national environmental justice campaign. Built in 1962 and located within the Abex Corporation Superfund Site, WPPHP drew NAACP's attention because of tenant Helen Person's advocacy for the relocation of the predominantly African American community due to the site's lead contamination (Whitt 1994).

Throughout the 1980s government officials gathered evidence of severe contamination resulting from the 50-year operation of a nearby foundry, but it was not until 1992 that EPA informed residents about their exposure. EPA's initial remediation plan called for the permanent relocation and compensation of affected private homeowners, with all residential properties – except WPPHP – rezoned for commercial or industrial usage.

In 1998 Person's coalition filed a class action federal lawsuit claiming that WPPHP was part of "a system of segregation" that exposed Black residents to disproportionate

DOI: 10.4324/9781003365259-18

environmental harms. The resulting settlement marked the first time the EPA modified an ongoing Superfund waste cleanup due to charges of racial discrimination. Portsmouth officials completed the settlement's mandated demolition in 2003, and eventually converted the vacant land into an extension of the PortCentre Commerce Park (Coffey et al. 2020).

Fifteen years after WPPHP's demolition, a similar episode of slow violence unfolded in East Chicago, Indiana, at the 346-unit West Calumet Housing Complex (WCHC), built in 1972 and located in the USS Lead Superfund Site. Similar to the EPA's actions at Washington Park, officials first discovered evidence of lead contamination at WCHC in the 1980s but did not inform residents until several decades later.

After conducting a feasibility study for cleanup options, EPA proposed returning WCHC to residential standards by removing the top two feet of soil, reasoning that "digging deeper is not meaningfully more protective of residential users and does not justify the additional cost" (USEPA 2018a). Local advocacy groups disagreed. In its 2018 public comment on EPA's proposed plan, the East Chicago Calumet Coalition Community Advisory Group (CAG) noted that the sub-24" contamination would endanger future generations due to ecosystem dynamics of flooding and erosion, projected to worsen with climate change. "Who wins if USEPA selects the less protective option?" CAG asked. "The companies, who profited off the land for decades, will pay less to address their pollution. Meanwhile, families whose lives have been permanently altered, at great economic and emotional cost, will remain in harm's way" (USEPA 2018b).

Despite WCHC residents' protests, EPA proceeded with demolition in 2018. Due to limited housing stock, the East Chicago Housing Authority relocated families to public housing sites in other contaminated northwest Indiana and Chicago areas. In 2020 the East Chicago City Council rezoned the former housing site to light industrial use, with plans to build a distribution center (Browning 2021) (Figure 1).

As the Superfund cleanup process coheres around erasure (defined here as a transformation of the landscape so as to prompt a forgetting of former contamination, but also, collaterally, the rupturing of community), what might toxic heritage mean for the memorialization of relocated public housing communities? Although Washington Park and West Calumet residents both fought for a more thorough remediation process, they had contradictory perspectives about the best way to fulfill this goal. Helen Person pushed for the demolition of WPPHP and residents' relocation, whereas many WCHC residents supported a comprehensive cleanup that would ensure the safety and longevity of the housing complex. Despite these different approaches, toxic heritage endures in both episodes of slow violence in two ways: first, through the poisonous bioaccumulated remnants of lead contamination within affected children and adolescents as they continue to encounter unforeseen health challenges from their exposure; and second, through the cascading social and emotional "root shock" effects resulting from this experience (Fullilove 2004). Toxic heritage, in these Superfund cases then, signifies forgotten heritage – with virtually all but the former residents

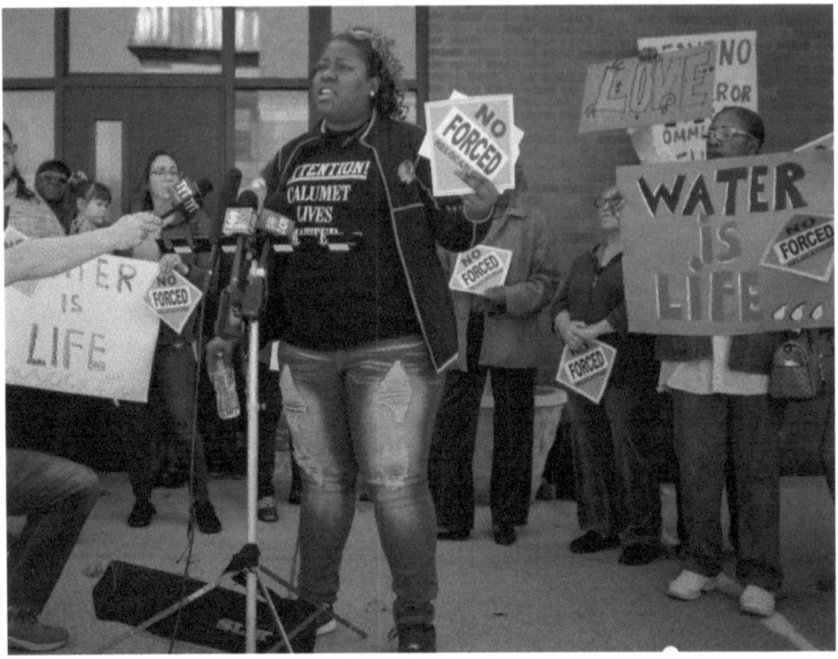

FIGURE 1 Former West Calumet Housing Complex resident Akeesha Daniels discussed the ongoing lead contamination crisis with the media in 2017 during a protest outside of East Chicago Mayor Anthony Copeland's office. (photograph by Alyssa Schukar).

themselves losing sight of the complex corporeal and social manifestations of this discrimination – as well as the uncertainty that persists beyond Superfund's containment efforts and follows former residents via health disparities and recurring patterns of social and emotional distress (Wollentz et al. 2020). As the Superfund remediation process effaced toxic heritage within the "reclaimed" physical landscapes of Portsmouth and East Chicago, so too did the process efface toxic heritage in official narratives that emphasized the transformed, capitalist-oriented nature of the sites – with both sites transitioning from residential to commercial and industrial purposes.

However, toxic heritage has persisted – not in the physical landscape itself, but through the migratory flows of the bodies of former residents and their long-term, embodied contamination. So too have former residents and their allies actively resisted the devastation of the communities' social networks and environmental health, by perpetuating creative expressions of toxic heritage through social media's virtual discourse in dedicated groups, such as the Calumet Lives Matter Facebook group, and through commemorative and advocacy-oriented activities such as toxic tours, public protests, digital mapping, and public art (Figure 2).

These forms of resistance and remembrance serve as memorializations of what Stacy Alaimo has called "trans-corporeality" – which foregrounds "the material

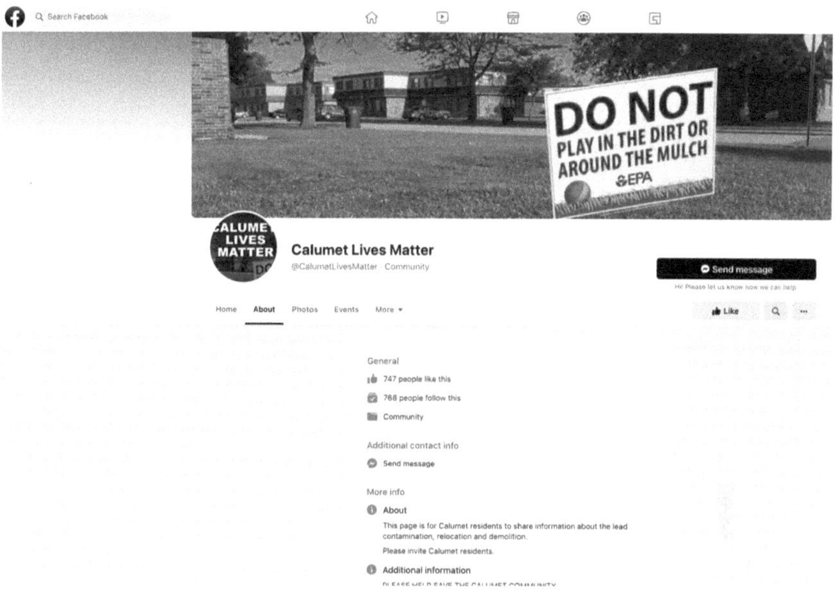

FIGURE 2 Members of the Calumet Lives Matter advocacy group, led by Sherry Hunter, formed a Facebook group in 2016 to organize protests and foster community-wide conversations about the Superfund remediation process at the West Calumet Housing Complex.

interconnections of human corporeality with the more-than-human world," and thus raises new questions about an ethics that is not just social, but material – "the emergent, ultimately unmappable landscapes of interacting biological, climatic, economic, and political forces" (Alaimo 2010, 2). Toxic heritage in fact makes some components of Alaimo's "unmappable landscapes" legible, by establishing informal archives that both foster resilience for those who lend their voices to the documentation of this toxic heritage and help ensure that future histories account for the complex, intertwined social and environmental histories behind long-standing environmental inequities.

References

Alaimo, Stacy. 2010. *Bodily Natures: Science, Environment, and the Material Self.* Bloomington: Indiana University Press.

Browning, Elizabeth Grennan. 2021. "Wastelanding and Racialized Reproductive Labor: 'Long Dyings' in East Chicago from Urban Renewal to Superfund Remediation." *Environmental History* 26: 749–775.

Coffey, Emily, Kate Walz, Debbie Chizewer, Emily A. Benfer, Mark N. Templeton, Robert Weinstock. 2020. *Poisonous Homes: The Fight for Environmental Justice in Federally Assisted Housing*. Chicago: Shriver Center on Poverty Law and Earth Justice.

Fullilove, Mindy Thompson. 2004. *Root Shock: How Tearing up City Neighborhoods Hurts America.* New York: One World/Ballantine Books.
HUD Office of Inspector General. 2021. "Contaminated Sites Pose Potential Health Risks to Residents at HUD-Funded Properties," Doc. 2019-OE-0003.
Nixon, Rob. 2011. *Slow Violence and the Environmentalism of the Poor.* Cambridge: Harvard University Press.
USEPA. 2018a. News Release No. 18-OPA067, November 7, 2018.
USEPA. 2018b. *Public Comments on Proposed Amendment to the Cleanup Plan for Zone 1.* https://semspub.epa.gov/work/05/946378.pdf.
Whitt, Toni. 1994. "NAACP's New Target: Toxic Waste." *The Virginian-Pilot*, March 25, 1994.
Wollentz, Gustav, Sarah May, Cornelius Holtorf, and Anders Högberg. 2020. "Toxic Heritage: Uncertain and Unsafe." In *Heritage Futures: Comparative Approaches to Natural and Cultural Heritage Practices*, Harrison et al., pp. 294–312, London: UCL Press.

VISUAL ESSAY 2

TRANSLATING AND TRANSFORMING TOXICITY: MOVING BETWEEN ETHNOGRAPHY AND GRAPHIC ART

Amelia Fiske and Jonas Fischer

One of the principal means through which outsiders have come to recognize histories of oil contamination in the Ecuadorian Amazon is through "Toxic Tours" in which a guide brings students, lawyers, environmental activists, journalists, and tourists to visit toxic sites. Inspired by the creative ways that the guides engage participants – inviting them to smell crude oil or to step out onto logs suspended over 3 meters of crude waste – the graphic novel *Toxic: A Tour of the Ecuadorian Amazon* immerses readers in the materiality of toxic contamination and struggles for environmental justice in everyday life. Based on 27 months of ethnographic research conducted by Fiske between 2011 and 2013, the novel narrates the discovery of crude oil in waste pits and the stories of Ecuadorian interlocutors living alongside industry through the journey of three participants on a "Toxic Tour."

Following the story of three fictional participants, readers visit the waste pits, gas flares, and homes of people living in this region. Along the way, the participants struggle with encounters with toxicity *in vivo*, seeking to make sense of the weight of histories of oil extraction in Ecuador that are woven through the air, water, soils, and the stories recounted by their guide, Donald and his young daughter, Leonela (Figures 1 and 2).

Amazonian toxic tours have been closely aligned with the plaintiffs in the longstanding *Aguinda v. Texaco* lawsuit, and rely on local figures to offer first-hand experiences with contamination and struggles for justice. By drawing on experience with oil operations – such as farm animals lost from falling into uncovered waste pits, or miscarriages from toxic exposures – Toxic Tours combine personal and anecdotal knowledge to convince visitors of the immediacy of environmental struggles with pollution.

Toxic Tours are a sensory engagement with contamination: participants pull oily soil from the ground with an auger, squish black and slippery leaves between their

DOI: 10.4324/9781003365259-19

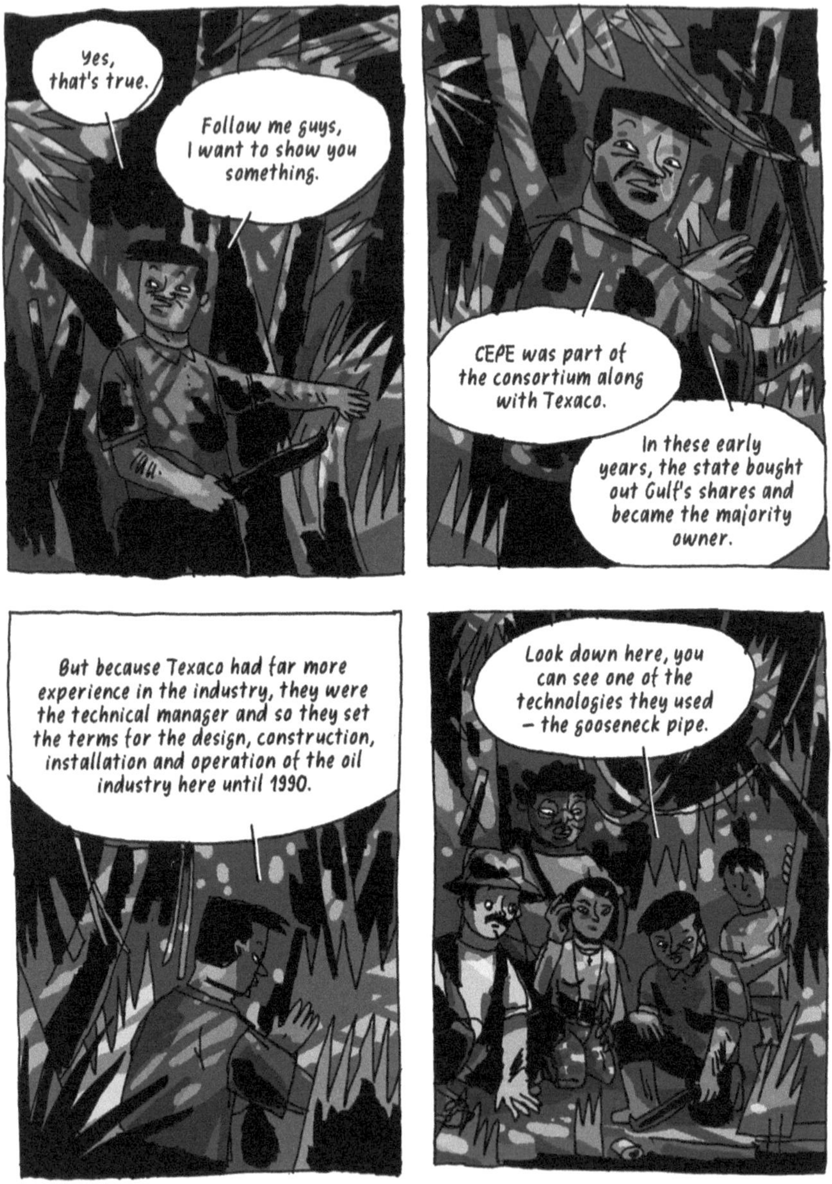

FIGURE 1 Donald leads the group through the forest around the edge of the Aguarico-04 waste pit while narrating how oil operations began in Ecuador. This Toxic Tour is the first up-close encounter for these participants with the realities of oil contamination.

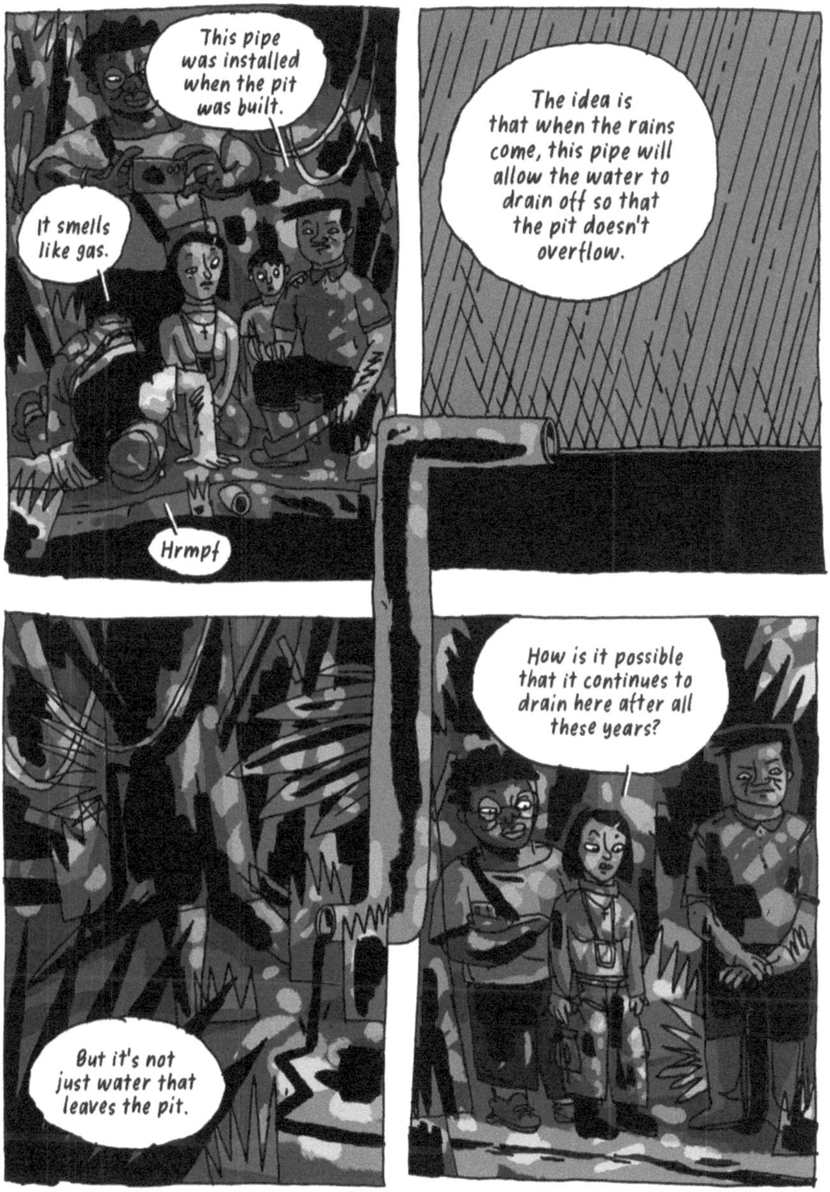

FIGURE 2 One opening of the gooseneck pipe sits inside the waste pit, where rainwater collects on the surface of old crude oil. When the pit becomes too full, the toxic runoff trickles through the pipe and drains down the bank and into the nearby waterways. Donald helps the participants to see aspects of the Amazonian landscape that they would not otherwise.

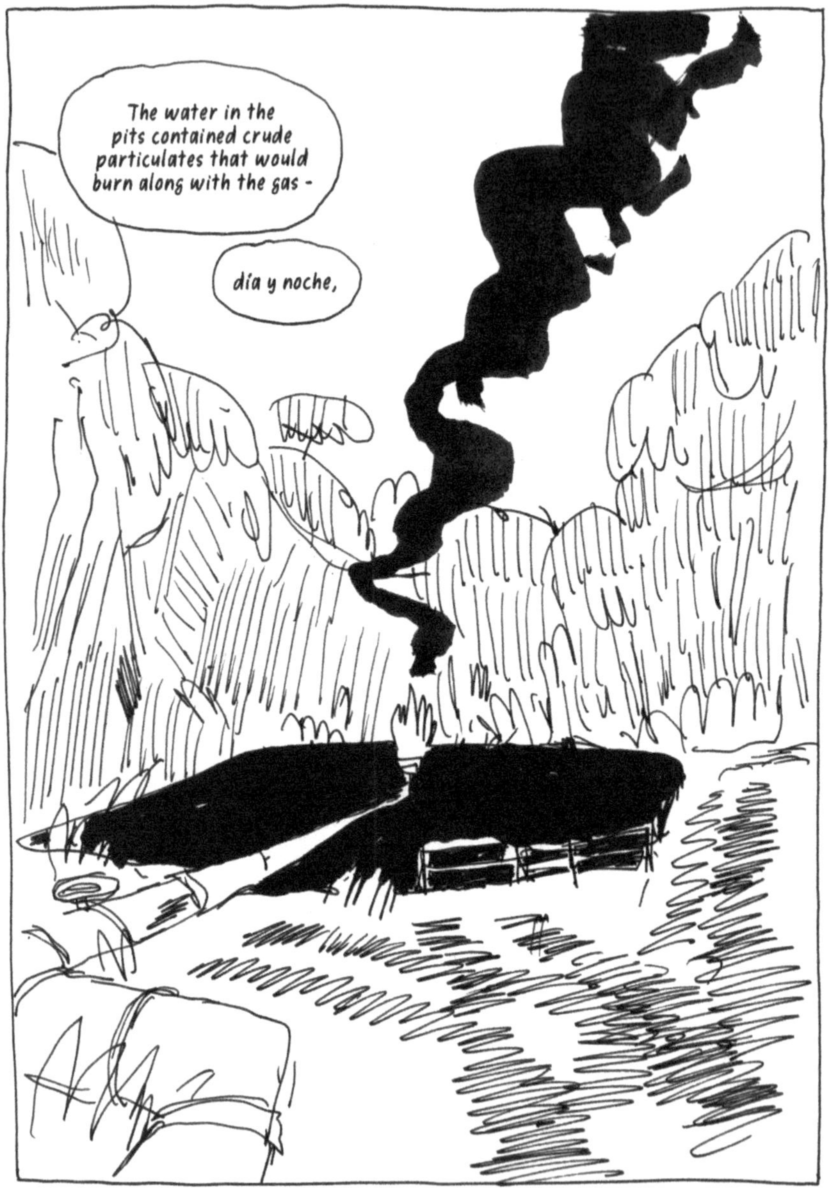

FIGURE 3 Billowing smoke offers readers a sense of the suffocating effects of gas flares for those in their midst. These are draft sketches that the final artwork for the *Toxic* graphic novel will be based upon.

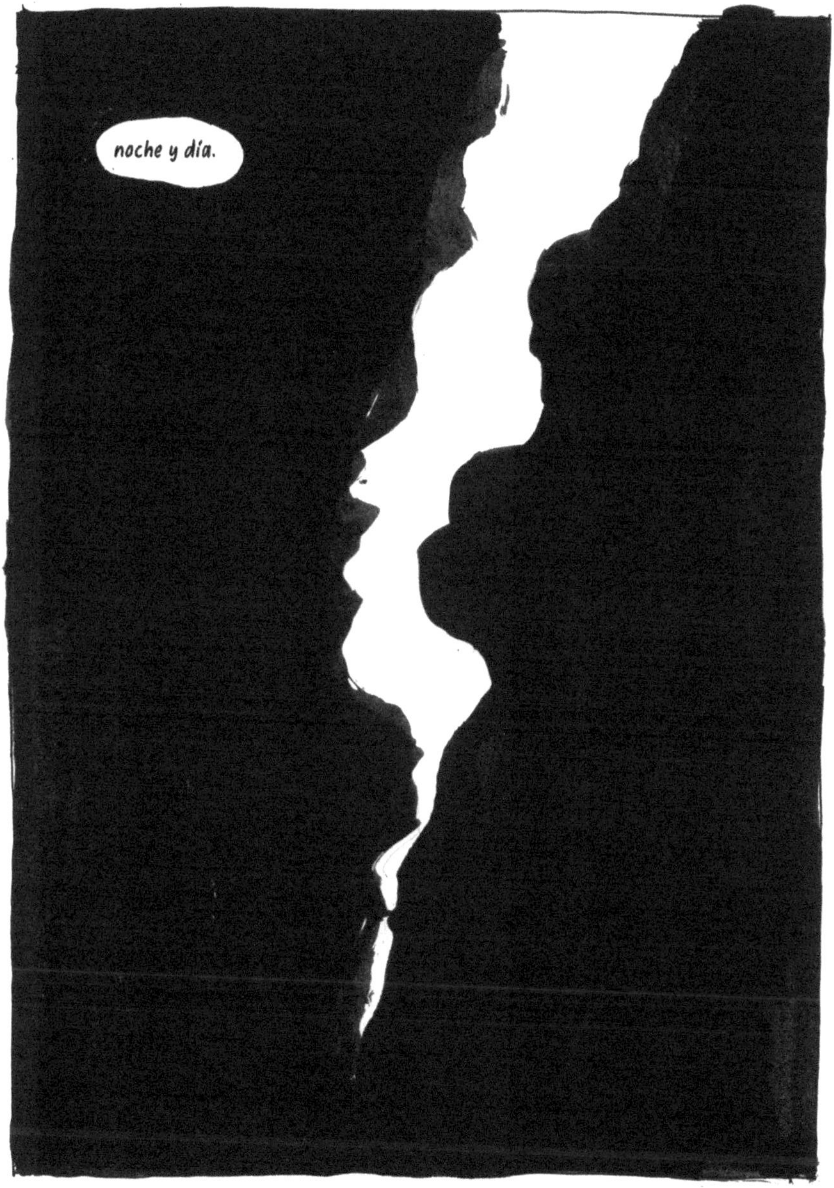

FIGURE 4 The inversion of smoke against the skies suggests the enduring effects of the chemicals released from the flares. These are draft sketches that the final artwork for the *Toxic* graphic novel will be based upon.

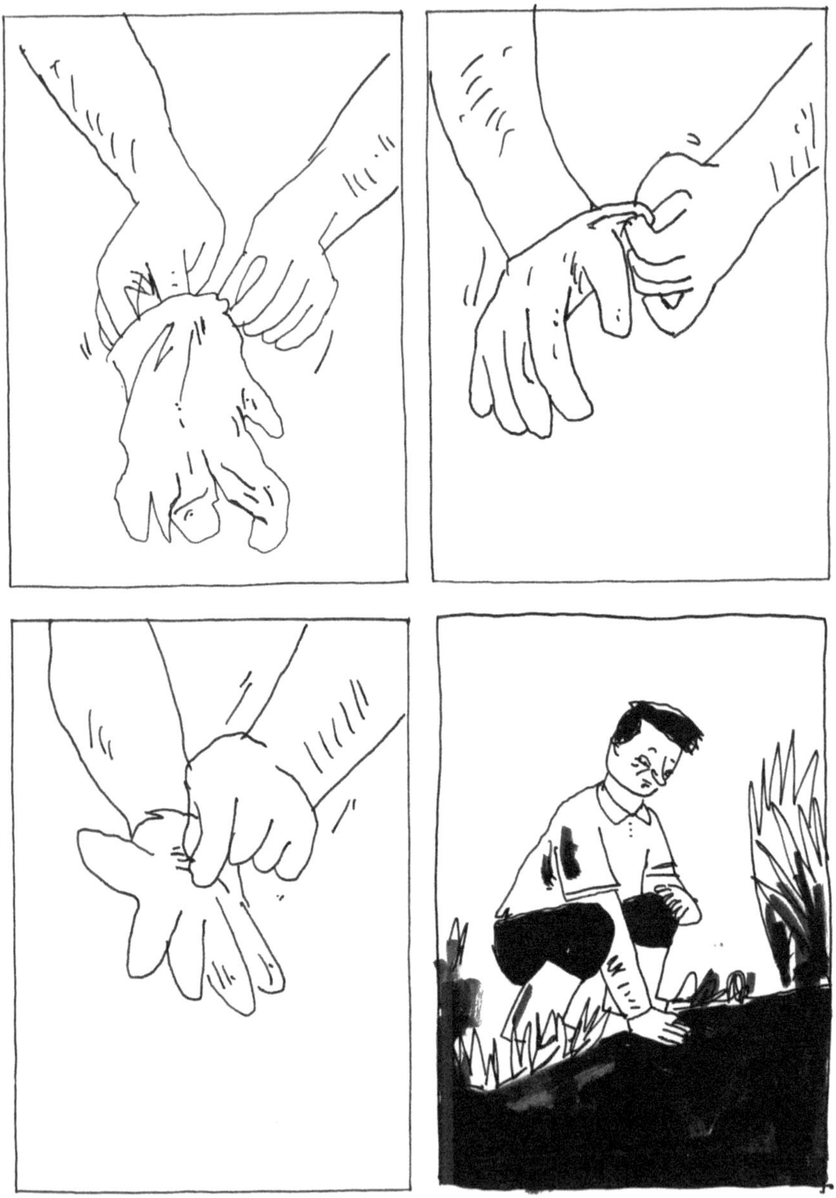

FIGURE 5 Using gloves, participants on the tour get to experience the feel and smell of contamination first-hand. The genre of the graphic novel is ideal for playing with the many sensory aspects of the Toxic Tour. These are draft sketches that the final artwork for the *Toxic* graphic novel will be based on.

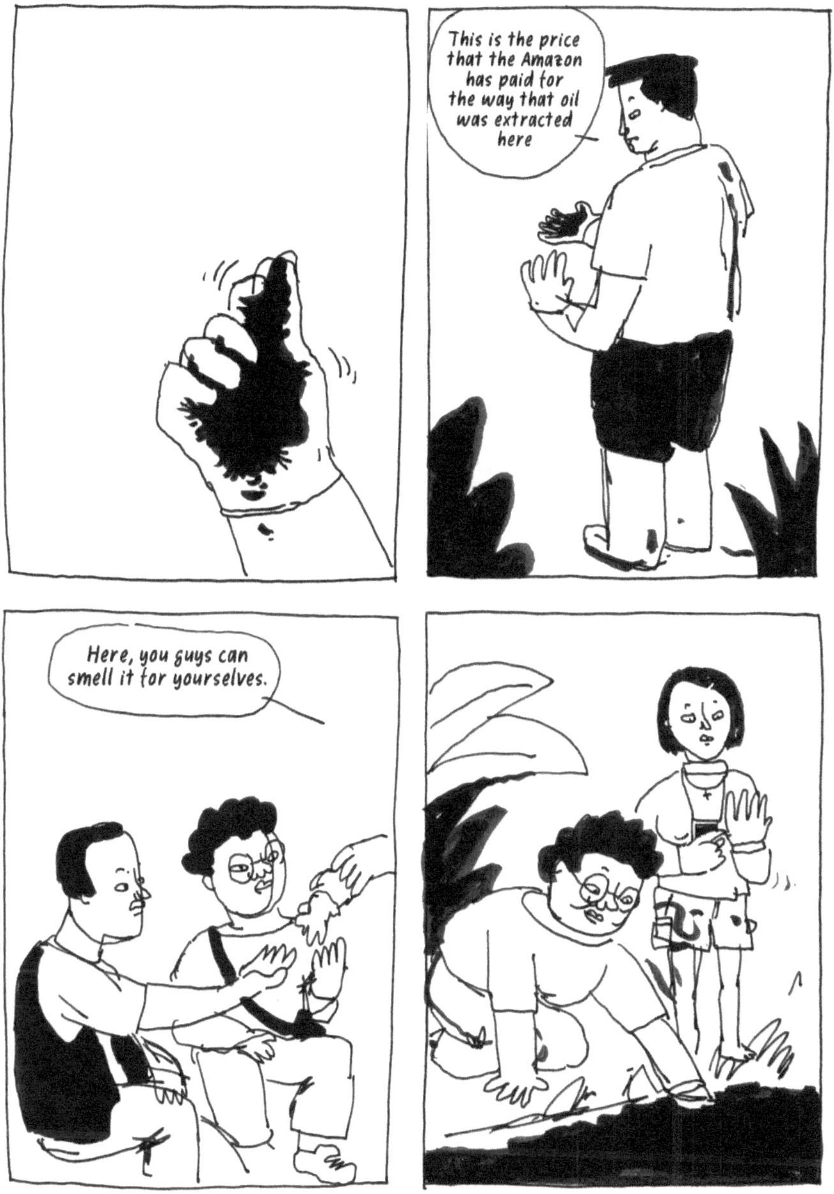

FIGURE 6 Feeling the squish of crude between their gloved hands is a powerful experience for the participants. These moments of toxic encounter in the graphic novel also help readers to relate to the experiential aspects of the Toxic Tour. These are draft sketches that the final artwork for the Tóxico graphic novel will be based upon.

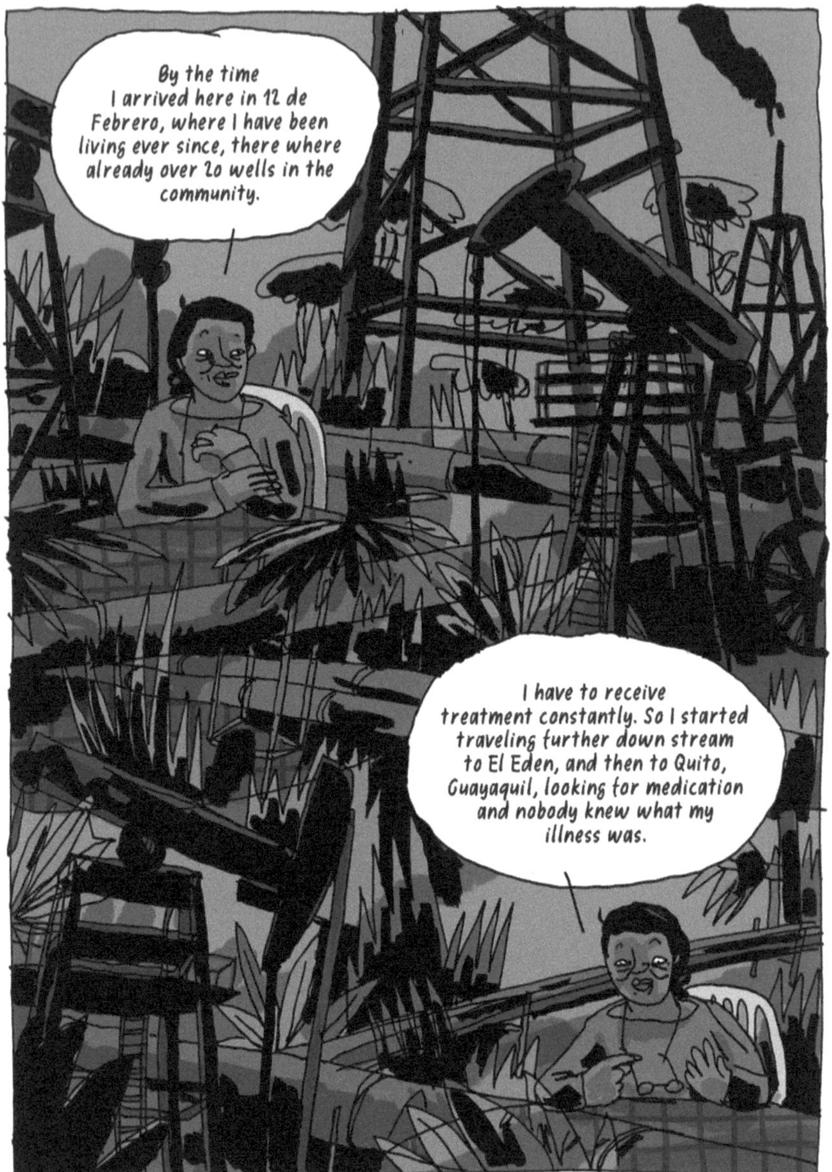

FIGURE 7 Personal narratives of Amazonian residents that are recounted to participants on the tour are visually overlaid with oil derricks and pipelines, illustrating some of the literal and metaphorical entanglements of extraction.

fingers, breathe the air beneath gas flares, or in some cases, have close calls with waste pits. One of the most striking parts of the Toxic Tour is the possibility for visitors – whose lives are often geographically and culturally distant from the Amazon – to relate to the consequences of oil production, forging a brief space of shared connection between outsiders and the lives of Amazonian residents (Figures 3, 4, 5 and 6).

The intention of this approach is that it helps to render toxicity within the oil complex palpable. The visual depiction of these events allows for the subtlety of the various, insidious, even contradictory ways that oil permeates life to emerge in all their difficulty. *Toxic* is a vivid reflection on the role of toxicants in our everyday lives, ultimately asking readers to reflect on the ways that we are each implicated in the petrochemical complicities of production, consumption, and exposure in the Amazon and at home (Figure 7).

The movement from ethnography to render toxicity in graphic form poses challenges in representing a kaleidoscope of partial, shifting, entangled knowledges of chemicals and their effects on bodies and places. Because the graphic form necessitates slower movements than text, it becomes more difficult to include a range of voices and experiences with toxicity in the story. In writing the story, we had to make choices about which stories to privilege, and which to leave out, as well as to find creative ways to make sure that the characters were not put in the position of speaking for the experiences of others.

The graphic form is best suited to conveying the effects of toxicity that are visible. Yet many effects of toxicity are not discernible by sight alone. Moments when bodies are affected by contamination – either in the flesh, or overwhelmed by fumes from a flare – need to be rendered legible for readers. The graphic form lends itself to imaginative explorations of what it feels like to be submerged in toxicity, or to be surrounded by losses of life – whether through the lives of neighbors or farm animals. Doing so translates toxicity from its chemical traces into its effects on daily life for those in the Amazon.

Inspired by the urgency of life in a moment where preventing permeation by the continual, residual, overwhelming onslaught of industrial toxicants is neither possible nor expected, the book explores the implications of conveying the sensory experiences of toxicity in graphic form, inviting readers to explore their own relationship to toxicity from where they sit.

10

PRESERVATION BY DEMOLITION: TOXIC HERITAGE IN CONTEMPORARY CHINA

Loretta I.T. Lou

Heritage preservation is usually understood as an act of preserving a human legacy for the benefit of future generations. Yet what this really means has seldom been examined (Harrison et al. 2020, 3). A deep dive into questions such as what should be preserved, why, and for whom has the potential to open up the notion of heritage as not only a tangible material legacy originated *from the past*, but also something that is intangible, processual, and discursive *for the future* (Harrison et al. 2020, 5). In this chapter, I explore how residents in a Chinese neighbourhood engaged in this future-making process by bargaining with their "toxic heritage." Here, toxic heritage is defined as the tangible and intangible legacies of living next to a toxic facility. Based on ethnographic fieldwork conducted between the spring and autumn of 2018, I show how the villagers of Samtilwei, a periurban neighbourhood adjacent to a petrochemical plant in southern China, tried to maximise their financial gains through the double act of destroying and remaking their toxic heritage. Such an attempt was made possible because the intangible heritage that the villagers wish to preserve, notably their land ownership and their peasant landlord status, would only be kept after the tangible forms of their toxic heritage are demolished. This tangible heritage includes not only the villagers' own farms and homes, but also several nineteenth-century buildings that are officially listed by the government as a cultural and historical heritage in the district. On the one hand, this deliberated reconceptualisation of heritage exemplifies the villagers' opportunism and the local government's development-above-all mentality. On the other hand, it highlights the significance of negotiation in contemporary heritage practice (Suntikul and Jachna 2013; Kidd and Cardiff 2017; Witte 2019). In exploring how villagers tactically mobilised "preservation" to retain their intangible connections to the villages while justifying the demolition of their tangible estates, this chapter

DOI: 10.4324/9781003365259-20

expounds the peculiar phenomenon of "preservation by demolition" in China and discusses its implications for heritage studies.

Resistance to relocation

Situated within 800 metres of the petrochemical zone, Samtilwei is an industrial neighbourhood consisting of three natural villages (New Village, Old Village, Happy Valley) that existed long before the government acquired their land to build the petrochemicals facilities in 1973. Although there has been no history of major accidents in this area, the local government has set out plans to demolish these three villages and relocate their residents in 2015. This is in response to the central government's call for strengthening the health and safety standards of petrochemical production in the aftermath of the Tianjin explosions[1] (He et al. 2018, 825) (Figure 10.1). After nearly two years of preparation, the relocation process officially began in March 2017. A year later, 80% of the villagers were said to have signed the Relocation and Resettlement Agreement. Those who remained in the villages, I was told, were either waiting for the keys to their new apartments or fighting against the relocation order until the last minute. At the time of my fieldwork in

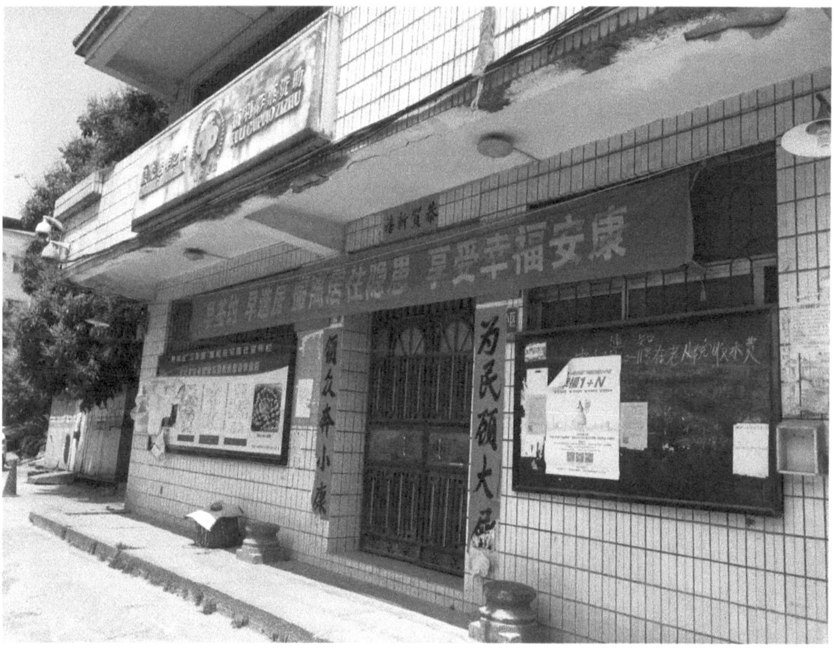

FIGURE 10.1 A typical Chinese socialist banner in yellow font and red background. The banner reads: "Sign the resettlement contract and choose your apartment as soon as possible. Move away from health and safety hazards. Enjoy peace and happiness."

early 2018, there were about 300 households and 1000 people living in these three villages, with a mix of local villagers and migrant workers.

Nostalgia for an industrial past is a well-researched subject across disciplines (Strangleman 2001; Mah 2009; Stephenson and Wray 2005; Mah 2012; Sherren et al. 2016; Emery 2020; Rhodes II, Walker, and Price 2020; Garrow 2021; Audin 2021). In her pioneering research on Newcastle-upon-Tyne in the UK and Highland, Niagara Falls, in the USA,, Alice Mah observed that "despite socioeconomic deprivation and material devastation in areas of industrial decline, houses and neighbourhood spaces can become invested with notions of family and community unity, nostalgia for a shared industrial past, and stability amidst socioeconomic change" (Mah 2009, 287). In Samtilwei, similar kinds of sentiments were also palpable among villagers who refused to leave their homes. Although the village houses looked dilapidated and the communal areas were inundated with filth and rats (Figure 10.2), the remainers defended the livability of the villages and insisted that their homes were far better than Ruidong Court, the urban apartment blocks that were designated for their resettlement. As Li Popo scorned, "Ruidong Court is low-cost housing for poor people! Our village houses are larger and are made of solid materials. How can the government force us to swap our detached houses with these tiny apartments in the city? It's not fair!"

The remainers were not the only people who had expressed nostalgia for village life. Those who had already left also felt homesick. Dan Dan and Yu-fang became

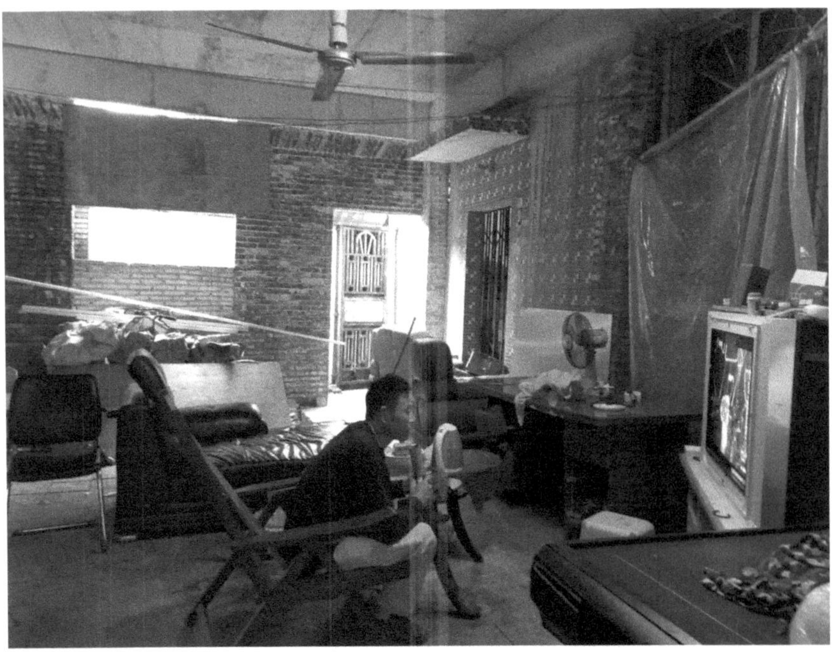

FIGURE 10.2 A villager enjoying his bamboo smoking pipe while watching TV.

friends when they married in the Old Village in their early 20s. Their husbands accepted the government's compensation offers and moved to Ruidong Court as soon as they were given the keys to the apartments. However, neither Dan Dan nor Yu-fang wanted to cut their ties with the villages. They came back to visit their friends and neighbours whenever they had time. "We miss living here," Dan Dan said, "people are closer to each other in the villages."

Villagers had also complained about the cramped condition of urban living. For example, it was particularly difficult for 91-year-old Chan Gongong to imagine living in the same block with hundreds of other residents. In his mind, a building block is a standard three-level detached home in which Samtilwei villagers lived (Figure 10.3). Normally, each block is occupied by no more than one or two households, often from the same kin. By contrast, Ruidong Court is made up of five 24-level blocks of apartments occupied by unrelated people of various origins. As such, Chan Gongong was convinced that Ruidong Court is inferior to the village houses. Being in a wheelchair, he was also apprehensive about living in a high-rise building. He said, "I don't trust the lifts. Lifts in China go out of order all the time!"

Although Ruidong Court was dismissed by these villagers as cheap, small, and cramped, this apartment complex is actually selected by the local government as a demonstration community to promote smart and intelligent property management in China. Not only is the gated community (*xiaoqu*) equipped with high technologies

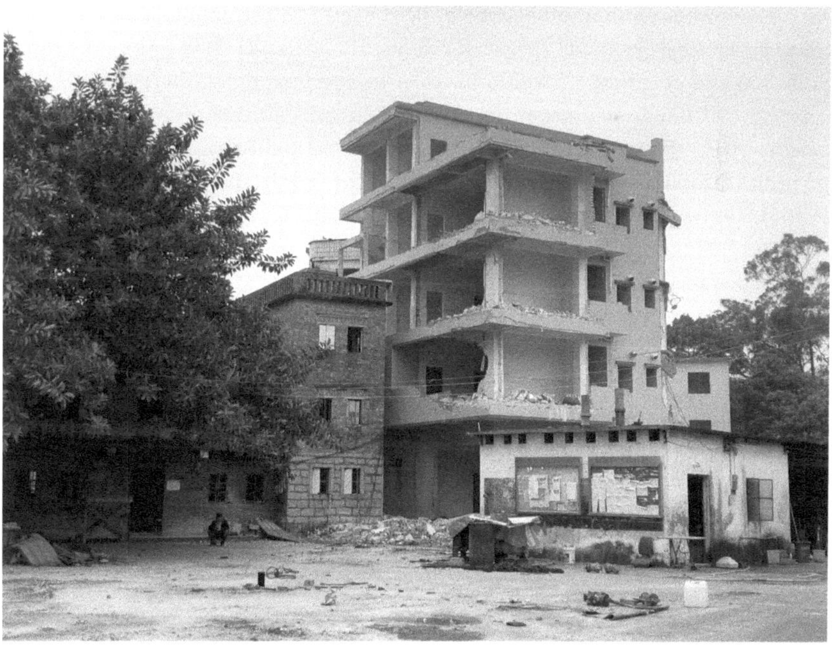

FIGURE 10.3 A standard three-level village house undergoing demolition.

like real-time air pollution monitoring and 5G wireless intelligent light poles, Ruidong Court is located within walking distance to many local amenities, including undergrounds, hospitals, schools, nurseries, and supermarkets. These benefits, however, did not seem to make a difference to the villagers' opinions.

However, what distinguishes the place attachment in Samtilwei from the place attachment observed in cases of deindustrialization (Strangleman 2001; Stephenson and Wray 2005; Emery 2020; Rhodes II, Walker, and Price 2020; Garrow 2021) and cases of environmental justice (Allen 2003; Davies and Mah 2020) is that toxic contamination has never been the driving force for changes and actions in Samtilwei. For one, people in China tend to downplay the health impact of toxic pollution (Tilt 2006; Lora-Wainwright 2009; Mah and Wang 2017; Lou 2022). Whenever I asked the villagers what they think about petrochemical pollution, they would brush it off as something minor and tolerable (Lou 2022). "There is no need to make a big deal of it! The pollution is not as bad as people said" Uncle Qin stated. Instead, villagers liked to brag about the sweet lychee they grew and the plump fish they kept in the pond. The rationale is that if humans can survive petrochemical pollution, the plants will certainly be alright. Another factor to keep in mind is that in view of the recurrent food-safety incidents, many farmers in China would rather rely on their homegrown food for subsistence than buy food from the market (Lora-Wainwright 2014, 661) as they perceived pesticides as far more dangerous than petrochemical contamination (Lou 2022).

As I have discussed in another article, petrochemical communities in China are adept to using what I coined "the art of unnoticing" (Lou 2022) to ignore the threat of pollution and chemical explosion in order to continue living there. This coping mechanism is not uncommon in (post-)industrial communities (Mah 2009; Jovanović 2016). Frequently, people in these affected communities would attribute their health problems to stress (Lora-Wainwright 2009) and ageing (Mah 2009, 303–304). Other times, they used longevity as evidence that pollution was overstated (Lou 2022). As Granny Ma of the New Village claimed, "I've lived next to the petrochemical plant for three decades already. What's the point of worrying now? Many people in our village live 'til their nineties. My mother-in-law died at age eighty-nine. My father-in-law died at ninety-two. There is even a nursing home nearby called the Village of Longevity!"

In short, these narratives not only reinforce people's confidence that Samtilwei was a liveable place. More importantly, they cast doubt on the necessity of the forced relocation. Although the local government insisted that the move was purely driven by health and safety concerns, nobody in the village really believed that. In fact, everyone knew that the ulterior motive behind the land grabs was to enable the local government to "secure lucrative land deals with outside investors" (Lora-Wainwright 2012, 8), from which the profits were not fairly distributed among people at the grassroots (Lora-Wainwright 2014). It is this perceived unfairness in compensation rather than toxic pollution that motivated some of the villagers to bargain with their toxic heritage.

Bargaining with toxic heritage

At first glance, the villagers' emphasis on family and community, and their longing for continuity and stability, resembles the sentiments of "devastation but also home" in areas of industrial decline (Mah 2009). However, a closer examination reveals that family, community, and nostalgia for village life only accounted for a small part of people's resistance to relocation. As I bore witness to how villagers bargained and negotiated for a better compensation and resettlement arrangement over the course of four years (2018–2022), I realised that the situation is much more complicated. Given the infrastructural dereliction and the potential hazards of the area, why were people reluctant to move? What value did their toxic heritage bring to individuals, families, and communities?

To make sense of this seeming contradiction, it is necessary to contextualise the villagers' perspectives in light of China's rapid urbanisation in recent decades. Since the 1980s, China has set off a wave of rural land expropriation and housing demolition in response to the demand for urbanisation. Under the current Constitution, urban land is owned by the state and rural land is owned collectively by the village communities (Peng 2018; Kan 2019). By this law, if the government wants to develop rural land, it must first convert them into state-owned urban land through the process of expropriation (Kan 2019, 636), which has become a major source of social unrest in China over the years (Sargeson 2013; Yuen 2014). In many cases, the conflict is inflicted by inadequate compensation or poor resettlement of land-lost farmers and rural residents. Thus, earlier research on land expropriation tended to focus on the state's coercive measures and the peasants' struggle (Yuen 2014; Lora-Wainwright 2012). More recently, however, scholars have challenged the binary view that villagers are either activists or victims in such processes. As Wang's research demonstrates, contrary to popular imagination, the state could not "simply expel existing residents and liquidate their homes" (Wang 2022, 504). Instead, the local government has to "calculate and allocate monetary and property compensations among relocated households based on a variety of factors ranging from the size of the family to the value of the estate to be demolished" (Wang 2022, 503).

As such, "the extent to which peasants benefited or suffered from land requisition was determined by multiple factors which differed region by region, village by village, and household by household" (Chen 2019, 79). Indeed, "compensation standards in developed coastal areas differ substantially from those of western regions" (Yang and Qian 2021, 502). While villagers in the west of China usually receive only minimal compensation (Lora-Wainwright 2012; Lora-Wainwright 2014), in more developed areas and in the region where I carried out my fieldwork, rural residents have received a large lump sum of monetary compensation (Kan 2019). For these villagers, demolition and relocation projects create opportunities for them to reap direct monetary benefits (Kan 2019; Wang 2022), resulting in the advent of a new social class known as *chai erdai* in Chinese, meaning "demolition parvenus" (Yang and Qian 2021, 502; Shi et al. 2019, 11) or "new rich-through demolition" (Steffen 2022).

Although the law does provide guidelines for land requisition in China, local governments have considerable flexibility in their handling of compensation (Chen 2019, 102). Generally speaking, rural residents are entitled to a combination of all or some of the following: a one-off monetary compensation for their land, lost crops, demolished houses; resettlement allowances; employment alternatives; or an urban household registration (a.k.a. an urban *hukou*, a status that is linked to a wide range of social benefits and services) (Kan 2019, 638; Yang and Qian 2021, 501). However, as the urbanisation of rural China is increasingly achieved not through physical land grabs but "the strategic enrolment of rural communities in the commodification of land via speculative rentiership" (Kan 2019, 633), local governments "have to constantly update and negotiate protocols of calculation with property-owning villagers to solve emerging issues of commensuration" (Wang 2022, 506). For example, the villagers of Samtilwei contended that compensation should be calculated based on their land's *future* value rather than its current value. And as I mentioned above, villagers did not believe that pollution was what motivated the government to relocate them. The real reason, they said, was that there would be a real estate boom in the area in the near future:

> What pollution? There is no pollution! Health and safety hazards are not the real reason that the government wants us to move. It's just an excuse to grab our land and turn it into something more lucrative. I tell you, this place is going to be more prosperous than the Central Business District in the future!
>
> (Quote from an interview in 2018)

Initially, the compensation offered to the villagers of Samtilwei was either one-off cash compensation *or* resettlement housing based on the size of their original properties. Either way, as soon as the villagers signed the agreements, they would permanently lose their legal connections to Samtilwei. Evidently, the villagers' reluctance to leave was not merely driven by the sentiment of "devastation but also home" (Mah 2009), but also the lucrative prospects of demolition, a once-in-a-lifetime opportunity to receive some money from the government (Lora-Wainwright 2012; Kan 2019; Wang 2022).

After several rounds of negotiation, the villagers agreed that instead of giving up their collective land ownership in exchange for one-off cash compensation, they would retain the collective ownership *and* the property rights of any to-be-built structures on their lands. In return, the government would be given the use rights to develop this neighbourhood, and the villagers would earn rental income from these newly developed properties. This new deal enables the government to proceed with its development plans while incorporating villagers as stakeholders by converting them into "peasant landlords" (Chu 2022). Although the villagers would be physically removed from their lands and their original houses, they are able to preserve their rural household registration, known as rural *hukou* in Chinese.[2] This is a significant victory, as rural *hukou* is increasingly considered a valuable intangible asset that farmers can pass on to their children, allowing them to also "participate directly in the appropriation of value from land as rentiers" (Kan 2019, 640).

The villagers gave three reasons for their rejection of the one-off cash compensation. First, the seniors in the village were worried that some less-educated villagers might end up gambling all the money away or squandering it on conspicuous consumption if they received a large lump sum. This is a social problem that has been widely reported in recent years (Bao et al. 2017). Second, although the villagers were concerned about currency depreciation, they knew very little about investing. Rental income was by far their most familiar form of investment option. Many villagers had already been renting out their spare rooms or houses to migrant workers in the area. Last but not least, the villagers of Samtilwei wish to preserve their rural *hukou* – a status that is increasingly seen as a financial asset thanks to the waves of urbanisation and land expropriation.

The negotiation process in Samtilwei reminds me of what Jovanović observed in an industrial town called Bor in Eastern Serbia because of the villagers' seeming self-contradiction. In Bor, Jovanović found that people disliked smoke, but they also celebrated smoke. Smoke, she wrote, was "a sign that the company was working well and that the whole town and its citizens, whether they worked for the company or not, depended on its production" (Jovanović 2016, 490). From there, Jovanović concluded that "it was not that people only adapted to risks while accepting them as inevitable. The risk was also seen as something that could be calculated and bargained with in relation to hopes for stable futures" (Jovanović 2016, 496). Such a conundrum was epitomised in a meeting between citizens of Bor and representatives of the local smelting plant, where villagers abruptly shifted from demanding compensation for health damage to asking the plant to employ them for "a stable personal and communal future" (Jovanović 2016, 496).

Like the citizens of Bor, the villagers of Samtilwei had also learned to see the bright side of living next to a petrochemical plant, but their articulation of *why* it was a positive thing changed over the course of the negotiation. During the early stage of the negotiation, most remaining villagers felt defeated and powerless. After all, nearly 80% of their fellow villagers had accepted the compensation offer and relocated to Ruidong Court. Given that there was no hope in sight, all they could do was express their reluctance and nostalgia – the feeling of "devastation but also home". Borrowing Mah's words, at that time there was an "idealized vision of community projected onto a turbulent social and economic reality" (Mah 2009, 295). But as the negotiation reached a stalemate in subsequent years, the villagers realised that the government could not evict them without their consent, and that relocation is a "messy process" of "value calibration and translation" (Wang 2022, 515). During this phase of their negotiation, the villagers justified their stay by highlighting the liveability of their homes and downplaying the threat of pollution and chemical explosions. Finally, when the profits from demolition were in sight, villagers shifted to emphasise the necessity of relocation for the sake of health and safety. This change in emphasis not only aligned them with the government's official objectives, but also made them appear less greedy and morally questionable. In other words, the toxic heritage of petrochemicals was simultaneously perceived as "a source of hazard, a threat, a risk, and a source of money and prospects, an opportunity" (Jovanović 2016, 498). No one seemed to be troubled by their own inconsistency. In the face of limited economic and political alternatives, opportunism was endorsed while health and heritage were commodified.

Preservation by demolition: Destroying and remaking toxic heritage

The story of Samtilwei not only demands us to rethink what we should preserve, why, and for whom, but it also challenges the very definition of heritage preservation. These questions became even more poignant when we take into account that all listed historical buildings in the neighbourhood were being actively ignored by the villagers and the local government during the demolition. According to government records, there are currently 22 heritage sites listed in the Historical and Cultural Sites Protected at the Level of Huangpu District (*Huangpu qu dengji baohu wenwu danwei*). Of these 22 sites, three of which are located within Samtilwei. The earliest site, situated at the back hill of the New Village, is a family tomb dated back to the Yuen Dynasty (1271–1368). The other two are an ancestral hall and a front entrance (*men lou*) of a classic Chinese house (Figure 10.4) dated 1823 and 1821, respectively.

Despite being listed as protected heritage sites, neither the villagers nor the local officials were interested in shielding them from demolition. From the villagers' perspectives, the "real heritage" that needs to be preserved is their ownership of the rural lands and their rural household registration. Their indifference to preserving the officially listed heritage forces us to rethink not only *what* heritage is, but also *who* decides what can be considered real heritage. Although the UNESCO Convention on Safeguarding of the Intangible Cultural Heritage (ICHC) (2003) has made an attempt to acknowledge non-Western manifestations and practices of heritage (Smith and Akagawa 2009, 1), in practice, it continues to "privilege colourful and exotic examples of intangible heritage; that represent nationally valued cultural events or performances, and which coincide with romanticised Western perceptions, while Indigenous works remain under-represented" (Smith and Akagawa 2009, 4). Indeed, international organisations are unlikely to classify China's collective land ownership and household registration system as intangible heritage, even though villagers of Samtilwei clearly privileged them over the tangible heritage that manifests their lineage (family tombs and ancestral halls) and their industrial past (the petrochemical sites). Such incongruence makes apparent the need to reconsider both the theory and practice of heritage preservation in order to address the complexity of the politics of recognition (Taylor 1994; Chakrabarty 2007). In light of this, previous studies have proposed to conceptualise heritage "as a cultural practice, rather than simply a site, place or intangible performance or event" (Smith and Akagawa 2009, 6). They argue that heritage is best understood as a cultural and political process of "remembering/forgetting" (Urry 1996; Dicks 2000;

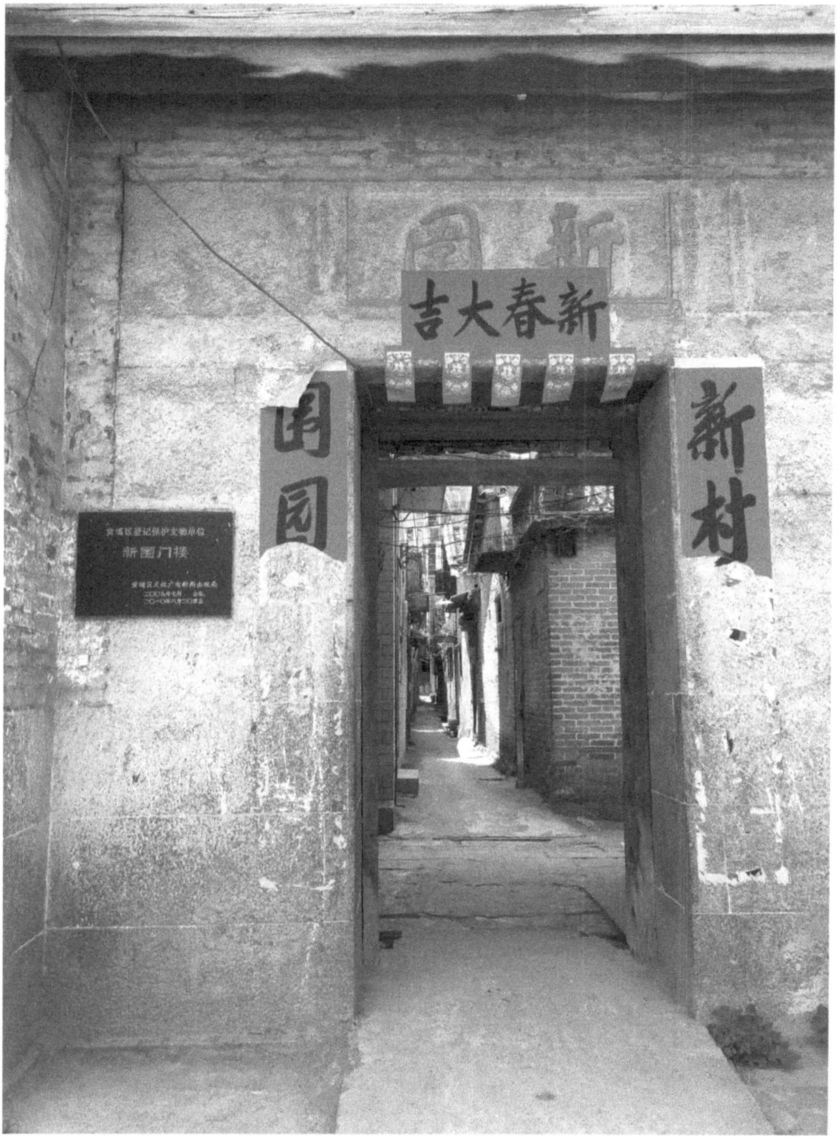

FIGURE 10.4 A *Men lou* in the New Village dated 1821. *Men lou* is the front entrance of a classic Chinese house. It is known as the 'face' of a family. The more elaborate the front entrance, the wealthier the family. The black plaque on the left records that the New Village *men lou* was officially listed as a heritage site in Huangpu District in July 2009.

Graham 2002; Peckham 2003; Smith 2006; Harrison 2013), and as a "verb" rather than a "noun" (Harvey 2001).

Building on these lines of thinking, my case study in China adds two nuanced dimensions to the ongoing re-theorisation of heritage preservation. First, the peculiar phenomenon of "preservation by demolition," whereby a heritage would only take form after it has been destroyed, may obscure the dialectics between heritage and preservation if the connections between the tangible (e.g. properties, lands, historical buildings) and intangible heritage (e.g. ownership, use rights, residential status) are not unravelled. While some scholars might attribute their seeming incongruence to the "inherent dissonant nature of heritage" (Smith and Akagawa 2009, 5), others would argue that there is no intrinsic nature in heritage, because heritage does not exist, but is made by people who imbue them with meanings, status, and values *for present purposes* (Kenny 2009, 151; Bendix 2009, 255). The case in Samtilwei proves the latter point, as land ownership and rural identity in themselves are not lucrative without the commodification of land.

Second, my ethnographic study has shown that the present wave of land expropriation in China offers rural residents a once-in-a-lifetime opportunity to become rich not just for themselves, but also for their future generations who would inherit their peasant landlord status. This "positive" case in China echoes the growing research on the possibilities of life in grim circumstances (Tsing 2015; McTighe and Raschig 2019; Murphy 2017; Ahmann and Kenner 2020) and challenges the widespread assumptions that toxic heritage is intrinsically undesirable for people who own and inherit it. While previous research has shed light on the intersection between the legacies of toxicity and the often celebratory interpretations of labour, prosperity, and place attachment to post-industrial sites, my case study contributes to the existing literature by elucidating the toxic site's financial promise for the current and future generations. It provides a case where the value of toxic heritage (in its material sense) derives not from the act of preservation, but from the act of demolition.

Finally, contrary to popular belief that heritage preservation is about the preservation of the past, I join other scholars to argue that heritage preservation is "deeply rooted in the present" (Kenny 2009, 151) and closely focused on the future (Harrison 2020). Heritagisation (Bendix 2009, 255) of toxicity is a process rife with calculation and contradictions in contemporary China, but for those who live the experience, it is also an opportunity to create a future in the brutality of neoliberal late-industrialism.

Acknowledgements

The research presented in this chapter is supported by the European Research Council under the European Union's Horizon 2020 research and innovation program (grant agreement no. 639583). It is part of the Toxic Expertise project led by Alice Mah at the University of Warwick. I would like to thank the Toxic Expertise team, especially Alice Mah and Thom Davies, for their encouragement and guidance. I also want to thank Elizabeth Kryder-Reid and Sarah May for their constructive feedback on the earlier drafts of this chapter. All remaining errors are my own.

Notes

1 According to an official statement published by the People's Government of Guangdong Province, the decision was jointly made by the inspection teams of the State Council of the People's Republic of China and the Ministry of Ecology and Environment.
2 According to the *hukou* system, holders of land use rights are still considered rural citizens (Chu 2022).

Bibliography

Ahmann, Chloe, and Alison Kenner. 2020. "Breathing Late Industrialism." *Engaging Science, Technology, and Society* 6 (0): 416–438. 10.17351/ests2020.673

Allen, Barbara L. 2003. *Uneasy Alchemy Citizens and Experts in Louisiana's Chemical Corridor Disputes.* Cambridge: MIT Press.

Audin, Judith. 2021. "Reconnecting Spatialities in Uninhabited Industrial Spaces: Ruination and Sense of Place in a Coal Town (Datong, Shanxi)." *China Perspectives* 4 (4): 41–49. 10.4000/chinaperspectives.12669

Bao, H., S. Deng, S. Xu, and Y. Peng. 2017. Conspicuous Consumption Behavior of Land-lost Farmers: A Perspective of Social Identity. *Cities* 66: 81–90. 10.1016/j.cities.2017.04.003

Bendix, Regina. 2009. "Heritage between Economy and Politics: An Assessment from the Perspective of Cultural Anthropology." In *Intangible Heritage*, edited by Laurajane Smith, and Natsuko Akagawa, pp. 253–269. London: Routledge.

Chakrabarty, Dipesh. 2007. "History and the Politics of Recognition." In *History and the Politics of Recognition*, edited by Alun Munslow, Keith Jenkins , and Sue Morgan. London; New York: Routledge.

Chen, An. 2019. "The Impact of Land Requisition on Peasant Life in China." *Modern China.* 10.1177/0097700419839638

Chu, Nellie. 2022. "Tu Er Dai Peasant Landlords and the Infrastructures of Accumulation in Guangzhou's Urban Villages." *Positions: Asia Critique* 30 (3): 479–499. 10.1215/10679847-9723685

Davies, Thom, and Alice Mah. 2020. *Toxic Truths: Environmental Justice and Citizen Science in a Post-Truth Age.* Manchester: Manchester University Press.

Dicks, Bella. (2000) *Heritage, Place and Community.* Cardiff: University of Wales Press.

Emery, Jay. 2020. "After Coal: Affective-Temporal Processes of Belonging and Alienation in the Deindustrializing Nottinghamshire Coalfield, UK." *Frontiers in Sociology* 5 (May). 10.3389/fsoc.2020.00038

Garrow, Hannah. 2021. "Living in and Loving Leith: Using Ethnography to Explore Place Attachment and Identity Processes." In *People-Centred Methodologies for Heritage Conservation: Exploring Emotional Attachments to Historic Urban Places*, edited by James Lesh. Abingdon, Oxon; New York, NY: Routledge. 10.4324/9780429345807

Graham, B., G. Ashworth, and J. Tunbridge. 2000. *A Geography of Heritage: Power, Culture and Economy.* London: Arnold.

Harrison, Rodney. 2013. *Heritage: Critical Approaches.* Milton Park, Abingdon; New York: Routledge.

Harrison, Rodney, Caitlin DeSilvey, Cornelius Holtorf, Sharon Macdonald, Nadia Bartolini, Esther Breithoff, Harald Fredheim, et al. 2020. "For Ever, for Everyone … ." In *Heritage Futures: Comparative Approaches to Natural and Cultural Heritage Practices*, 3–19. UCL Press. 10.2307/j.ctv13xps9m.7

Harrison, Rodney. 2020. "Heritage as Future-making Practices." In *Heritage Futures: Comparative Approaches to Natural and Cultural Heritage Practices*, 20–50. UCL Press.

Harvey, David C. 2001. "Heritage Pasts and Heritage Presents: Temporality, Meaning and the Scope of Heritage Studies." *International Journal of Heritage Studies* 7 (4): 319–338.

He, Guizhen, Chunci Chen, Lei Zhang, and Yonglong Lu. 2018. "Public Perception and Attitude towards Chemical Industry Park in Dalian, Bohai Rim." *Environmental Pollution* 235: 825–835. https://doi.org/10.1016/j.envpol.2017.12.105

Jovanović, Deana. 2016. "Prosperous Pollutants: Bargaining with Risks and Forging Hopes in an Industrial Town in Eastern Serbia." *Ethnos* 83 (3): 489–504. 10.1080/00141844.2016.1169205

Kan, Karita. 2019. "Accumulation without Dispossession? Land Commodification and Rent Extraction in Peri-Urban China." *International Journal of Urban and Regional Research*, February. 10.1111/1468-2427.12746

Kenny, Mary L. 2009. "Deeply Rooted in the Present: Making Heritage in Brazilian Quilombos." In *Intangible Heritage*, edited by Laurajane Smith, and Natsuko Akagawa, pp. 151–168. London: Routledge.

Kidd, Jenny, and Rosie Cardiff. 2017. "'A Space of Negotiation': Visitor Generated Content and Ethics at Tate." *Museum and Society* 15 (1): 43–55. 10.29311/mas.v15i1.661

Lora-Wainwright, Anna. 2009. "Of Farming Chemicals and Cancer Deaths: The Politics of Health in Contemporary Rural China."*Social Anthropology* 17 (1): 56–73. 10.1111/j.1469-8676.2008.00057.x

Lora-Wainwright, Anna. 2012. "Rural China in Ruins: The Rush to Urbanize China's Countryside Is Opening a Moral Battleground (Respond to This Article at Http://Www.therai.org.uk/At/Debate)." *Anthropology Today* 28 (4): 8–13. 10.1111/j.1467-8322.2012.00884.x

Lora-Wainwright, Anna. 2014. "Grassroots Perspectives on Relocation: Threats and Opportunities." *Positions: Asia Critique* 22 (3): 661–689. https://muse.jhu.edu/article/552124

Lou, Loretta. 2022. "The Art of Unnoticing: Risk Perception and Contrived Ignorance in China." *American Ethnologist* 49 (4): 580–594.

Mah, Alice. 2009. "Devastation But Also Home: Place Attachment in Areas of Industrial Decline." *Home Cultures* 6 (3): 287–310. 10.2752/174063109x12462745321462

Mah, Alice. 2012. *Industrial Ruination, Community, and Place: Landscapes and Legacies of Urban Decline*. Toronto: University of Toronto Press.

Mah, Alice, and Xinhong Wang. 2017. "Research on Environmental Justice in China: Limitations and Possibilities." *Chinese Journal of Environmental Law* 1 (2): 263–272. https://doi.org/10.1163/24686042-12340016

McTighe, Laura, and Megan Raschig. 2019. "Introduction: An Otherwise Anthropology." Theorizing the Contemporary, *Fieldsights*, July 31. https://culanth.org/fieldsights/introduction-an-otherwise-anthropology

Murphy, Michelle. 2017. "Alterlife and Decolonial Chemical Relations." *Cultural Anthropology* 32 (4): 494–503. https://doi.org/10.14506/ca32.4.02

Peckham, R. S. (ed.) 2003. *Rethinking Heritage. Cultures and Politics in Europe*. London: I. B. Tauris.

Peng, Chun. 2018. *Rural Land Takings Law in Modern China: Origin and Evolution*. S.L.: Cambridge Univ Press.

Rhodes II, Mark Alan, Amy Walker, and William R. Price. 2020. *Geographies of Post-Industrial Place, Memory, and Heritage*. Milton: Taylor & Francis Group.

Sargeson, Sally. 2013. "Violence as Development: Land Expropriation and China's Urbanization." *Journal of Peasant Studies* 40 (6): 1063–1085. 10.1080/03066150.2013.865603

Sherren, Kate, Thomas M. Beckley, John R. Parkins, Richard C. Stedman, Kristina Keilty, and Isabelle Morin. 2016. "Learning (or Living) to Love the Landscapes of Hydroelectricity in Canada: Eliciting Local Perspectives on the Mactaquac Dam via Headpond Boat Tours." *Energy Research & Social Science* 14 (April): 102–110. 10.1016/j.erss.2016.02.003

Shi, Jiangang, Xinru Min, Hongyun Si, Daizhong Tang, and Wei Miao. 2019. "The Transition from Housing Demolition to Conservation and Renovation in Shanghai: Challenges and Countermeasures." *Land* 8 (11): 175. 10.3390/land8110175.

Smith, Laurajane. 2006. *Uses of Heritage*. London: Routledge.

Smith, Laurajane, and Natsuko Akagawa. 2009. "Introduction." In *Intangible Heritage*, edited by Laurajane Smith, and Natsuko Akagawa, pp. 1–10. London: Routledge.

Steffen, Megan. 2022. "The Fruits of Demolition: Generative Neglect in Zhengzhou's Urban Villages." *Positions: Asia Critique* 30 (3): 571–594. https://doi.org/10.1215/10679847-9723737

Stephenson, Carol, and David Wray. 2005. "Emotional Regeneration through Community Action in Post-Industrial Mining Communities: The New Herrington Miners' Banner Partnership." *Capital & Class* 29 (3): 175–199. 10.1177/030981680508700111

Strangleman, Tim. 2001. "Networks, Place and Identities in Post-Industrial Mining Communities." *International Journal of Urban and Regional Research* 25 (2): 253–267. 10.1111/1468-2427.00310

Suntikul, Wantanee, and Timothy Jachna. 2013. "Contestation and Negotiation of Heritage Conservation in Luang Prabang, Laos." *Tourism Management* 38 (October): 57–68. 10.1016/j.tourman.2013.02.005

Taylor, Charles. 1994. "The Politics of Recognition." In *Multiculturalism: Examining the Politics of Recognition*, edited by Amy Gutman. New Jersey: Princeton University Press.

Tilt, Bryan. 2006. "Perceptions of Risk from Industrial Pollution in China: A Comparison of Occupational Groups." *Human Organization* 65 (2): 115–127. 10.17730/humo.65.2.69dufcr1hatnpu5v

Tsing, Anna. 2015. *The Mushroom at the End of the World: On the Possibility of Life in Capitalist Ruins*. New Jersey: Princeton University Press.

Urry, J. 1996. "How societies remember the past." In *Theorising Museums*, edited by S. Macdonald, and G. Fyfe. Oxford: Blackwells.

Wang, Mengqi. 2022. "Calculating Needs and Prescribing Properties: *Chaiqian* and the Commensuration of Value at Nanjing's Urban Edge." *Positions: Asia Critique* 30 (3): 501–521. 10.1215/10679847-9723698

Witte, Marleen. 2019. "Black Citizenship, Afropolitan Critiques: Vernacular Heritage-Making and the Negotiation of Race in the Netherlands." *Social Anthropology* 27 (4): 609–625. 10.1111/1469-8676.12680

Yang, Chen, and Zhu Qian. 2021. "'Resettlement with Chinese Characteristics': The Distinctive Political-Economic Context, (In)Voluntary Urbanites, and Three Types of Mismatch." *International Journal of Urban Sustainable Development* 13 (3): 496–515. 10.1080/19463138.2021.1955364

Yuen, Samson. 2014. "China's New Rural Land Reform? Assessment and Prospects." *China Perspectives* 2014 (1): 61–65. 10.4000/chinaperspectives.6403

11

UNWANTED LEGACY AND MEMORY OF THE MILIEU: TOXIC MATERIALS, REMEDIATION, HABITUATION (ESTARREJA, PORTUGAL)

Fabienne Wateau, Carmem Regina Giongo, Daniela Figueiredo, Johnny Reis, and Manuelle Lago

Introduction – Damaged World

Between Land and Water, the Place is Magnificent

On the Portuguese Atlantic coast, south of Porto, a large lagoon of brackish water has been designated as a Special Protection Area (SPA) by the Natura 2000 network (Natura 2000). It offers a rich habitat favourable to the reproduction of a large number of migratory birds; the landscape has become very popular for birdwatchers and tourists because of the layout of its promontories, pastoral paths, and cyclable bridges. Between land and water, the place is magnificent. Nature is as if preserved there, some of it wild, totally unspoiled by contamination.

And it is this particularly positive image that local politicians and institutions go to great lengths to defend and protect. On the periphery of Estarreja city, which has been home to an industrial chemical facility since the 1950s and also has its feet in the water of this lagoon, one area in particular – which is called BioRia today, and was once a spillway for toxic products from the facility's factories – won the 2008 prize for best environmental rehabilitation and biodiversity preservation on the Ria de Aveiro. BioRia is the new calling card brandished by the city council, in association with PACOPAR, a multi-actor consortium organised around the factories, which itself also won a prize from the 2005 European Responsible Care Awards.[1] It is therefore between Aveiro and Estarreja that this study is geographically situated; and it is between toxic materials, remediation and habituation that it takes its sociological context.[2] In response to this book's theoretical argument, which encourages us to question the idea of toxicity as heritage, the article will begin by reviewing some of the anthropological literature on recovery, then consider the proposed concept of toxic heritage, and finally, present the chosen case study in connection with memory and heritage.

DOI: 10.4324/9781003365259-21

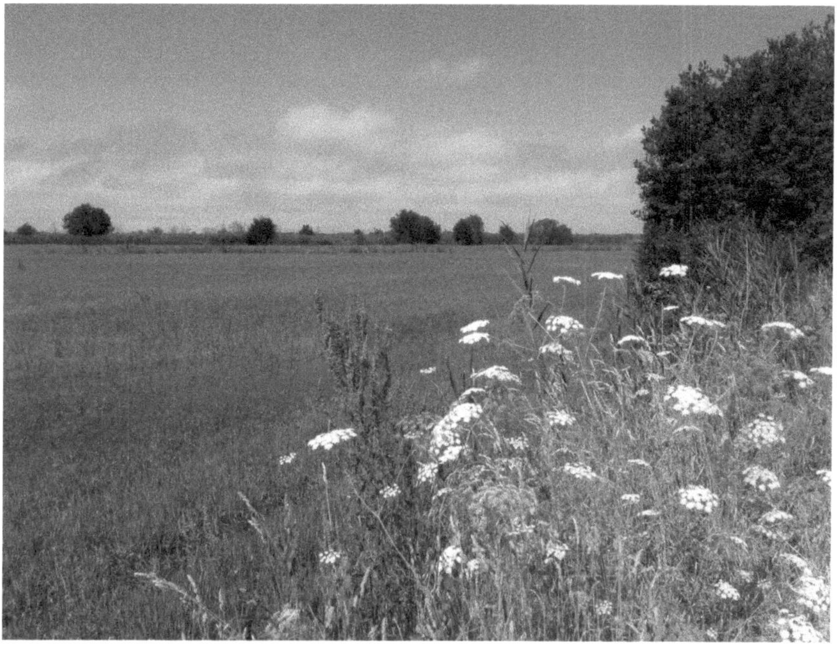

FIGURE 11.1 The BioRia, biodiverse wetland. Photograph by Manuelle Lago 2016.

Among Optimism and Utopia

Rehabilitating, restoring, and remedying. Because we didn't manage to prevent, contain, or protect. Rachel Carson's book *Silent Spring* certainly attempted to alert populations and decision-makers when she asserted that 'along with the possibility of the extinction of mankind by nuclear war, the central problem of our age has therefore become the contamination of man's total environment with such substances of incredible potential for harm – substances that accumulate [...]'. (Carson 2002, 7). In addition to this 'obligation to endure', it seems that we have subsequently been imposed the obligation of 'staying with the trouble'. Donna Haraway (2016), Ana Tsing (2015) and Bruno Latour (2017, 2021) – to name a few – are part of this optimistic, benevolent trend of raising awareness about a better world, advocating possible forms of earth resurgence (Terra/Gaïa): 'The task is to become capable, with each other [...], of response. Staying with the trouble does not require such a relationship to times called the future. Staying with the trouble requires learning to be truly present, not as a vanishing pivot between awful or edenic pasts and apocalyptic or salvific futures [...]'. She sets out another model, of 'oddkin [...] unexpected collaborations and combinations [...] We become-with each other or not at all [...] Sympoiesis—making-with' is the way to go. Haraway's complex theoretical book abounds in ideas and interpretations. It also outlines concrete actions, at local and regional levels that are situated, 'partial', that is to say, contextualised – as proof of the possible actions of 'art/science activisms [...]' aimed at 'multispecies resurgence [...] a recuperation [...]' that would be part of response through our abilities, 'response-abilities' (Haraway 2016, 1; 5; 16). In Latour's view, faced with the 'reaction of the earth system' that began after the 70 post-war years that specialists called the 'Great Acceleration' (Latour 2017, 76), 'we need to reinvent everything all over again – the law, politics, the arts, architecture, cities. But – and this is stranger still – we also need even to reinvent movement, the vector of our actions. We need not forge ahead into the infinite, but learn to step back, to unplug, in the face of the finite. [...]'. It is a quite clear political position, where the 'violence of the power struggle between the Extractors' (over-exploiters of the earth's resources) 'and the Menders' (the 'terrestrials' or repairers) is at play (Latour 2021, 155–157). So of course, if 'recuperation is still possible, but only in multi-species alliance, across the killing divisions of nature, culture, and technology and of organism, language, and machine' (Haraway 2016, 118), then it is nevertheless very difficult to go from philosophy to the field, and especially to find among our interlocutors in civil society the forces and means to concretely take up these challenges of 'assemblages' (Tsing 2015) that almost smack of utopia.

The Pessimistic View

Soraya Boudia and Nathalie Jas's vision is much more pessimistic (2019). After having sketched the history of the governance of the toxic world, which they divide into three forms (governance through control, governance through risk, and governance through adaptation), they rather condemn this policy of rehabilitation, maximum adaptability, and resilience. Because 'it is no longer a matter of rehabilitating damaged sites—something that has hardly ever been done satisfactorily—but rather to rehabilitate life possibilities on permanently contaminated lands'. And it is the 'populations living on contaminated lands [that] must learn to show resilience, to reinvent themselves and move forward despite the problems they face'. This government of adaptation 'contributes to normalising, or more precisely "naturalising", not only serious accidents but also ongoing contaminations that are often less immediately perceptible. This naturalisation process makes it possible not to single out those responsible for pollution bringing about visible and less visible catastrophes, and not to determine the deep causes of these'. The axe abruptly falls on this 'staying with' policy. Because by 'overemphasising the adaptability of individuals and populations [...] these policies are not devoid of cynicism' (Boudia and Jas 2019, 89–92). The portrait they sketch of the situation is therefore rather sombre, as they explain: 'Sombre because of the scale of the environmental pollutions described by many scientific texts [...]; Sombre also because of the cynicism of certain actors who try to ignore or cover up the damaging effects of their activities. Sombre finally because of the inability of various committed mobilisations and implemented regulations to curb this movement over the long term' (Boudia and Jas 2019, 95). In her quite chilling book, Lucie Taïeb (2020) tells the story of the human remains of the attack on the Twin Towers, mixed with garbage at the former Staten Island dump, which is in the process of becoming a large public park with recreational space for children. She describes her guided tour of the site, where the process of confining and capping the dump, through successive layers of materials possessing specific properties, is described to visitors. Depressed, Taïeb adds: 'Now everything is repaired, and this is the only consolation for all of the evil we know. The wounded earth is infinitely resilient, and memory is plastic. Shame and blemishes are erased. [...] Everything is repaired and everything is erased; yet nothing disappears'[3] (Taïeb 2020, 76).

Unwanted Legacy in Estarreja

A Logical Continuum of Heritage Studies

Since it has been recognized that 'heritage comes also in diverse 'negative' forms, such as environmental pollution, social inequalities, and poverty, inherited from the past and having a myriad of impacts in the present,' other definitions of heritage have been included in 'a logical continuum [...] of this subfield of heritage studies' (Thomas et al. 2019). Wollentz et al., in 2020, first proposed a definition of toxic heritage: 'we use the term "toxic heritage" to describe instances in which forms of difficult heritage come to endanger certain core values in society (such as equal rights and opportunities, peaceful coexistence, freedom of speech, health, and well-being), most often because of a lack of responsible management' (Wollentz et al. 2020, 299). In this book, Kryder-Reid and May (2024) are using the term toxic heritage to refer to the materiality of toxic substances, focusing on literal meanings; and connecting that history of harm with both formal heritage institutions and informal memory practices. Can toxic material be heritage? What does framing toxic sites as heritage offer? How are toxic sites remembered or, conversely, erased in the memory practices of communities and in official heritage narratives such as museums and historic markers? Are some of the questions asked in this volume (Kryder-Reid and May 2024).

Toxic Materials

Estarreja[4] falls within the definition of toxic heritage explored in this collection (Kryder-Reid and May, 2023) and also partly in the other framework (Wollentz et al. 2020). A chemical facility has existed there since the 1950s, one that discharged tons of toxic chemical products into the environment (earth, water, air) without oversight until the 1990s. Among the toxic materials found there, 'the most common pollutants are sulphur oxides and nitrogen oxides, ammonia, mononitrobenzene, suspended solids and vestiges of heavy metals, particularly arsenic and mercury. Liquid effluents include aniline, ammonium, benzene, mononitrobenzene, arsenic, iron, zinc and lead. In general, the liquid effluents originating from the chemical facility's factories (whether or not they

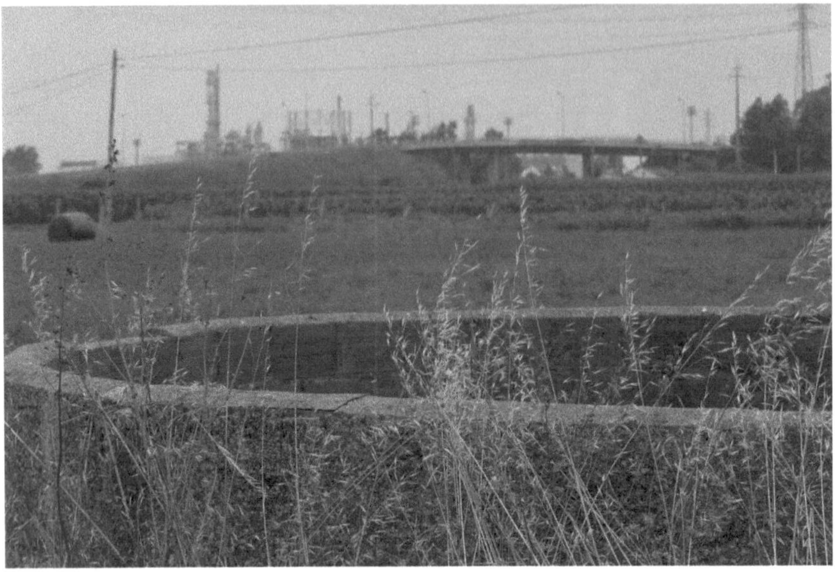

FIGURE 11.2 A well for irrigation next to the factories. Photograph by Carmem Giongo 2019.

have undergone pre-treatment) end up in two ditches (those of São Filipe and Breja) leading to their final destination (the Esteiro de Estarreja) [...]' (Barrosa 1985 quoted by Azevedo 1999). The superficial and groundwater are contaminated and will remain so, apparently due to a lack of possible treatments and their costs. But this water is still used to irrigate maize (and adjacent plants like gourds and beans in backyard gardens), being easily accessible through pumping from wells.[5] In response to our questions, the city health department asked us to systematically remind inhabitants not to drink the well water – implicitly revealing a habit acquired by some people.

Danger to human (and animal) health, therefore, seems real and existent – yet research into miscarriages, for example, has not succeeded in proving a higher incidence in Estarreja than elsewhere (Valente 2016).

The Materiality of Toxic Substances

From the institutions, there exists a certain kind of silenced information and a choice about which type of past to promote. On the developed bank of the *Esteiro de Estarreja*, whose underwater sludge has been proven to be contaminated with arsenic and mercury, a sign recalls the history of salt shipping by boat from the sea to the city. It provides information about the lagoon's pre-1950s social and trade history. But nothing is said about the toxic sediments it has retained, which were discharged by factories after the 1950s and are still dangerous; nor about the ban on swimming there. This lack of warning signs, which shocked us in 2016, was viewed by some members of the team as deliberate concealment from tourists, who are very numerous in this region. This was also observed near the *Largo do Laranjo*, another area heavily polluted by mercury, where the banks have been attractively developed into cyclable paths, without any signs warning about the dangerousness of the sludges in the deep water. But according to the biology teachers at the high school in Estarreja, it seems that a lack of information is notorious in that area, where (illegal) shellfish divers take risks that pose a threat not just to themselves, but also to the environment, by digging up sludge full of heavy metals. The town council prefers to emphasise a rich history of salt and of algae harvesting (*moliço*) instead of stressing the more recent unwanted legacy, which is damaging to its image.[6] All the more so since it actively intervenes to repair the environment. It also seems that the contamination of the environment is a legacy that can cross generations, regardless of the attitude of the population and their point of view on this contamination. The unwanted legacy seems accepted, with resilience (Boudia and Jas 2019).

Remediation and Memory of the Milieu

Legacy versus Heritage

Actually, in Estarreja, the fieldwork did not lead us to heritage or toxic heritage but remediation. Through sponsored interventions at different scales from local to international, the physical and environmental recovery of contaminated spaces is ongoing. In the past, this has already been applied to two solid waste dumps. In the older one (*Parque de Lamas*), mercury sludge dumped on permeable sands has been covered over, using only lime. This space remains open and easily accessible to the public. Motocross bike tracks are visible on it. The other one (*Aterro de Cinzas brancas*) retains other toxic sludges under a waterproof plastic film. Covered in the soil, it forms a sloping hill surrounded by a fence, access to which is prohibited or restricted. Both are adjacent to the factories, far from the flagship remediation operation on a former liquid waste spillway, a wetland that is today called BioRia. Through phytoremediation – controlled cooperation between species, the basic rule of 'making-with' (Haraway 2016) – and *Phragmites australis*, a plant that is already abundant in that lagoon, the lands caught in the roots are gradually being decontaminated. A refuge for birds and hikers with its wild and/or cultivated lands interspersed with water channels, some of which are navigable, the BioRia is today the pride of the city council and people of Estarreja, a *natural heritage*[7] whose interpretation centre and mascot have been set up as symbols of clean, sustainable renewal. The interpretation centre stresses biodiversity and suggested walks. No link is made with the area's industrial history.

Recovery of a Contaminated Ditch

The last regeneration programme concerns the removal of contaminated soil from one of the drainage ditches, carried out in 2021. This ditch (*Vala de São Filipe*)

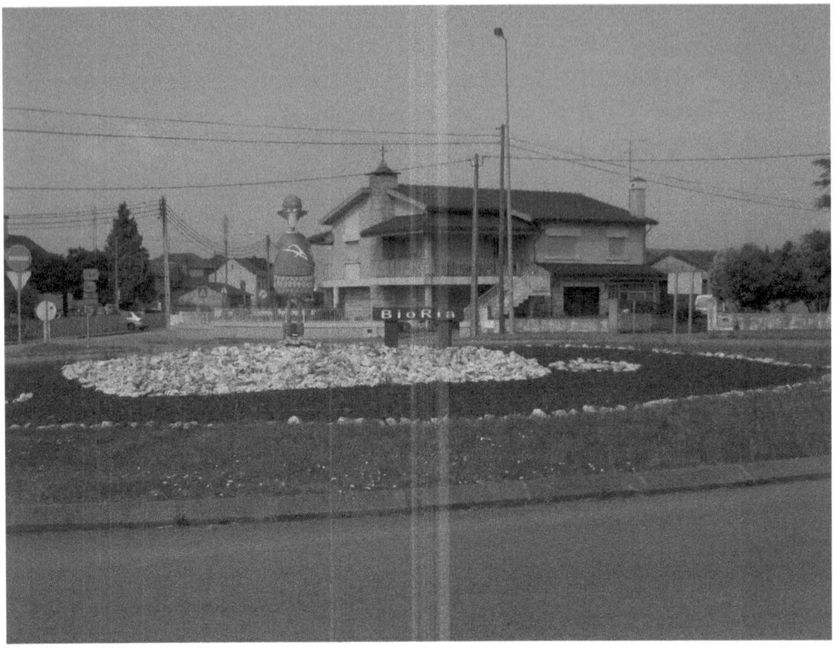

FIGURE 11.3 BioRia mascot. Photograph by Fabienne Wateau 2016.

crosses woodland, land cultivated with maize, and a hamlet. Its course ends in the Esteiro de Estarreja, which itself belongs to the water network of the whole of the Ria de Aveiro. In 100 days the ditch was cleaned out; its soil was removed to a depth of 40 cm and transported by truck to a decontamination centre near Lisbon; then a geotextile fabric and heavy rocks were placed on the bottom. Contaminated soil was also removed from the verges, up to a width of 5 metres, and arable replacement soil was added for cultivated fields. The ditch is now paved, and this is supposed to prevent all possible contact with pollution, especially for the people who had the routine of scrubbing it after winter. It will prevent neither the grasses from growing there again nor the still-contaminated groundwater from overflowing during heavy rain. But it contains and represents the space, it warns through its restoration, it informs, it protects.

A Border Case

On the area's memory, traces and a few scars, therefore, remain: old abandoned factories; still contaminated groundwater; an unmissable white mountain of lime; a sloping hill almost forgotten; a crossing line that will fade into the vegetation over time; a mended and maintained walking space set up as natural heritage. Underneath or inside, the materials remain (Storm 2014). The toxic sites of Estarreja are neither remembered nor erased. Belonging to the landscape, they are gradually decontaminated or confined.

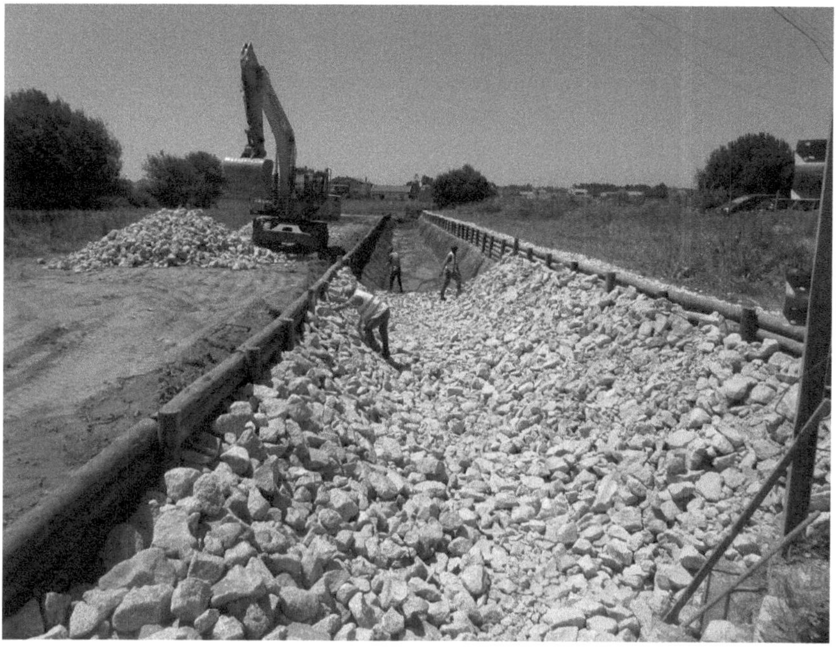

FIGURE 11.4 Paving of the ditch in the hamlet. Photograph by Fabienne Wateau 2021.

As things stand, they are not subject to any interpretation, or any reflexivity on history, nor do they serve as strategies of resistance or restorative justice. Estarreja appears as a border case, a situation where the question of toxic heritage is not yet asked by the institutions. Only a natural heritage is praised and protected.

Habituation and Activatable Memory

The City's Image

The reconstruction of Estarreja is also societal. In 2001, the PACOPAR was created, a multi-actor consortium grouping the five active chemical factories (Air Liquide, AQP, Cires, Bondalti and Dow), to establish connections with the main entities of civil society, including the city council, the firemen, the health centre, the schools, the University of Aveiro, and the ecological Cegonha association. This institutional and political coming-together supports the factories' desire for transparency and goodwill, and its members' common aim of applying Responsible Care principles, for a 'sustainable development of Estarreja and the world, by linking economic development with the protection of the environment and health' (CME 2010). In a sense, the PACOPAR is trying out a societal and environmental *making-with*, and it was awarded a prize for this in 2005. During its regular meetings, it also distributes grants for the city's educational and cultural activities, as well as for environmental research. It is in this context of physical and political reconstruction that Estarreja is managing the present time of its still-active chemical facility. And this might be the reason why, unlike other industrial cities, Estarreja does not seem prone to turn its factories into a potential social or cultural heritage that can be valorised, nor to repurpose some of them (Berger 2019). Having a cinema/theatre, a library, several museums, a jazz festival and an urban art festival, its rich artistic-cultural program somewhat systematically bypasses any reference to the factories. There is no mural painting representing the industrial part of the city, nor any existing book about the history of the chemical facility. The unwanted legacy is not a source of pride. It remains a heavy burden. Its valorisation is not on the agenda. Neither is its patrimonialisation. In the booklet for the 2020 ESTAU Urban Art Festival, Estarreja is presented as the hometown of Egas Moniz (1949 Nobel Prize in Medicine) and as a unique biodiversity zone with the BioRia.[8]

The Worrying Comes from Others

On the other hand, the neighbouring University of Aveiro just 20 kilometres away has an obvious interest in the chemical facility. Since its creation in 1973, much research has been conducted at Estarreja, and academic theses and publications in biology, geoscience, toxicology and other fields abound. Since 2010, an Observatoire Hommes-Milieux receiving funding from France (Labex Driihm-CNRS-INE) has been reinforcing this research laboratory. Yet, very few of the results get back to the local population – the requirement that academics write in English being one of the reasons. A book *The Factory and the Life* should soon be published in Portuguese. This collection of interviews with former workers of the chemical facility shows that a social and rather kindly memory is accessible and activatable.[9] The film we created (Giongo and Wateau 2021) also features workers with fairly happy memories associated with economic prosperity and pride in not having been obliged to emigrate or live in agriculture. The relationship with the dangers is more complex. Unsurprisingly, however, the worrying comes from others: the closer one stands to the epicentre, the less one's awareness of danger is activated – or activatable (Beck 1999). Without any denial of the existence or futile resistance, there is something like necessary habituation to an evil accepted as necessary. The film also presents the citizen science project CITAQUA[10], which, by linking the university, the locals and the schools, succeeded in raising awareness, as well as teaching (particularly adolescents) about other contamination sources in the water of private wells, with a focus on nitrate levels. The monitoring of 27 wells by their owners using measurement kits distributed by this academic project made some users realise that they could no longer use well water for domestic purposes. Awareness among citizens of the groundwater quality and its vulnerability was increased. The fact that political decision-makers could now seize upon that successful action to distribute kits to well owners would magnify that principle of *making-with*, and also ensures prevention and information, quite obviously. Schoolchildren also

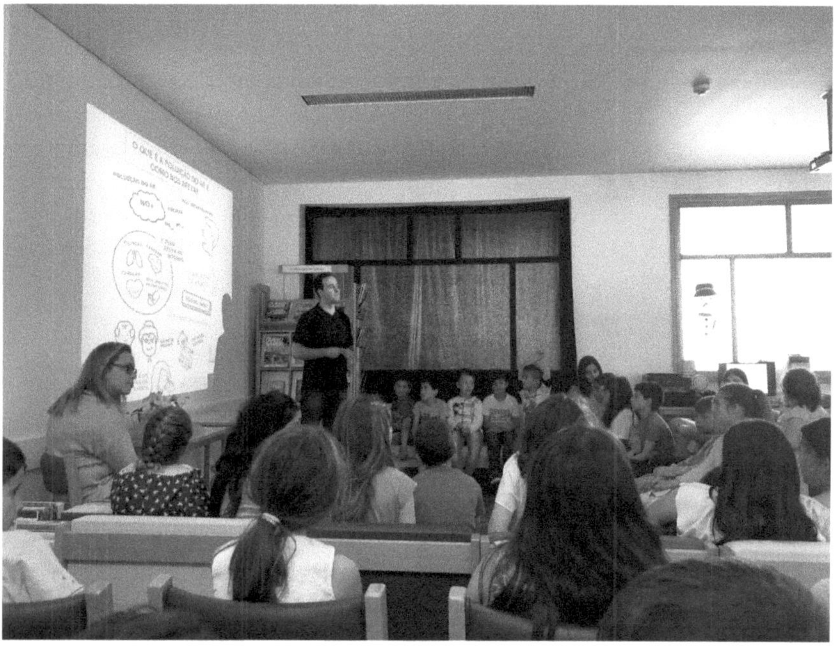

FIGURE 11.5 Air quality awareness-raising with school children. Photograph by Fabienne Wateau 2019.

benefit from a few air quality awareness-raising sessions, in which they are asked fun questions for example.

At the same time, thanks to university funding, extra air-measurement sensors have been installed, able to supply tangible indicators for research.[11] It is not yet a matter of scientific-artistic activism, like *Pigeon Blog*, which involves civil society, artists, and pigeons equipped with air quality sensors that can take readings on the ground and at various heights in the sky (Haraway 2016, 37), but maybe it should get to that point someday.

A Kind of Balance

Because although actions are being carried out with a civil society receptive to awareness-raising programmes, these people prove to be more resistant to the possibility of self-organising to create an opposition force, or simply a monitoring force. Estarreja's only ecological association, created in the 1990s to (successfully) fight the government's decision to locate a very polluting waste incinerator in Estarreja has almost no members today. Therefore, whether sufficiently informed or not, the population seems as if it is 'staying with' the chemical facility. There is no silenced memory, nor any memory or recollections activated by the municipality (at the risk of losing them) with a view to possible patrimonialisation. No one seems to have taken possession of a certain history, neither the city council, nor the factories, nor civil society, nor even a historian. Therefore, it seems that some kind of balance has been established in Estarreja, through habituation that takes comfort in turning around the city's image through concrete environmental regeneration actions, and through the energy invested in sociocultural activities that make the city attractive and pleasant to live in. A *rotten peace*?[12] Maybe. But between acceptance of a toxic heritage that – in a certain way – attributes an active recognition to pollution, and the gradual recovery of an unwanted legacy of which no one wants to be the standard-bearer, the choice seems to have been made.

Conclusion

Estarreja is neither *Flammable* in Argentina (Auyero and Swistun 2009) nor even Lubrizol in France (Izoard 2020). Fortunately, no serious accident occurred at this location. Estarreja is only one case among many others where, after the euphoria of full employment thanks to factories, the damage to the environment and health now has to be repaired. Estarreja is also not in a post-industrial context, where a form of nostalgia for the past could unfold (Storm 2014; Berger 2019). Quite the contrary. In August 2022, one of the complex's powerful entrepreneurs announced in the press that the contract between the chemical plants had just been renewed for 15 years because their competitiveness in hydrogen production technologies was ensured (*Expresso*, 19/08/2022). Industrialists, city councils and people continue to rely on the chemical and industrial complex for the future, in terms of dynamic economic production and jobs.[13] What is perceived today in Estarreja is a toxic legacy, which leads to environmental repair operations. What is not formalized is the existence of a toxic heritage, which from the past to the future is irreversibly inscribed in this place: no local historian, citizen association or municipal actions seems (yet?) to think of the place in terms of memory or heritage. The case of Estarreja shows that material toxicity is not synonymous with negative associations and that 'habituation' can be useful in thinking about how toxic heritage intersects memory practices. Unfortunately, conceptualising toxicity as heritage does not tend to resolve real problems, like spatiotemporal, political and institutional differences. While on the one hand, there is recovery, generally in advanced economies, on the other hand, there is our own recovered and ever-accumulating waste (toxic materials, technology, clothing …) in emerging countries, sent there by land, air and sea – that social and environmental injustice that is much evoked, but that we are only maintaining and reinforcing. While discourse here is in favour of green environmental protection technologies, over there, where others live, discourse concerns the 'extremely polluting extraction of those many metals that those green technologies need for their production, carried out under conditions that are catastrophic for workers and residents living near the production sites, and will soon end up as waste for which we have no solution available – nor is any being studied' (Boudia and Jas 2019, 96). There are also gaps between the university and the insufficient return of research to locals; between philosophical thought and actors in the field. Our contribution appears here as a border case for this volume devoted to Toxic Heritage and Heritage Studies. It responds more directly to research on repair and species alliances – as suggested in the introduction. With success, BioRia has become that dynamic environmental and ecological *patch* evoked by Tsing (2015). Are we 'capable, with each other in all of our bumptious kinds, of response?' (Haraway 2016, 1). Let us hope that operations to minimize damage to health and the environment continue; that no chemical accident ever occurs; and that one day, perhaps, once all chemical and industrial activities have ceased, a human memory – not just a memory of the milieu – is developed there.

Notes

1 Respectively: Regional Prize in the 'Environment' category of the first edition of the Annual Tourism Prizes, awarded by Turismo da Rota da Luz; 'For the example of innovation, transparency and success in relations between the chemical industry and the community' (CME 2010).
2 This ongoing research has benefited from several sources of financing: from the Labex Driihm (Dispositif de Recherche Interdisciplinaire sur les Interactions Hommes Milieux) CNRS-INE, in 2016 and 2019; from the Labex PP (les Passés dans le présent) CNRS-INSHS in 2020; from the MSH-Mondes (for the film SAFE) in 2019; from SMI (Soutien à la mobilité internationale) of the INSHS CNRS in 2021. It has been conducted collectively by all of the co-authors, under the supervision of the first author of the article. The team unites several fields: social anthropology, psychology, geography, biology and environmental engineering.
3 On the positive role of forgetting see this volume (May 2024).
4 A mixed territory with 27,000 inhabitants, including the chemical facility, an urban centre, hamlets, and rural areas with breeding, horticulture, and cereal farming.
5 Research conducted by the University of Aveiro proved that mercury was retained in the maize roots but did not get into the grain. On the other hand, nothing has been confirmed about the associated plants (Cabral-Pinto et al. 2020).
6 However, very high financial damages that are not without importance in many ways are repaid by the factories. In fact, every industrial site pays compensation to the municipalities. We do not yet have these figures for Estarreja.
7 Title of the book published by the city council of Estarreja (CME 2010).
8 See also Silva (2015).
9 Elisabeth Figueiredo (Dir.), Labex Driihm 2015. See Figueiredo (2018).
10 Daniela Figueiredo (Dir.), Labex Driihm 2015. See the first results in De Figueiredo et al. (2017).
11 Myriam Lopes (Dir.), Labex Driihm 2018. See the first results in Reis et al. (2020).
12 Criticism made by PACOPAR through one of the people interviewed in 2016. See the verbatim statement in Wateau (2018).
13 It would be 3,000 direct and indirect jobs (*Expresso*. August 19, 2022). De Mello.

References

Auyero, Javier, and Débora Alejandra Swistun. 2009. *Flammable: Environmental Suffering in an Argentine Shantytown*. Oxford University Press.

Azevedo, Maria Celeste. 1999. "Contribuição para a caracterização ambiental da área envolvente do complexo químico de Estarreja." University of Aveiro. Master thesis.

Beck, Ulrich. 1999. *World Risk Society*. Cambridge: Polity Press.
Berger, Stefan. 2019. "Industrial Heritage and the Ambiguities of Nostalgia for an Industrial Past in the Ruhr Valley, Germany." *Labor. Studies in Working-Class History* 16 (1): 37–63.
Boudia, Soraya, and Nathalie Jas. 2019. *Gouverner Un Monde Toxique*. Versailles: Éditions Quae.
Cabral-Pinto, Marina, Manuela Inácio, Orquídia Neves, Agostinho Almeida, Edgar Pinto, Bárbara Oliveiros, and Eduardo Ferreira da Silva. 2020. "Human Health Risk Assessment Due to Agricultural Activities and Crop Consumption in the Surroundings of an Industrial Area." *Exposure and Health* 12 (4): 629–640. 10.1007/s12403-019-00323-x
Carson, Rachel. 2002. *Silent Spring* 40th ed. Boston: Mariner Books.
CME - Câmara Municipal De Estarreja. 2010. *Estarreja, Património Natural*. Estarreja: BioRia.
De Mello, João. 2022. "Somos quem mais sabe de tecnologia de produção de hidrogénio em Portugal." *Expresso*. August 19, 2022.
de Figueiredo, Daniela, Pedro Marquinhos, and Luís Gonçalves. 2017. "Groundwater Quality Monitoring by Local Citizens at a Small Town." In *Book of Abstracts of the SciCom PT 2017 - Congresso de Comunicação Em Ciência, 12–13 October*. Coimbra, Portugal.
Figueiredo, Elisabete. 2018. "Não Queriam Que Ninguém Soubesse Que Era Uma Fábrica Perigosa» – Narrativas Dos Trabalhadores Do Complexo Químico de Estarreja Sobre Os Riscos No Trabalho e a Sua Evolução." In *Ambiente e Direitos Humanos*, 709–721. Aveiro: Universidade de Aveiro.
Giongo, Carmem, and Fabienne Wateau. 2021. *SAFE – Social Awareness for Environment in Estarreja*. Documentary, 24m. Paris, France: Iiac-Cnrs audiovisual department (Labex Driihm/Msh Mondes). 10.34972/driihm-a125h8
Haraway, Donna. 2016. *Staying with the Trouble. Making Kin in Chthulucene*, Durham and London. Durham: Duke University Press.
Izoard Celia. 2020. "L'effet Lubrizol"."Rouen - Fumées Noires et Gilets Jaunes." *Z - Revue Itinérante d'enquête et de Critique Sociale* 13: 18–25.
Kryder-Reid, Elizabeth, and Sarah May, co-editors. 2024. *Toxic Heritage: Legacies, Futures, and Environmental Injustice*. London: Routledge.
Latour Bruno. 2021. *Où suis-je? Leçons du confinement à l'usage des terrestres*. Paris: La Découverte.
Latour, Bruno. 2017. *Où Atterrir? Comment s'orienter en politique*. Paris: La Découverte.
Natura. 2000. Network website: https://natura2000.eea.europa.eu/Natura2000/SDF.aspx?site=PTCON0061
Reis, Johnny, Carlos Ribeiro, Daniel Rodrigues, and André Rodrigues. 2020. "Accident Scenario Modelling with Hazardous Materials in Transportation Infrastructures." *Proceedings of the Institution of Civil Engineers in Forensic Engineering* 172 (4): 133–149. 10.1680/jfoen.20.00015
Silva, Adriana. 2015. "O BioRia e a Construção da Imagem de Estarreja." University of Aveiro. Master thesis.
Storm, Anna. 2014. *Post-Industrial Landscape Scars*. New York: Palgrave MacMillan.
Taiëb, Lucie. 2020. *Freshkills. Recycler La Terre*. Monts: La Contre allée.
Thomas, Suzie, Vesa-Pekka Herva, Oula Seitsonen, and Eerika Koskinen-Koivisto. 2019. "Dark Heritage" C. Smith (ed.), *Encyclopedia of Global Archaeology*. Cham: Springer.
Tsing, Ana Lowenhaupt. 2015. *The Mushroom at the End of the World: On the Possibility of Life in Capitalist Ruins*. Princeton: Princeton University Press.
Valente, Sandra, Celeste Coelho, Cristina Ribeiro, Anne-Marie Guihard-Costa, Eduardo Ferreira da Silva, and Manuela Inácio. 2016. "Environmental Pollution and Human Health: Exploring Adverse Pregnancy Outcomes and Food and Water Consumption in

Estarreja Municipality, Portugal." In *9th ISEG – International Symposium on Environmental Chemistry, 15–22 July*. Aveiro: University of Aveiro.

Wateau, Fabienne. 2018. "Entre Terre et Eau Contaminées: Populations, Institutions et Perceptions Du Risque." In *X Congresso Ibérico de Gestão e Planeamento Da Água, University of Coimbra, 6–8 September*, AT4-C30: 1–12.

Wollentz, Gustav, Sarah May, Cornelius Holtorf, and Anders Högberg. 2020. "Toxic Heritage: Uncertain and Unsafe." In *Heritage Futures: Comparative Approaches to Natural and Cultural Heritage Practices*. UCL Press.

12

ENVIRONMENTAL AND EMBODIED AGRO-TOXIC HERITAGE IN RURAL URUGUAY: FROM RECOGNITION TO TRANSITION TO SUSTAINABILITY AMONG DAIRY FARMERS

Victoria Evia, Santiago Alzugaray, and Javier Taks

Introduction

Life, as a process of development, involves continuous exchanges between organisms and their environments. The Anthropocene has been recognized as a new geological era in which human activity becomes a dominant force in the transformation of the globe, with important effects on the conditions that sustain life as we know it. In particular, "the saturation of the biosphere with toxic substances is the most notable reminder of the lasting human impact on Earth" (Hendlin, 2021; Nading, 2020). Likewise, the emerging knowledge in epigenetics provides evidence that the human genome responds to the environment we are altering and that the effects of environmental toxicity can be transmitted between generations, configuring "toxic local biologies" (Lock, 2020:33). This entails the collapse of the distinction between natural history and human history and the recognition of the importance of histories lived in both natural and social environments. As Nading (2020) argues, toxicity emerges in situated biologies since the effects of chemicals in the body depend on doses, the presence of other substances, genetic and nutritional factors, among others. In turn, these particulars are framed in global patterns of colonialism, racism and economic exploitation, exposing some people to much more toxic substances and doses than others.

Environmental toxicity can be understood as "heritage". Toxic heritage includes both, the history of the substances and processes that create physical harm or threats of harm to environments (soil, water, air, etc.), the life supported within them (plants, animals, humans, etc.), and the intersections of that history of harm with both formal heritage institutions and informal memory practices (Kryder-Reid, 2021).

The field of heritage studies is concerned with how we manage human and natural legacies; how one's sense of place, social and cultural experiences and memories' are

DOI: 10.4324/9781003365259-22

legitimised or delegitimised and a responsibility towards the future with future generations' quality of life (Wollentz et al., 2020). Important for anthropological studies on toxicity is the understanding of how "toxic risks" and "toxic experiences" are constructed; how meanings are formed and negotiated from interactions between different types of actors (lay, experts, decision-makers), the role of the experience of dwelling and embodied knowledge about exposure, as well as the processes of uncertainty and frustration regarding knowledge gaps (Auyero & Swistun, 2009; Little, 2016; Shapiro, 2015; Singer, 2011). In both fields, social and cultural memories and experiences are central elements for understanding the different present and future narratives and practices regarding toxicity.

Most social studies on the experience of toxicity and its socio-health impacts, including toxic heritage studies, have been developed mainly in urban contexts, linked to industrial or deindustrialization processes and/or industrial and nuclear emergencies (Auyero & Swistun, 2009; Fortun, 2012; Little, 2016; Mah, 2012; Nading, 2020; Wollentz et al., 2020). These works, among others, have shown that although people may recognise the damage that environmental deterioration processes have on the material conditions of existence of their communities as well as on their health, there are also complex ties of affectivity and attachment that bind them to the territories they inhabit which condition its propensity to move away.

Fewer studies address the experiences of toxicity linked to the dominant agro-industrial regime, despite growing criticism regarding its environmental and health negative externalities (Levidow, 2015; Melby & Mauger, 2016). The technoeconomic package of conventional agribusiness dependent on fossil-based inputs such as agrochemicals, and the abuse of medicines for production animals has shown systemic limits: prevalence of hunger in the world, nutritional problems caused by the global food regime and environmental and health externalities, manifested among others in soil and water contamination and the release of viruses and bacteria that impact human health (Svampa & Viale, 2020). Anthropological studies (Gamlin, 2016; Nading, 2017; Saxton, 2015; Tsing et al., 2019) show that in industrial agricultural systems pesticides and other toxic substances affect racialized communities of the global south, migrants, indigenous groups, workers and inhabitants of agro-industrial enclaves to a greater extent. They also show that pesticide toxicity experience among vulnerable social groups is plagued by uncertainties about the potential damages that these substances cause in their bodies and in their environment, recognising acute signs and symptoms whilst minimising or denying chronic outcomes.

In the Latin American Southern Cone (Argentina, southern Brazil, Paraguay and Uruguay), since the early 2000s, there has been a process of expansion of industrial agriculture under the agribusiness model, led by the expansion of transgenic soybean cultivation. This model has been strongly criticised for its environmental, social and health consequences, such as the intensification of the use of natural resources, the displacement of small family farmers and the dependence on transgenic cultivars, pesticides and other petroleum derivatives (Catacora-Vargas, G. et al., 2012; Gras & Hernández, 2013). A growing social literature addresses the socio-environmental and health impacts of this model and the emerging alternatives among affected populations. These studies have focused on documenting the lived experiences of the rural population and waged workers on the increasing environmental degradation, impacts on individual and community health as well as socio-environmental conflicts between urban actors, rural producers and government agents.

Uruguay is an agro-exporting country. In 2019 the country's main export product was beef, followed by cellulose, soybeans and dairy products (Uruguay XXI 2019). During the last 20 years, the consolidation of hegemonic production models, highly dependent on synthetic chemical inputs and large-scale machinery, has been consolidated, with increasing concentration of productive area in fewer owners, intensification of natural resources usage and higher use of water, with its consequential impact on quality (Alonso et al., 2020; García Préchac et al., 2010; Gómez Perazzoli, 2019). These models have been promoted by government policies in recent decades, supported by multilateral financial institutions, and have great support from the majority of farmers and other rural actors.

Hegemonic narratives that sustain the agribusiness regime-originated not only in academy, industry and state agencies but also among local folks-have historically minimised the toxicity of the dominant way of production, which is intensive in the use of pesticides, synthetic fertilisers, and veterinary medicinal products. Nevertheless, for the last decade, there has been increasing visibility and recognition by inhabitants in both rural and suburban territories of consequential environmental and embodied effects (Abbate et al., 2017; Alonso et al., 2020; Chiappe, 2020; Evia, 2018, 2020, 2022). For example, events such as the eutrophication of watercourses and the consequent blooming of cyanobacteria, or the spraying of people and population centres as a result of pesticide applications, have brought agricultural technologies to the forefront of public discussion in the country. The increasing recognition of this toxicity has also become a claim that mobilises reactions -still fragmented- facing multiple pressures and dependencies determined by agribusiness as a way of life. The socio-ecological transition in food production has become a central concern in international forums and local technical conferences (Svampa & Viale, 2020, pp. 248 et seq.).

In a context of the growing concern about the environmental limits of the dominant production model, some dairy farmers and agricultural technicians are questioning the consequences of environmental toxicity in terms of personal and collective health; moreover, they link these hazards to 'agribusinesses'. Based on ethnographic fieldwork and collaborative research among agriculture scientists, rural extensionists and family dairy farmers searching for a transition towards sustainability in milk production, in this article, we want to discuss how agricultural toxicity is understood and experienced by farmers who are part of family and community histories linked to this production. The questioning can be problematized as a heritage struggle: is frequently signified as an affront by farmers; as inherited farming practices are questioned, their whole legacy is under the spotlight.

Toxic Heritage and the Search for Alternatives to the Conventional Dairy Production Model

Dairy production in Uruguay is mainly based in the South and South-western region (comprising the departments of Canelones, Colonia and San José). The most traditional dairy basin has been located, since the beginning of the 20th century, in the territory of the Santa Lucía river basin (Figure 12.1). This river is also the main source of raw water for drinking water for almost half of the country's population, concentrated in the metropolitan area of Montevideo. The southwest region of the

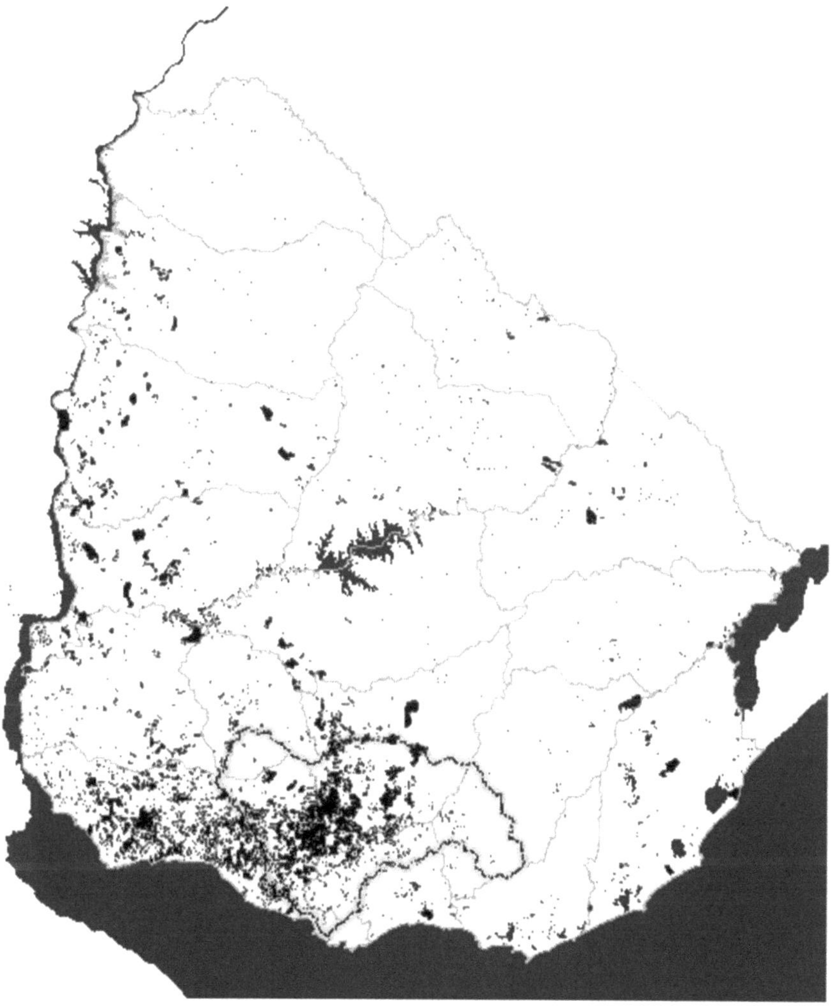

FIGURE 12.1 Territorial distribution of dairy farms in Uruguay, signalling the Santa Lucía river basin.

Source: Mvotma, 2018: 22.

country has a high intensiveness of land use and a relatively low concentration of land ownership (Riella & Mascheroni, 2011). It is characterised by a greater presence of family producers on medium and small farms, mainly dedicated to horticultural, vineyards, and dairy production. Likewise, extensive agriculture, mainly of grains and oilseeds, has advanced (MGAP, 2014).

In the last two decades, this region has seen an increase in socio-environmental controversies and conflicts due to the coexistence of different production systems. In particular, the Santa Lucía river basin has been the subject of proposals for agriculture land use planning and the application of restrictive regulations, with limited progress in terms of transitions to sustainability among cattle and dairy farmers (Schelotto et al., 2015).

In this context, a group of six dairy farms grouped around the NGO Emmanuel Centre (ecumenical social organisation that promotes agroecology), involving around 40 farmers and waged workers, are working to transition from conventional production systems to agroecological production systems (Centro Emmanuel, 2020). The area of influence of the NGO, at the centre and Eastern territories of the province of Colonia, is characterised by a strong presence of populations belonging to the Waldensian Church. These migrants were established in the area at the end of the 19th and beginning of the 20th century, linked to colonisation policies of the Uruguayan state that promoted agricultural production and settlement in the countryside (Geymonat, 2004). Although not all the producers involved in the transition project profess the Waldensian religion, they share a concern for the common good and care for the environment that allows life to be sustained. Furthermore, in all six cases, their families have been involved in agricultural production for several generations, and the lands where they produce are inherited and/or shared with close relatives. This implies a strong weight of the material and immaterial legacies of their respective families, both in terms of production technologies and the values associated with them.

Before the start of the mentioned project, most of the producers in the group had previous links with the Centre because they had participated in other social and/or religious activities in the region, or because of personal friendships with the technical experts. These previous links of affiliation and mutual knowledge contribute to the construction of a common framework in which agroecology is seen as a horizon that transcends technical relations and includes a dimension of care for the communal and spiritual "quality of life".

Personal and community experiences regarding the toxicity of the chemical products used in the dominant production model appear as a common denominator of the motives to begin the transition to more sustainable production practices. Agricultural toxicity begins to be recognised and named from stories that imply a concern for the effects on the health of the people who work in the production systems, who are their own relatives, workers and/or neighbours, and/or for the environmental effects on the territories they themselves inhabit. For example, at one of the group's first meetings in 2020, when asked about their motivations for coming together, one of the farmers shared a motivation that had led her and her husband to rethink their traditional method of dairy farming and to look for alternatives: their concern about the effects of fumigating crops – used as fodder for the cows – on the health of a close family member linked to the farm. "He has health problems that we associate with his work because he has always dedicated himself to 'that' [fumigating]" (...) "He was very reluctant. These are things that go back a long time. Today there is more openness to speak".

Another farmer also highlighted that among his main motivations for transit is the "caring of health, of those of us who work, and care of the communities", the "care of the environment" and "giving coherence to life", coherence between his and the community beliefs and values, and productive practices. He became aware of the toxicity of agrochemicals because there is a rural school on the farmstead, on a piece of land donated to the State by his ancestors who arrived in Uruguay from the Waldensian region. He stated: "My mother used to work in the school's orchard. At one point I saw that we were spraying [fumigating] 20 metres from the school garden. So I started not to spray in that field". The decision to stop fumigating there is related to a change in perception about the danger of pesticides, particularly glyphosate: "They [technicians, researchers] lied to us. They told us that glyphosate degrades by itself and it doesn't".

A third dairy farmer also pointed out the increasing use of glyphosate and other pesticides, and their consequent generation of waste, referring to plastic containers, as one of the first indicators of problems with the production model. "The drums of glyphosate are piling up. And it wasn't just that. You had to pour [apply] more and more [product] and the problems [with weeds] accumulate". During a personal health recovery that gave him "more time to think", he remembers: "I thought about the soil and I realised that many things were not what we had been told. [For instance], that glyphosate degraded, it wasn't like that either".

As evidenced in the testimonies above, the process of signifying the ways in which they are traditionally produced as potentially harmful or toxic legacies centres on concerns regarding the impact of pesticides on human and environmental health. The herbicide glyphosate is the most commonly named product, as for years it has been a substance widely promoted as a technical solution to the problem of soil erosion by the hegemonic agronomic discourses (technological package of no-till or minimum soil tillage) (Cáceres, 2018). At the same time, it is a total-action herbicide (it acts on all plant species), which is why it is massively adopted thanks to its ease of use. However, in recent years, more and more controversy has arisen regarding its alleged safety (Arancibia & Motta, 2015), and more evidence of its harmful effects on soil, watercourses, environmental toxicity and chronic toxicity has been disseminated (Antoniou et al., 2012). Although a popular belief persists that it is a less dangerous product than other pesticides, a belief rooted in the hegemonic discourses of agribusiness actors, and scholars linked to agricultural production (Arancibia, 2013; Cáceres, 2018) and generally found among rural workers (Evia, 2020), dairy farmers involved in this project began to question such a belief. From the reception and appropriation of new information, producers begin to allude to the hegemonic paradigm as "lies".

Recognising the toxicity in this legacy can be a difficult experience, as it implies tensions and contradictions with one's own family inheritance and/or with the technical models that have been taken as valid references up to now. Socialisation in the daily life of family work is fundamental for learning the dairy farming trade, in the production and reproduction of the sexual division of labour and in the mechanisms and criteria that construct and make intergenerational succession possible (Camors, 2016; Courdin et al., 2010; Malán, 2013). For example, in the case of two brothers who, together with their parents, manage a prosperous dairy farm in the area, one of them states that for years he had been eager to "do something different", because "it has not made much sense to think one thing [to have an idea about best practices] and then get involved in a [productive] system that leads to another". However, "As it is a family business it is not so easy to change. My father always worked that way".

Decision-making on productive changes in family businesses can be difficult because not only rational economic-productive variables are at stake, but also affective relationships and alliances marked by kinship and emotional ties. Questioning toxic heritage, as part of inherited practices, is interpreted by those being questioned as an affront to the legacy as a whole; a questioning of a way of life that was intended to be bequeathed and which is signified positively as a whole. Dialogues and agreements regarding changes in management paradigms and forms of production can become conflictive because they confront not only technical differences, but also differences in social evaluations of technology, its potential effects and, eventually, worldviews (Geels, 2005). In most of the testimonies, it is discernible that concerns about the increasingly evident contradictions of the dominant model go back several years, but taking the step towards the new can take time.

Responsibility towards present and future generations also acts both as a motivator and as a barrier for productive transitions. As a motivator, concerns about the quality of life, the environment they live in, and the health of their offspring, workers, and future generations -whom they do not want to continue exposing to new sources of agrochemical toxicity- are drivers for change. As a barrier, the uncertainties regarding the economic and technical viability of these changes are often the main obstacles to advancing concrete modifications in practices within the framework of redesigning production systems. There are cultural legacies and imperatives to be honoured by dairy landlords. We identified the following main ones: the responsibility of not losing inherited land, maintaining and creating new sources of employment for wage earners coming from their local community, and reproducing and transmitting a patrimony (soil fertility, water sources, productive cattle, machinery, etc.) for their offspring that allows the latter to continue a lifestyle in communities heavily dependent on agricultural production.

"This [farm] has to be economically profitable. (...) We have to look for a model that is competitive with what we have today (...) Technology leads us to extract, extract, extract. In our case, thanks to my father who worked his ass off, we have a larger area. (...) We want to do it out of conviction, for life itself. But, if you are not competitive [you] lose the land. How do we make a change in the production system without putting at risk what we do have? We want a change. How do we start?"

Although there is a model, agroecology, which is proposed as a shared horizon towards which to move in order to reverse this agro-toxic heritage, there are also many doubts about how to do it. Unlike the conventional model, there is no agreed "recipe" either on the part of technicians or producers. There are also technical limitations that make it difficult to substitute key technologies in the hegemonic production model, and there is a lack of scientific knowledge to support different production practices. So far, transitional trajectories have been based on trial-and-error dynamics – sometimes very costly – at the individual farm level.

Final remarks

Toxic heritage, for the people involved in our research, is part of a larger heritage coming from their ancestors who built a way of life in the region. Only recently, dairy farmers started to question that problematic heritage, as they live on their farms and have a situated and embodied experience of the toxic heritage. The shared ideas about the common good, the care of the creation, and the continuity of the legacy of a way of life linked to the land and strongly communal are drivers for change, away from unsustainable productive practices.

The grandparents, first and second generations, brought from Europe the farm, the effort, and ethics and experienced scarcity. The fathers and mothers (third generation) rode the green revolution, increased productivity and improved their standard of living in economic terms, and the possibility of upward social mobility.

The next generation, who are now taking the reins on the farms, became aware of the toxic heritage and want to eliminate toxicity from their legacy, without losing their material well-being and rootedness in the rural milieu.

They face a heritage struggle as, on the one hand, it is very difficult to accuse the green revolution generation of leaving a negative heritage that causes environmental unsustainability and human health problems. For several reasons: a) they were not individual decisions, but the decision-making takes place in a framework of institutional, economic-financial and cultural pressures. The dominant values and goals were increasing production, taking on debt for growth, investing capital in new technologies, and modernising social relations which, among other things, represents the separation between business and family domains. The technological imperative (Mumford, 1974) was strong; b) the results of the application of the toxic technological package are far from being only negative; it allowed the professional and family fulfilment of those who managed to build up a "patrimony" in land, animals and machinery. It has not only provided a material base, but is also accompanied by positive identity aspects and the configuration of a "community of practice" proud of itself in the south-west region of the country; and c) although a business transfer is taking place, with increasing decision making by the fourth generation, the green revolutionary fathers (and mothers) continue to strongly influence the processes of food and milk production, either because they still control a large part of the means of production or for moral reasons of respect for those who have developed the dairy farms.

The concept of agro-toxic heritage provides a temporary dimension to the crisis of agrochemicals in national agriculture, which transcends individual responsibility for the present and includes, on the one hand, previous generations who transformed the environment on their farms and beyond, constituting an immediate ecological heritage (Foladori, 2005, p. 122) which descendants received and had to adapt to. On the other hand, this temporality of the agro-toxic heritage concept involves future communities and generations of persons, some not yet born that should be taken into account when evaluating current practices within dairy environments.

In another sense, agro-toxic heritage, contests the idyllic image and superior morality of country people, due to their close relationship with nature and their supposed care for resources. Well, the discovery of the toxic legacy of the last twenty years, manifested in the contamination of waters, the degradation and loss of soils and the effects on human and animal health -not yet fully acknowledged- relativizes the "countryside" as a moral reservoir of the national society. However, the recognition of this not-so-desirable heritage gives rise to a deeper critique of the dominant production and reproduction model and to practise an "anticipatory awareness" (Payne & Phillips, 2012), that reality can be defined in terms other than those of the mainstream development paradigm and that, consequently, people and social groups can act on the basis of these different definitions of good living. A modest beginning for the detoxification of life.

References

Abbate, Silvana, M. Colazzo, Andrés Fonsalía, Horacio Heinzen, Jimena Heinze, Santiago Niell, and Nicolás Rodríguez. 2017. *Agroquímicos, salud laboral y ambiental: diálogo de saberes y búsqueda de alternativas en una comunidad urbana del litoral del país*. Montevideo: CSIC.

Alonso, Jimena, Federico Quintans, Javier Taks, Daniel Conde, Guillermo Chalar, Silvia Bonilla, Rafael Arocena, and Signe Haakonsson. 2020. «Calidad del agua en Uruguay: Actualidad y desafíos». En *Water Quality in the Americas: risks and Opportunities*, editado por Vammen. Mexico: Interamerican Network of National Academies of Sciences-IANAS/ UNESCO. https://www.ianas.org/images/books/wb09.pdf

Antoniou, M., M. E. M. Habib, C. V. Howard, R. C. Jennings, C. Leifert, R. O. Nodari, C. J. Robinson, and J. Fagan. 2012. «Teratogenic Effects of Glyphosate-Based Herbicides: Divergence of Regulatory Decisions from Scientific Evidence». *Journal of Environmental & Analytical Toxicology* 4 (006): 2161-0525. doi: 10.4172/2161-0525.S4-006

Arancibia, Florencia. 2013. «Challenging the bioeconomy: The dynamics of collective action in Argentina». *Technology in Society* 35 (2): 79–92.

Arancibia, Florencia, and Renata Motta. 2015. «Health Experts Challenge the Safety of Pesticides in Argentina and Brazil». En *Medicine, Risk, Discourse and Power*, editado por John Martyn Chamberlain, 187–214. New York: Routledge. 10.4324/9781315658742

Auyero, Javier, and Debora Alejandra Swistun. 2009. *Flammable: Environmental Suffering in an Argentine Shantytown*. Oxford University Press.

Cáceres, Daniel M. 2018. «Biotecnología y poder. ¿Usan los cultivos transgénicos menos agroquímicos?» *Revista Interdisciplinaria de Estudios Agrarios* 48: 29–56.

Camors, Verónica. 2016. «Hacia la construcción de una política de acceso a la tierra y desarrollo rural con perspectiva de género; atendiendo las desigualdades existentes». En *Aporte a las políticas públicas desde la perspectiva de género. 10 años de FLACSO Uruguay*, editado por Silvana Darré, FLACSO Programa Uruguay. Montevideo.

Catacora-Vargas, G., P. Galeano, S. Agapito-Tenfen, D. Aranda, T. Palau, and R. Nodari. 2012. *Soybean production in the Southern Cone of the Americas: Update on land and pesticide use*. GenØk/UFSC/REDES-AT. Cochabamba.

Centro Emmanuel. 2020. «Validación de la herramienta de co-innovación para la transición hacia la agroecología en predios lecheros en el Uruguay». PROYECTO PROGRAMA DE BIENES PÚBLICOS SECTORIALES PARA LA COMPETITIVIDAD 2020 ANDE. MODALIDAD: REACTIVACIÓN.

Chiappe, Marta. 2020. «Conflictos por uso de agroquímicos: el papel de las mujeres rurales en Uruguay». *Agrociencia Uruguay* 24: 1–15.

Courdin, Virginia, Annie Dufour, and Benoit Dedieu. 2010. «Las mujeres en las explotaciones familiares lecheras: análisis de situaciones francesas y uruguayas». *Agrociencia Uruguay* 14 (1): 55–63.

Evia, Victoria. 2018. «Saberes y experiencias sobre la exposición a plaguicidas entre mujeres que residen en contextos agrícolas en Soriano, Uruguay». *Revista Trama* 9 (diciembre): 13–35.

Evia, Victoria. 2020. «Venenos, curas y matayuyos. Trabajadores agrícolas y saberes sobre plaguicidas en Uruguay». *Revista de Ciencias Sociales* 34 (48): 67–92. 10.26489/rvs.v34i48.3

Evia, Victoria. 2022. «Entre denunciar y aguantar. Sojización, plaguicidas y participación en salud ambiental en Uruguay». *Saúde em Debate* 46 (2): 62–74. 10.1590/0103-11042022E204

Foladori, Guillermo. 2005. «Una tipología del pensamiento ambientalista». *Sustentabilidad*, 83–136.

Fortun, Kim. 2012. «Ethnography in Late Industrialism». *Cultural Anthropology* 27 (3): 446–464. 10.1111/j.1548-1360.2012.01153.x

Gamlin, Jennifer. 2016. «Huichol migrant laborers and pesticides: structural violence and cultural confounders». *Medical anthropology quarterly* 30 (3): 303–320.

García Préchac, Fernando, Osvaldo Ernst, Mario Bidegain, Clara Pristch, Alejandra Ferenczi, and Mercedes Rivas. 2010. *Intensificación agrícola: Oportunidades y Amenazas para un país productivo y natural*. Montevideo: CSIC.

Geels, Frank W. 2005. *Technological Transitions and System Innovations: A Co-Evolutionary and Socio-Technical Analysis*. Cheltenham, UK; Northampton, MA: Edward Elgar.

Geymonat, Roger. 2004. «El elemento religioso como factor identitario en el caso de los valdenses en el Uruguay». En *Las religiones en el Uruguay. Algunas aproximaciones*. Montevideo: La gotera.

Gómez Perazzoli, Alberto. 2019. «Uruguay: país productor de alimentos para un sistema alimentario disfuncional». *Agrociencia (Uruguay)* 23 (1): 92–100. 10.31285/agro.23.1.8

Gras, Carla, and Valeria Hernández. 2013. *El agro como negocio. Producción, sociedad y territorios en la globalización*. Buenos Aires: Biblos.

Hendlin, Yogi Hale. 2021. «Surveying the Chemical Anthropocene: Chemical Imaginaries and the Politics of Defining Toxicity». *Environment and Society* 12 (1): 181–202. 10.3167/ares.2021.120111

Kryder-Reid, Elizabeth. 2021. Toxic Heritage: Post-Industrial Sites and the Unspoken Stories of Environmental Harm. https://vimeo.com/653300388

Levidow, Les. 2015. «European Transitions towards a Corporate-Environmental Food Regime: Agroecological Incorporation or Contestation?» *Journal of Rural Studies* 40 (agosto): 76–89. 10.1016/j.jrurstud.2015.06.001

Little, Peter C. 2016. «New Toxics Uncertainty and the Complexity Politics of Emerging Vapor Intrusion Risk». En *A Companion to the Anthropology of Environmental Health*, 281–301. Malden: Wiley-Blackwell. 10.1002/9781118786949.ch14

Lock, Margaret. 2020. «Toxic Environments and the Embedded Psyche». *Medical Anthropology Quarterly* 34(1): 21–40. 10.1111/maq.12545

Mah, Alice. 2012. *Industrial Ruination, Community, and Place: Landscapes and Legacies of Urban Decline*. University of Toronto Press.

Malán, Inés. 2013. «La sucesión generacional de la lechería familiar». *Revista Plan Agropecuario* 147 (agosto): 48–53.

Melby, Melissa K., and Megan Mauger. 2016. «Effects of Agriculture on Environmental and Human Health». En *A Companion to the Anthropology of Environmental Health*, 44–67. Malden: Wiley-Blackwell. 10.1002/9781118786949.ch2

MGAP. 2014. «Censo general agropecuario 2011. Resultados definitivos.» Ministerio de Ganadería, Agricultura y Pesca. Estadísticas agropecuarias DIEA. https://www.gub.uy/ministerio-ganaderia-agricultura-pesca/sites/ministerio-ganaderia-agricultura-pesca/files/2020-02/censo2011.pdf

Mumford, Lewis. 1974. *Pentagon of Power: The Myth of the Machine, Vol. II: 002*. New York: Harcour, Brace Jovanovich.

MVOTMA. 2018. «Plan de acción para la protección de la calidad ambiental de la Cuenca del Río Santa Lucía. Medidas de Segunda Generación.» https://www.gub.uy/ministerio-ambiente/sites/ministerio-ambiente/files/documentos/publicaciones/PLAN_DE_ACCION_RIO_SANTA_LUCIA_-_MEDIDAS_DE_2da_GENERACION.pdf

Nading, Alex M. 2017. «Local Biologies, Leaky Things, and the Chemical Infrastructure of Global Health». *Medical Anthropology* 36 (2): 141–156. 10.1080/01459740.2016.1186672

Nading, Alex M. 2020. «Living in a Toxic World». *Annual Review of Anthropology* 49 (1): 209–224. 10.1146/annurev-anthro-010220-074557

Payne, Anthony, and Nicola Phillips. 2012. *Desarrollo*. Madrid: Alianza.

Riella, Alberto, and Paola Mascheroni. 2011. «Desigualdades sociales y territorios rurales en Uruguay». *Pampa (Santa Fe)*, n.o 7 (diciembre): 39–63.

Saxton, Dvera I. 2015. «Strawberry Fields as Extreme Environments: The Ecobiopolitics of Farmworker Health». *Medical Anthropology* 34 (2): 166–183. 10.1080/01459740.2014.959167

Schelotto, Salvador, José Freitas, Hugo Gilmet, Javier Taks, and Álvaro Soba. 2015. *Atlas de la Cuenca del Río Santa Lucía*. Montevideo: Ministerio de Vivienda, Ordenamiento Territorial y Medio Ambiente.

Shapiro, Nicholas. 2015. «Attuning to the Chemosphere: Domestic Formaldehyde, Bodily Reasoning, and the Chemical Sublime». *Cultural Anthropology* 30 (3): 368–393. 10.14506/ca30.3.02

Singer, Merrill. 2011. «Down Cancer Alley: The Lived Experience of Health and Environmental Suffering in Louisiana's Chemical Corridor». *Medical Anthropology Quarterly* 25 (2): 141–163. 10.1111/j.1548-1387.2011.01154.x

Svampa, Maristella, and Enrique Viale. 2020. *El colapso ecológico ya llegó. Una brújula para salir del (mal)desarrollo*. Buenos Aires: Siglo veintiuno.

Tsing, Anna Lowenhaupt, Andrew S. Mathews, and Nils Bubandt. 2019. «Patchy Anthropocene: Landscape Structure, Multispecies History, and the Retooling of Anthropology: An Introduction to Supplement 20». *Current Anthropology* 60 (S20): S186–S197. 10.1086/703391

Uruguay XXI. 2019. «Informe de Comercio Exterior de Uruguay - 2019». https://www.uruguayxxi.gub.uy/es/centro-informacion/articulo/informe-de-comercio-exterior-de-uruguay-2019/

Wollentz, Gustav, Sarah May, Cornelius Holtorf, and Anders Högberg. 2020. «Toxic heritage: Uncertain and unsafe».

SECTION 4

INTRODUCTION: NARRATIVES OF TOXIC HERITAGE

Elizabeth Kryder-Reid and Sarah May

This section focuses on the stories and discourse practices around toxic legacies. Heritage is often said to be 'about stories', but the structure of the narrative arc is often under-examined and its silences ignored. In an era of political inertia and widespread denial of environmental crises, there is value in rooting contemporary contamination in a historical narrative. Different phases of environmental harm in the same place may produce different narratives which compete, support, or undermine each other. It is also productive to examine how toxic heritage rhetoric, discourse, and narrative navigate tropes like loss, heroism, redemption, and reversals. Particularly when there is a yearning for a simpler, somehow more pristine past and when both industrial and state interests benefit from obscuring the causes of environmental damage, toxic heritage narratives are positioned to offer a counterpoint to nostalgia as well as to speak truth to power. As the authors in this section demonstrate, the narratives of toxic heritage can complement, undermine, or act as alternatives to existing heritage narratives.

In this section, Kryder-Reid, Dwyer, and Filippelli examine the hidden discourses of dry cleaning heritage in contrast to industry and environmental science-authored sources and share an example of participatory heritage as a counter-narrative. Hubé and Bausinger's study of the 'Place a gaz' shows how the valorised, commemorative narratives associated with WWI can deflect or overwhelm the stories of toxic heritage associated with the conflict. Muniz' account of rubber production in Brazil highlights how the authorized heritage associated with colonial and corporate industry marginalises the intangible heritage of indigenous knowledge and the often invisible embodied toxic heritage of labourers producing rubber. McIvor explores community environmental justice activism associated with the chemical industry in Glasgow through creative modes of resistance such as advocacy, campaigning, the deployment of memories and storytelling. These

DOI: 10.4324/9781003365259-23

heritage practices, he argues, offer a counternarrative to the erasure of the toxic environment in the city's formal heritage institutions. Shackel's chapter on coal mining in Pennsylvania lays out three narrative tropes which sit uncomfortably with one another. Benussi finds a similar pattern at Chernobyl where distinct narratives are associated with different spatial scales of memory practices from the local to the national.

13

DIRTY LAUNDRY: THE TOXIC HERITAGE OF DRY CLEANING IN INDIANAPOLIS, INDIANA

Elizabeth Kryder-Reid, Owen Dwyer, and Gabriel Filippelli

Introduction

In September 2012, the Environmental Protection Agency (EPA) began remediating what was known as "The Tuchman Cleaners Site" in Indianapolis, Indiana (USA).[1] The process involved excavating ruptured underground storage tanks and removing contaminated soil, assessing and mitigating vapor intrusion in the surrounding residential neighborhood, and transporting and disposing excavated hazardous materials to an "off-site" location. (EPA Tuchman Site Profile n.d.). As the EPA report states, this dry cleaning contamination site is in a mixed commercial and residential area with approximately 10,000 people living within a mile. It is also less than a quarter mile (0.4 km) from a well field that supplies drinking water for the city and approximately 500 feet from Fall Creek, a tributary of the White River that flows through central Indianapolis. The Tuchman plant serviced most of its retail stores throughout the city from 1952 to 2008 when its parent company declared bankruptcy. IDEM, the state environmental agency, had noted contamination on the site during inspections in 1989, 1993, 2002, and as early as 1994 identified contamination in a nearby aquifer. In 2011 at the request of the state (Groves 2011), the EPA got involved and the Tuchman site became one of the US's largest EPA Superfund[2] sites associated with dry cleaning contamination (Figure 13.1). Yet, based on conversations with residents of the city and on official and public discourses, few people in Indianapolis are aware of it.

Unlike highly visible sites such as mines, major industrial plants, and sprawling dumps, dry cleaning contributes to the pervasive but largely invisible toxicity that characterizes modern cities. As sociologists Scott Frickel and James R. Elliott have argued, itinerant, small-scale, and largely unregulated and undocumented enterprises have left a legacy of accumulated environmental hazards across American

DOI: 10.4324/9781003365259-24

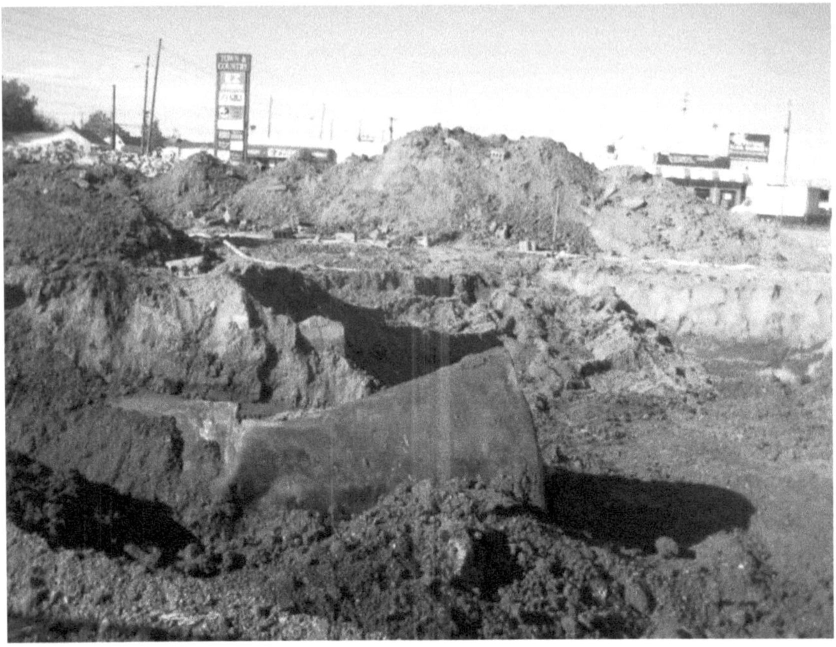

FIGURE 13.1 Underground storage tanks being excavated by the EPA in 2012 in the first phase of mitigation at the Tuchman Cleaners site, Indianapolis, IN (Photo: Environmental Protection Agency).

cities (Frickel and Elliott 2018). Unlike operations confined to industrial areas, dry cleaners (along with gas stations) are distributed throughout cities and often even integrated within residential areas. Due to their neighborhood locations, dry cleaners, particularly wholesale plants, contribute to negative health impacts in urban environments, particularly in marginalized neighborhoods, through fumes, spills, and dumping (Gelobter 1993, 850). Dry cleaners are also part of a pattern of frequent turnover. Frickel and Elliott (2018) argue that cities' accumulated environmental burdens are created by three interrelated processes: industrial churning (frequent relocations and closing of urban industry), residential churning (changing residential patterns and demographics), and risk containment (politically expedient and highly selective governmental management of legacy risks). Significantly for this examination of toxic heritage, Frickel and Elliott contend that one result of this rapid spatial and temporal turnover combined with public policy inattention is a loss of "public memory" and a process of "collective forgetting" (Frickel and Elliott 2018, 6, 84). They argue that the three processes "reinforce the lost knowledge of neighborhood history and legitimize institutional forgetting in ways that perpetuate the accumulation of industrialized urban lands and the hazards they may still contain" (Frickel and Elliott 2018, 7). Humans have long relegated waste to less obtrusive places, whether dumping household trash in pits or siting municipal dumps away from population centers (Melosi 1973). But the inescapable exposure to modern pollution has often necessitated what Loretta Lou calls the "art of unnoticing" as a coping strategy, such as in Guangzhou, China, where people live in the shadow of petrochemical plants with their attendant risks of pollution and chemical explosions (Lou 2022). Politicized narratives of silence and denial have been documented in situations in which allegiances with or challenges to sources of contamination "operate as strategies of normalization, of exclusion and inclusion" and that resonate at an emotional, as well as symbolic and conceptual level (Hart 2022).

Beyond polluted urban contexts, scholars across biology, environmental science, theology, and psychology have explored the phenomena of environmental or ecological amnesia. The term describes the suppression or absence of knowledge of ways of engaging with the natural world and also a loss of affective connection to the places one inhabits. As Forrest Clingerman argues, "This amnesia is not merely a forgetfulness of how to encounter environments in general; it is equally a loss of home and place" (Clingerman 2018, 186). The idea of environmental amnesia also references the phenomenon that each generation takes their lived experience of the environment to be the norm and that people are habituated to whatever conditions they were born into (Kahn 2002; Buell 2017). But understanding affected communities' responses to environmental harm and risk requires recognizing that amnesia and habituation are entangled with the power dynamics of economic and political structures that produced the pollution (Cahill and Pain 2019; Chagnon et al. 2022; Lou 2022, Parks 2021; Wateau et al.; Little and Akese in this collection). As the editors of *Arts of Living on a Damaged Planet* state, people face a "barrage of messages asking us to *forget* – that is, to allow a few private owners and public officials with their eyes focused on short-term gains to pretend that environmental devastation does not exist … . Our era of human destruction has trained our eyes only on the immediate promises of power and profits. This refusal of the past, and even the present, will condemn us to continue fouling our own nests" (Gan et al. 2017, G1–2). A reckoning with toxic heritage requires recognizing, therefore, that habituated or suppressed awareness of toxicity is integral to extractivist processes in which resources (minerals, human labor, water, soil, air) are consumed and/or contaminated in the pursuit of profit, while the ongoing costs of despoiled communities, negative health impacts, and mitigation are borne by those who inherit the pollution. An extractivist framework acknowledges, as Justin Parks has argued, that this is not simply an industrial practice of "large-scale, profit-driven operations for the removal and processing of natural resources such as hydrocarbons, minerals, lumber, and other materials," but it is also the habitus of modern society "in which our agency as subjects is thoroughly energy-dependent, underwritten by narratives of abundant resources available for the taking" (Parks 2021, 353).

For many Indianapolis residents, environmental contamination – dry cleaning, air pollution, lead in their soil and water, *E. coli* in the public waterways, and a host of other issues – is neither a conscious concern nor a common topic of public conversation. This chapter explores narratives of the social and environmental history of dry cleaning in Indianapolis as a lens into the ways in which toxic heritage, much like the dry-cleaning chemical contamination itself, has been channeled in hidden discourses and archives. While it has largely disappeared from public memory, it circulates nonetheless, and it can be activated with intentional curation. Specifically, the study highlights the tensions between, on the one hand, the amnesia and habituation of accumulated environmental harm and, on the other, the narratives of pollution as a central heritage of modern human history. Indianapolis offers examples of how mobilizing toxic heritage storytelling can engage people in public conversation, make connections across communities, and amplify residents' long-standing activism and advocacy. The city's legacy of "dirty laundry" demonstrates the power of participatory heritage not only to reflect but to shape understandings of the relationship between past and present in a time of ecological and climate crises.

A Brief History of Dry Cleaning

While the techniques have changed over time, people have cleaned textiles without water since Romans used ammonia (derived from urine) to clean wool without shrinking it (Bradley 2002), and laundries applied clay to leather gloves in order to remove oils (*Greenfield Evening Republican*, 1 January 1895, 3). In the United States in 1821, a patent was granted to Thomas L. Jennings, the first African-American patent-holder, for a process he called "dry scouring" (Bellis 2021). In France, industrialist Jean Baptiste Jolly developed a similar method using petroleum-based solvents such as kerosene and gasoline in the mid-19th century. This appears to have been at least one source for the adoption of the process in Indianapolis, as evidenced by an 1880 advertisement for Walker Chemical Dye Works offering dry cleaning "by the new French process" (*Indianapolis Leader,* Oct. 2, 1880). Products were also available for in-home dry cleaning. An 1895 Indiana newspaper advised, "If the article to be cleaned cannot be put in water, it is dipped in a vat of benzine" (*Greenfield Evening Republican*, Jan. 1, 1895), and a 1947 ad for Renuzit stated "All you do is dip and rinse. So easy a child can do it!" (Indianapolis Times, April 24, 1947).

With the growth of commercial laundries in the early 20th century (Mohun 1999) and the development of the synthetic chemical industry following World War I, the technology for dry cleaning became more complex. Sophisticated equipment and the introduction in the 1930s of perchloroethylene or tetrachloroethylene, a chlorinated solvent commonly known as PERC, enhanced dry cleaning's popularity (Doherty 2000, 70). Chlorinated solvents were highly effective in removing stains, evaporated more quickly and with less odor, and were less flammable than petroleum distillates. PERC was more expensive than petroleum solvents, however, so it was not until 1962 that chlorinated solvents became more common than petroleum distillates. The problem, however, is that PERC (CTC, PCE, TCE, and TCA) is a reproductive toxicant, neurotoxicant, carcinogen, and persistent environmental pollutant, and it was unregulated until the 1977 Clean Water Act (Ceballos et al. 2021; EPA 2016; CDC 1997; Morrison 2003). By the 1990s some dry cleaners had adopted safer, more environmentally friendly solvents, and the industry has also seen a significant market share decline over the past 30 years. But PERC is still the most commonly used chemical (Ceballos et al. 2021), and its historical use has left a legacy of slow violence (Nixon 2013, Davies 2022) in the form of contaminated sites, particularly in urban areas.

Dry cleaning is significant for toxic heritage not only because workers are exposed to toxic solutions and fumes, but also because spills, dumping, and leaking underground storage tanks contaminate soils. Prior to 1970s regulations, many plants had drains in the floor so that spills were channeled directly into the ground (Lohman 2002). Contamination creates underground plumes that continue to spread over time through the transfers from a subsurface source zone of residual dense nonaqueous phase liquid (DNAPL) to flowing groundwater (Frind 1999). These spreading underground plumes of chemical contamination contaminate groundwater, but can also enter open waterways posing a direct risk to ecosystems. Significantly for health risks, the vapors can seep upward from volatile sub-surface plumes and enter basements and crawl spaces. If not ventilated, the invisible and largely odorless gas can become concentrated and pose a health risk. The EPA's cleanup proposal for the Tuchman site noted, "Exposure to very high levels of TCE or PCE can cause dizziness, headaches, sleepiness, imbalance, confusion, nausea, unconsciousness, and even death" (EPA 2020). The odorless gas and its underground spread mean that dry cleaning's environmental harms, as well as their health risks, are largely invisible.

Narratives of Dry-Cleaning Heritage in Indianapolis

Environmental amnesia and the invisibility of urban industrial contamination offer productive starting places for considering Indianapolis' dry-cleaning heritage, particularly the way its narratives intersect with public memory. As Samuels (2015) has argued, heritage rhetoric, including text and images, mobilizes and motivates as it is used to persuade and codify. As such, heritage rhetoric is a social practice and a dynamic part of creating social change. In this light, Indianapolis' dry-cleaning heritage narratives are a lens into the fields of power that produced, perpetuate, and navigate urban pollution. These narratives circulate in discrete discourses that may be characterized by their authorship: industry-produced, dry cleaning workers, environmental regulatory and advocacy organizations, cultural heritage institutions, and activist voices including public history, journalism, community members, and participatory heritage. Each offers insights into the meaning and the politics of dry cleaning heritage as they intersect with the history of the industry's environmental harms.

Industry-produced narratives reveal both consumer-focused messages and the contours of social inequalities and segmentation. Specifically, Indianapolis dry cleaners operated in a highly racialized landscape that shaped the city's social fabric for much of the 20th century, and, many would argue, continues to be a significant force in the city.[3] For example, a 1938 story in the city's Black newspaper *The*

Indianapolis Recorder reported the opening of a United Laundries store at Lockefield Gardens, the new public housing project along Indiana Avenue, a center of African-American cultural life in the city. The story notes, "Having perfected a new system of dry cleaning, the company offers a non-odor method to its customers that is taking the city by leaps and bounds." It also reports that "Believing in equal representation of all groups, the United Laundries. Inc., operators of fourteen retail stores, employs more than 40 per cent of our group in their plants" (*Indianapolis Recorder* 1938).[4]

The racialized customer base and workforce are evident in other records as well. *The Fiery Cross* newspaper published by the white supremacist Klu Klux Klan, which was a powerful political and social force in Indiana through much of the first half of the 20th century, included dry-cleaning advertisements that subtly coded their affiliations by using the KKK initials, such as in the Arcade Garment Cleaners' promise of being Kareful, Klothes, Kleaners (Figure 13.2). Business Directory records

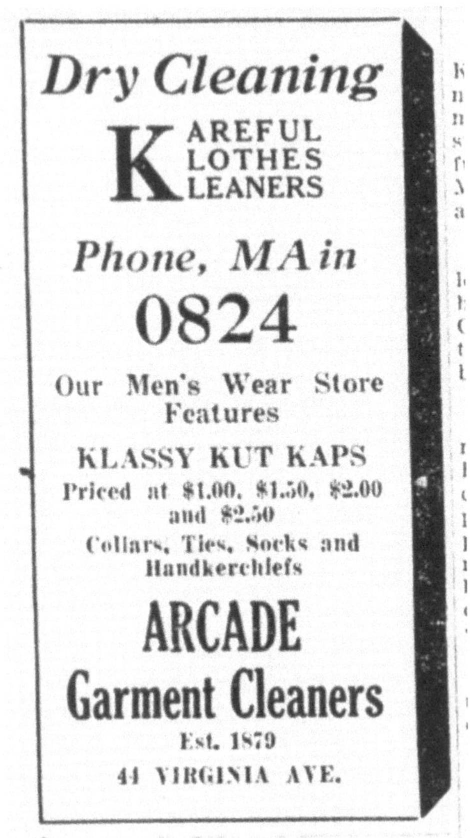

FIGURE 13.2 Arcade Garment Cleaners advertisement in The Fiery Cross, May 2, 1924.

document the location of cleaners and laundries which were dispersed across the city but concentrated along commercial thoroughfares. Many were in HOLC "Redlining" maps' red (D) and yellow (C) areas which represented classifications of higher risk and therefore excluded from mortgage lending based on racial criteria (Figure 13.3).

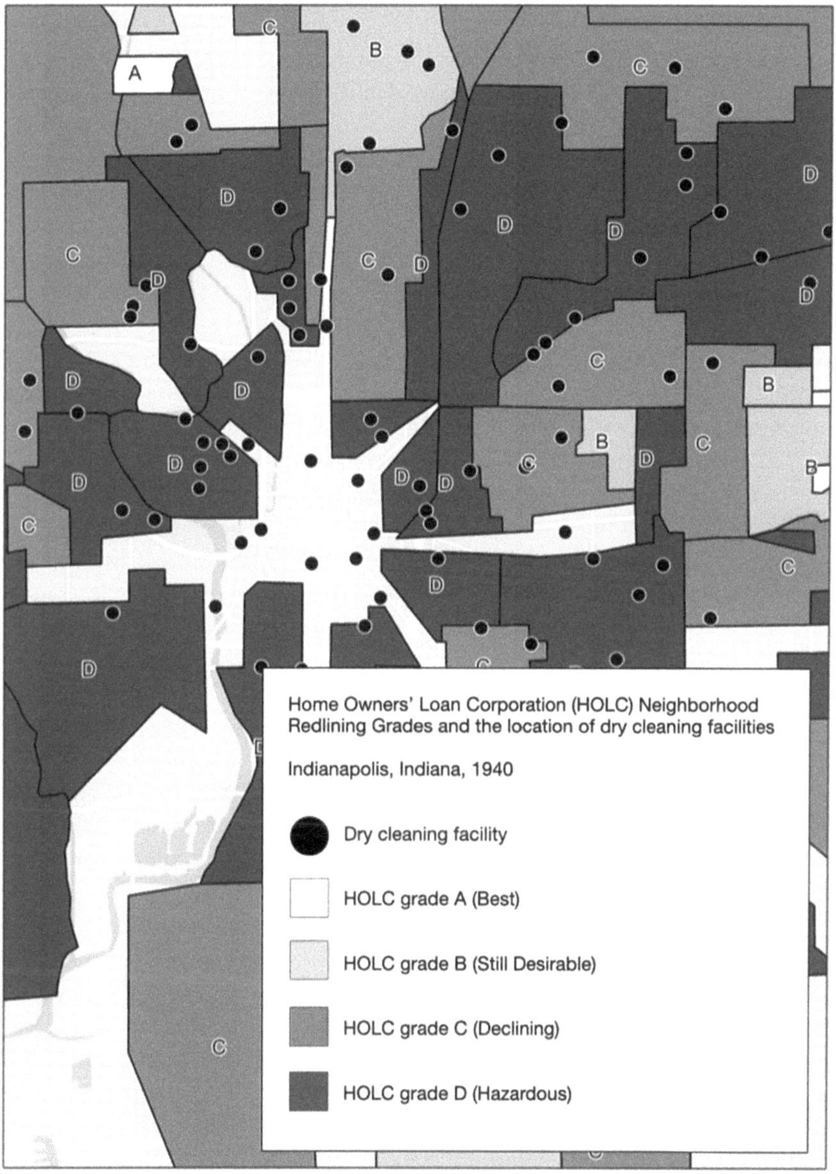

FIGURE 13.3 Distribution of Indianapolis dry cleaning facilities in 1940 overlaid on the Home Owners Loan Corporation (HOLC) Risk Assessment Categories map. (Map by Owen Dwyer).

Dry cleaning advertisements promoted a service that entails caring for clothing, one of the more intimate forms of material culture, and it was presented not only as convenient but an essential service for the "modern consumer." The persuasive language used in these ads reflects ideas about race and laundry in ways that elided racial superiority and white supremacy with notions of health and cleanliness. As with the broader commercial laundry industry (Mohun 1999; DeArmond 1950), dry cleaning was associated with modernity, professionalism (and the subtext of socio-economic status), progress, and the health benefits that an allegedly more sanitary process afforded. The central metaphor of commercial laundries and dry cleaning of "clean and white" taps into, as Carl Zimring (2015) has established, an essentialized racial metaphor of American culture and is critical to the construction of white identity and white supremacy. For example, industry-produced rhetoric such as a Swiss Cleaner's 1920 advertisement extolled the cleanliness and sanitation standards of their facility and process, stating

> In the absolutely fireproof cleaning room (the entire plant is of brick, concrete and steel construction), distilled gasoline is gently flushed through your garments in a clean cascade flow, fed into the cleaning machine from pipes underground. The gasoline is extracted by centrifugal force; no twisting by hand, no wringer to stretch delicate garments out of shape. A steam-heated drying tumbler deodorizes and dries simultaneously. Besides the usual pressing machines, special irons for velvets and puff-irons for puffs, ruffles, etc., insure correct and finished work. A corps of expert needlewomen attend to repairs. Four large, clean autos call for your garments and deliver them in sanitary, wardrobe bags (*Indianapolis News,* August 24, 1920).

The architecture of dry-cleaning establishments similarly reflected this metaphor of modernity and cleanliness, particularly when viewed across Indianapolis' racialized landscape. In communities of color, the cleaners were typically in modest brick and cinderblock structures or adapted residential dwellings as with Baird Cleaners on West Street (Figure 13.4). Larger businesses, which often had multiple locations, invested in storefront facades that conveyed their modern approach to clothing care such as Progress Laundry's streamlined art deco architectural motifs (Figure 13.5).

These storefronts belied the dangers of exposure for workers and nearby residents to chemical contamination and airborne pollution, as well as the risk of frequent fires inherent in volatile chemical processes. These risks highlight the complexities of workers' experiences and the ways dry cleaning worker narratives have circulated in "hidden archives," much as Lisbeth Haas (2014) has argued that "Indigenous archives" preserve Native Californian stories and memories outside of official archives and public memory. The heritage of dry cleaning has been similarly preserved in the traditions, craft, and knowledge of the industry's workers as well as in the biological legacy borne by their bodies. Interviews with dry-cleaning workers and owners indicated that there are rich memories and deep knowledge of the complex industry,

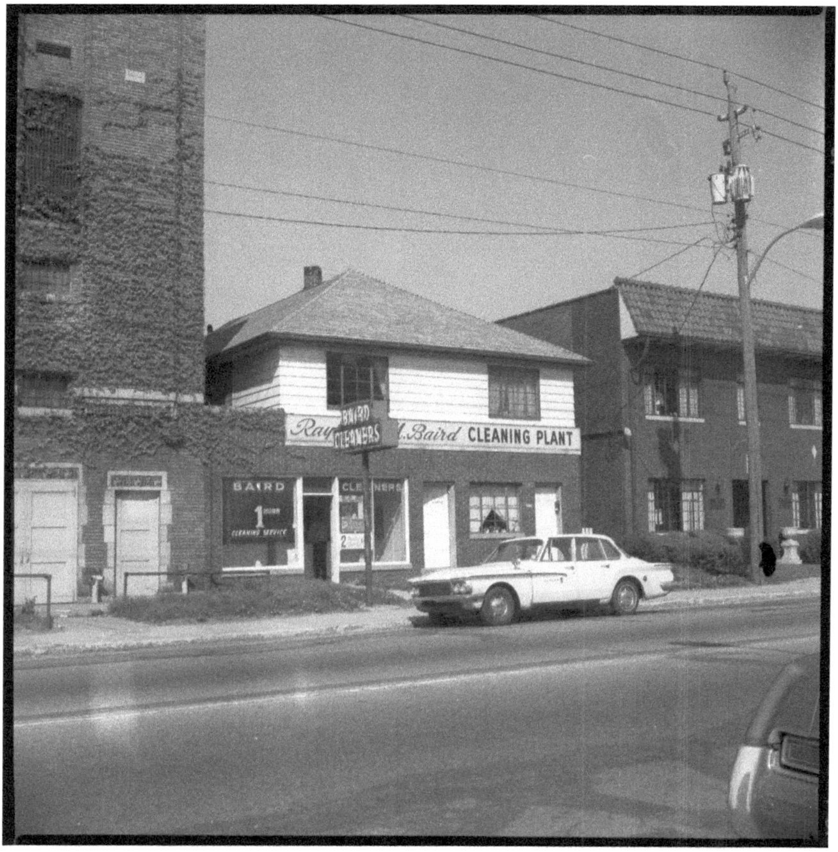

FIGURE 13.4 Baird Cleaners, 628 West Street, Indianapolis, c.1975. (Indiana Landmarks).

as well as a candid understanding of both the personal and environmental risks involved. Dry-cleaning workers, both retired and currently working, spoke with pride about the knowledge required to clean stains on different fabrics and the skills to provide a perfect finish to the garment. They spoke of masters of their craft, including one described as a "dry-cleaning magician" who knew all the old formulas and another who was a "master spotter." At the same time, the workers were aware of the risks and spoke about the changing awareness of the hazards of the industry over time. They knew who ran a "clean shop" and who cut corners by improperly disposing of dirty fluid even after the industry became more strictly regulated. This varying compliance was confirmed by a conversation with a former environmental inspector who commented that "if you want to find the hot spots at a dry cleaner's, check outside the back door" and by reports of a lawsuit over environmental cleanup costs with accusations of negligence among members of a family-owned dry-cleaning business.

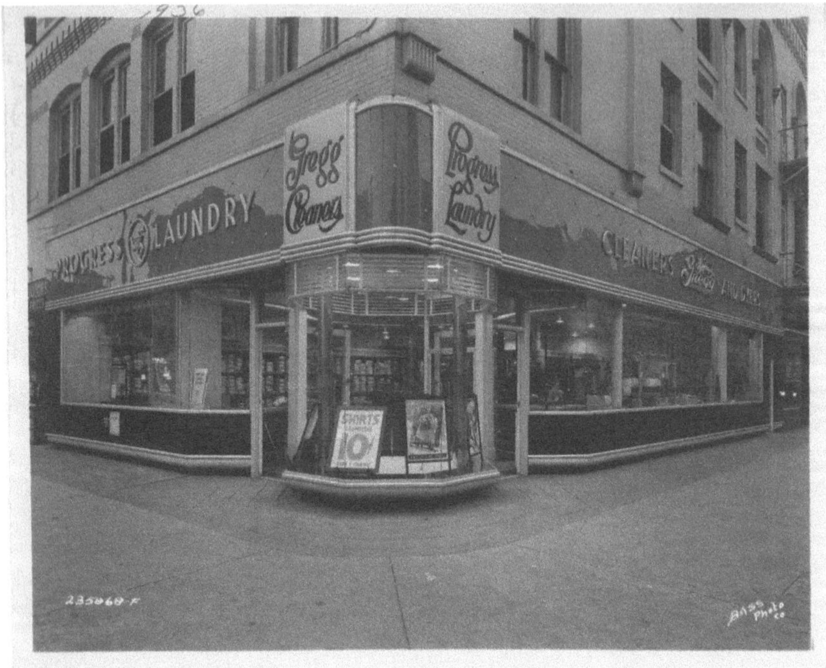

FIGURE 13.5 Progress Laundry, Gregg Cleaners, 1936 (Bass Collection, Indiana Historical Society).

There is clear evidence of links between exposure to dry cleaning chemicals and disease among industry workers (Vaughan et al. 1997; Zielhuis, Gijsen, and van der Gulden 1989), and dry cleaning laborers spoke candidly about the risks of working with chemicals, although they said they never talked about it while working. One retired owner commented, "I'm in dry cleaning. I know there are chemicals," and he described times he was "up to my elbows in PERC." Another recalled that when trying to fix a malfunctioning machine he had to reach his arm far into the tank. He added, "I looked down and saw my face was inches from the fluid." While the workers spoke of colleagues who had died of cancer, every person interviewed also brought up the pervasive smoking among dry-cleaning workers, and commented that it would be hard to determine the cause of those illnesses.

In contrast to industry and worker narratives, environmental regulatory and advocacy organizations' language is dispassionately scientific and largely presentist. Regrettably, Indiana currently ranks 42nd-46th out of 50 for environmental health, eco-friendliness, pollution, and air quality, while it is first for coal ash ponds and sixth for toxic chemical releases (Hopkins 2018). Governmental and environmental advocacy organizations typically focus on current conditions and plans for future mitigation or management. Non-profit organizations, such as Friends of the White

River, Hoosier Environmental Council, and Reconnecting Our Waterways, are strong advocates for legislation, education, and funding to restore and protect the future of Indiana's natural environment. The futurist focus of their narratives includes both the positives if action is taken (eg., people once again swimming in now polluted waters, return of healthy ecosystems in ponds now choked by agricultural runoff) and predictions of the dire consequences if nothing is done. And yet, these regulatory agency and advocacy organization narratives have failed to account for the complexity of pervasive urban pollution and its effects on human and non-human lives (Fiske 2020; Murphy 2017). Their messages are often muted in the methodological imperatives of the scientific process and suppressed by economic and political agendas (Boudia and Jas, eds. 2014; Murphy 2017), culminating in what Fiske has called the "disjunctures between scientific models and the entanglements of toxicants and injustice" (Fiske 2020: 9).

A fourth discourse arena is the city's thriving formal heritage network of museums and cultural organizations. Indianapolis is the 18th largest city in the US by area and 15th by population, and it houses the world's largest children's museum, a historical society founded in 1830, and one of the country's largest statewide historic preservation organizations. As the capital of Indiana, it also hosts the state archives, state library, and state museum. It is notable, therefore, that the work on environmental harm is evident only in a few ephemeral programs, in contrast to the more visible temporary or permanent exhibits. For example, the Indiana State Museum mentions industrial pollution and biodiversity loss briefly in their grade 9–12 online video "Human Impact on the Earth," screened "The Last Glaciers" at their IMAX theater, and hosted a guest lecture on environmental racism at an Underground Railroad site (Clark 2022). The silence of heritage organizations on the topic may be understood as part of a pattern of museal silence described by Mason and Sayner (2019), who enumerated eight types or explanatory factors in museum silences, including gaps in the collection, external pressures, collusion in society's silences, and use of silence as oblique or ambiguous messaging. It is clear that addressing the history of environmental harm is challenging in a state that recently passed legislation to disincentivize transitions to renewable energy and to weaken industrial pollution regulation and wetland protections (Bowman 2022). As an example of the political climate, a few years ago the Indiana State Museum hung a banner about climate change on its building, which is visible from the State Capitol. A state legislator objected to it, and it was removed. Furthermore, when approaching a local museum about the possibility of an exhibit on the state's toxic heritage, the authors were referred to the biology curator in the natural history division. Environmental issues were seen as part of the scientific discourse, rather than a social history topic, let alone an important part of the "Hoosier story." The museums' silence may be explained, therefore, by a combination of external pressures, complicity in society's silences, and a disjuncture in thinking about "environment" as a scientific rather than social history topic.

Activist Voices: Archives, Journalism, and Participatory Heritage

In this context of museal silence and environmental amnesia, it is significant that the most prominent narratives of Indiana's toxic heritage are from activist voices outside the formal heritage sector. The example first is a project developed by Ryan T. Schwier who combined his public history, library science, and law training to create the Indiana Legal Archive, "a digital platform for exploring the state's rich legal heritage." (Schwier 2016). The main project was a legal history of the environmental justice movement in the USA focusing on the ecological implications of residential segregation in Indianapolis. With images, maps, and case law history, Schwier traced the historical processes and consequences of environmental racism in the city through practices such as redlining, the Combined Sewer Overflow system, and urban development policies. Another significant contribution to Indiana's toxic heritage narrative has been a sustained project at the *Indianapolis Star*. Reporters[5] funded by the nonprofit Nina Mason Pulliam Charitable Trust have covered the state's environmental issues, integrating history, contemporary politics, and science, as well as the lived experiences of Indiana residents through articles, photo essays, multi-media long-form journalism, and social media.

Another response to silences about the state's environmental harm has been to mobilize participatory heritage in an activist scholarly mode. The projects share Roued-Cunliffe and Copeland's definition of participatory heritage as "a space in which individuals engage in cultural activities outside of formal institutions for the purpose of knowledge sharing and co-creating with others" (2017, xv). The projects are also part of a broader trend in the cultural heritage field to democratize knowledge, share authority, and engage a wide range of participants in active meaning-making (e.g. Adair, Filene, Koloski eds. 2011; Byrne et al. 2018; Kryder-Reid ed., 2018; Ševčenko 2023; Simon 2010) and deploy heritage to help people imagine and act for change, particularly around environmental issues (e.g. Harrison et al. 2020; the Climate Heritage Network, and the Coalition of Museums for Climate Justice).

A catalyst for participatory heritage research on environmental issues in Indianapolis was the invitation to IUPUI Museum Studies and Public History faculty to participate in the Humanities Action Lab's (HAL) collaborative project "Climates of Inequality" (Sevcenko 2023, 259–273). IUPUI students and faculty partnered with the Kheprw Institute (KI), a grassroots community development non-profit, and other community partners to research the history of Indianapolis waterways and to investigate the city's environmental racism and environmental justice issues. They produced exhibits (Figure 13.6), public programs, social media sites, and digital humanities, including StoryMaps and a Twitter thread (for details see the Toxic Heritage website and Kryder-Reid et al. 2021a, 2021b, 2022).

The projects were an intervention into the silence around environmental harm, and their goal was to raise awareness and provoke dialogue about Indianapolis' toxic

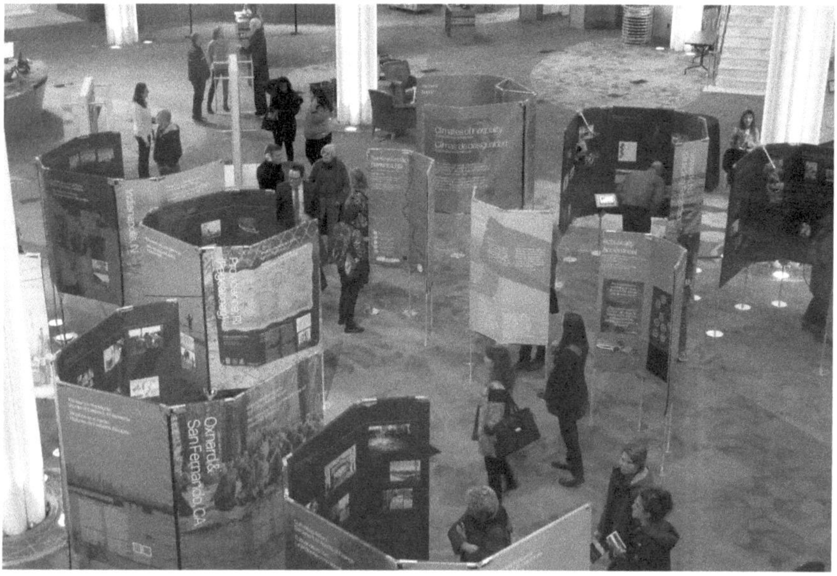

FIGURE 13.6 Climates of Inequality exhibit at the Indianapolis Central Library, January-February 2020.

heritage, as well as to amplify the voices of community advocates and activists. Bringing the history of the city's environmental harm into public discourse spaces sought both to raise awareness of the contamination of past industries and to illuminate the structures and policies that perpetuate the practices.

Conclusion

The paradox of dry-cleaning toxic heritage is that it is materially pervasive and yet largely invisible, manifested predominantly in silence, forgetting, avoidance, and denial. The historic processes of industrial and residential churning combined with governmental interests in denying or deferring risk are important factors (Frickel and Elliott 2018). There is also a clear record of the benefits for polluting industries to avoid accountability for their toxic legacies, whether accounting that displaces environmental damage costs or the practices of declaring bankruptcy, transferring assets, or successive purchases by parent companies that shield the perpetrators from responsibility. But the practices of silence and suppression go beyond the processes of urban development and the self-interested avoidance of businesses. It also speaks to the broader failure to reckon with the current state of the planet and to locate that harm squarely as our heritage. Yet heritage, particularly collaborative, participatory heritage, can be a tool to reclaim those silenced voices and hidden stories. It can be used to give voice to the waterways and soils and plants and animals that have been impacted, as well as the human communities. It can help blur the artificial distinctions of nature and culture, science and history, and name pollution, climate change, and biodiversity loss as a central heritage of our time.[6]

Notes

1 The Tuchman site was later designated as the Keystone Corridor Ground Water Contamination (INN000510399) and placed on the National Priority List in 2013 due to the actual and potential contamination in municipal wells.
2 Federal legislation (CERCLA) was passed in 1980 creating a program where sites, known as "superfund sites" are eligible for EPA funding and management for decontamination.
3 Indianapolis has also long had Asian-American-owned laundries and cleaners which tend to be family-owned and operated single-location stores (Mullins 2016).
4 The exposure for people of color continues. A 2017 national study identified c. 20,600 dry cleaners employing nearly 160,000 workers, with approximately 80% identifying as a racial or ethnic minority (Ceballos et al. 2021).
5 Reporters in these positions have included Emily Hopkins, Sarah Bowman, and London Gibson.
6 This research was supported by a grant from the IUPUI Arts and Humanities Institute.

Bibliography

Adair, Bill, Benjamin Filene, and Laura Koloski, eds. 2011. *Letting Go?: Sharing Historical Authority in a User-Generated World.* Philadelphia: The Pew Center for Arts & Heritage.

Bellis, Mary. 2021. "Biography of Thomas Jennings, First African American Patent Holder." ThoughtCo, Feb. 16. https://www.thoughtco.com/thomas-jennings-inventor-1991311

Boudia, Soraya anNathalie Jas, Eds. 2014. *Powerless Science?: Science and Politics in a Toxic World.* New York, NY: Berghahn Books.

Bowman, Sarah. 2022. "Indiana One of Most Polluted States. New Report says Legislature Gets a D+ on Green Bills." *Indianapolis Star*, March 22. https://www.indystar.com/story/news/environment/2022/03/22/indiana-environmental-bills-lawmakers-flunking-state-pollution/7094262001/

Bradley, Mark. 2002. "'It All Comes out in the Wash': Looking Harder at the Roman Fullonica." *Journal of Roman Archaeology* 15: 20–44. 10.1017/S1047759400013829

Buell, Lawrence. 2017. "Uses and Abuses of Environmental Memory." In *Contesting Environmental Imaginaries: Nature and Counternature in a Time of Global Change.* pp. 93–116. Series: Studies in Environmental Humanities, Volume: 4. Leiden, Netherlands: Brill. doi 10.1163/9789004335080_007

Byrne, John et al. eds. 2018. *The Constituent Museum: Constellations of Knowledge, Politics and Mediation: a Generator of Social Change.* Amsterdam: Valiz.

Cahill, Caitlin, and Rachel Pain. 2019. "Representing Slow Violence and Resistance: On Hiding and Seeing." *ACME: An International E-Journal for Critical Geographies* 18(5): 1054–1065.

Ceballos, Diana M. et al. 2021. "Perchloroethylene and Dry Cleaning: It's Time to Move the Industry to Safer Alternatives." *Frontiers in Public Health* 05 March 2021, Sec. Environmental health and Exposome, 10.3389/fpubh.2021.638082

Center for Disease Control (CDC). 1997. "Control of Exposure to Perchloroethylene in Commercial Drycleaning." DHHS (NIOSH) Publication no. 97-154. https://www.cdc.gov/niosh/docs/hazardcontrol/pdfs/hc16.pdf?id=10.26616/NIOSHPUB97154 (accessed 2-5-2020).

Chagnon, Christopher W. et al. 2022. "From Extractivism to Global Extractivism: the Evolution of an Organizing Concept." *The Journal of Peasant Studies* 49(4): 760–792, 10.1080/03066150.2022.2069015

Clark, Ben. 2022. "White Supremacy and Environmental Racism." Public Presentation, Levi and Catharine Coffin State Historic Site, March 5. Climate Heritage Network. https://www.climateheritage.org/

Clingerman, Forrest. 2018. "Conclusion: The Aesthetic Roots of Environmental Amnesia: The Work of Art and the Imagination of Place." In *Arts, Religion, and the Environment: Exploring Nature's Texture*, edited by Sigurd Bergmann, and Forrest J. Clingerman, 186–212. Studies in Environmental Humanities, Volume: 6. Leiden, The Netherlands: Brill. doi: 10.1163/9789004358980

Coalition of Museums for Climate Justice. https://cmcj.ca/

Davies, Thom. 2022. "Slow Violence and Toxic Geographies: 'Out of Sight' to Whom?" *Environment and Planning C: Politics and Space* 40 (2): 409–427. 10.1177/2399654419841063

DeArmond, Fred. 1950. *The Laundry Industry*. New York, Harper.

Doherty, Richard E. 2000. "A History of the Production and Use of Carbon Tetrachloride, Tetrachloroethylene, Trichloroethylene and 1,1,1-Trichloroethane in the United States: Part 1-- Historical Background; Carbon Tetrachloride and Tetrachloroethylene." *Journal of Environmental Forensics* 1: 69–81.

EPA. 2016. "Dry Cleaning Sector." EPA Region 2 *Insights Bulletin* September. https://www.epa.gov/p2/dry-cleaning-sector-insights-bulletin

EPA. n.d. "Tuchman Cleaners Site Profile." https://response.epa.gov/site/site_profile.aspx?site_id=7130 (accessed 9-15-2021).

EPA. 2020. Proposes Cleanup Plan For Underground Water: Keystone Corridor Ground Water Contamination Site Indianapolis, Indiana. 958132. May.

Fiske, Amelia. 2020. "Naked in the Face of Contamination: Thinking Models and Metaphors of Toxicity together." *Catalyst: Feminism, Theory, Technoscience* 6(1): 1–30. 10.28968/cftt.v6i1.32093

Frickel, Scott, and James R. Elliott. 2018. *Sites Unseen: Uncovering Hidden Hazards in American Cities*. New York: Russell Sage Foundation.

Frind, E. O., J. W. Molson, M. Schirmer, and N. Guiguer. 1999. "Dissolution and Mass Transfer of Multiple Organics under Field Conditions: The Borden Emplaced Source." *Water Resources Research* 10.1029/1998WR900064

Gan, Elaine, Anna Lowenhaupt Tsing, Heather Swanson, and Nils Bubandt. 2017. "Introduction: Haunted Landscapes of the Anthropocene." In *Arts of Living on a Damaged Planet: Ghosts and Monsters of the Anthropocene*, edited by Elaine Gan, Anna Lowenhaupt Tsing, Heather Swanson, and Nils Bubandt, G1–G14. Minneapolis: University of Minnesota Press.

Gelobter, Michel. 1993. "The Meaning of Urban Environmental Justice." *Fordham Urban Law Journal* 21: 841–856.

Groves, Dawn. 2011. Email - Site Referral: Former Tuchman Cleaners (IND016450561), 4401 North Keystone, Indianapolis, IN From IDEM to Shelly Lam, US EPA, July 14, 2011 [EPA Records 437138].

Haas, Lisbeth. 2014. *Saints and Citizens: Indigenous Histories of Colonial Missions and Mexican California*. Berkeley: University of California Press.

Harrison, Rodney et al. 2020. *Heritage Futures: Comparative Approaches to Natural and Cultural Heritage Practices*. London: University College of London.

Hart, Laura B. 2022. "Emotion and Othering in a Contaminated Community." *Ethnography* Vol. 0(0): 1–19 published online https://doi.org/10.1177/14661381221145812

Hopkins, Emily. 2018. "Indiana Has a Polluted Reputation. Here's How Bad it is." *Indianapolis Star* April 19. https://www.indystar.com/story/news/2018/04/19/how-bad-indianas-environment-actually-according/525003002/

Humanities Action Lab. n.d. "Climates of Inequality." https://climatesofinequality.org/

Kahn, Peter H. Jr. 2002. "Children's Affiliations with Nature: Structure, Development, and the Problem of Environmental Generational Amnesia." In *Children and Nature: Psychological, Sociocultural, and Evolutionary Investigations*, edited by Kahn and Stephen Kellert, 93–116. Cambridge MA: MIT Press.

Kryder-Reid, Elizabeth, ed. 2018. Tools for a Critical Heritage: Exploring Shared Authority and Stakeholder-Defined Values of Heritage. Special issue. *International Journal of Heritage Studies* 24 (7).

Kryder-Reid, Gabriel Filippelli, with Phyllis Boyd, Paula Brooks, Aghilah Nadaraj, Alvin Sangsuwangul, and Leah Humphrey. 2021a. "EJ Stories: Exploring University-Community Collaborations." *Engage!* 3(1) https://journals.iupui.edu/index.php/ENGAGE/article/view/25724. 10.18060/25724

Kryder-Reid, Elizabeth, Laura Holzman, Aghilah Nadaraj, and Leah Humphrey. 2022. "An Environmental Justice Lens on Indianapolis's Urban Ecosystem: Collaborative Community Curation." In *The City as Ecosystem: Sustainable Education, Policy, and Practice.* edited by Deborah Mutnick, Margaret Cuonzo, Carole Griffiths, Timothy Leslie, and Jay M. Shuttleworth. pp. 10.4324/9781003217442-20

Lohman, John H. 2002. "A History of Dry Cleaners and Sources of Solvent Releases from Dry Cleaning Equipment." *Environmental Forensics* 3: 35–58.

Lou, Loretta Ieng Tak. 2022. "The Art of Unnoticing." *American Ethnologist.* published online 17 September.

Mason, Rhiannon and Joanne Sayner. 2019. "Bringing Museal Silence into Focus: Eight Ways of Thinking about Silence in Museums." *International Journal of Heritage Studies,* 25(1): 5–20. 10.1080/13527258.2017.1413678

Melosi, M. 1973. "'Out of Sight, Out of Mind' The Environment and Disposal of Municipal Refuse, 1860-1920." *The Historian* 35 (4): 621–640.

Mohun, Arwen. 1999. *Steam Laundries: Gender, Technology, and Work in the United States and Great Britain, 1880-1940.* Baltimore, MD: The Johns Hopkins University Press.

Morrison, Robert. 2003. "PCE Contamination and the Dry Clean Industry." *Environmental Claims Journal* 15(1): 93–106. 10.1080/10406020390832471

Mullins, Paul. 2016. "The Landscapes of Chinese Immigration in the Circle City." Blog Post, Invisible Indianapolis. https://invisibleindianapolis.wordpress.com/2016/10/16/the-landscapes-of-chinese-immigration-in-the-circle-city/

Murphy, Michelle. 2017. Alterlife and Decolonial Chemical Relations. *Cultural Anthropology* 32(4), 494–503. 10.14506/ca32.4.02

Nixon, Rob. 2013. *Slow Violence and the Environmentalism of the Poor.* Cambridge: Harvard University Press.

Parks, Justin. 2021. "The poetics of extractivism and the politics of visibility." *Textual Practice* 35 (3): 353–362. 10.1080/0950236X.2021.1886708

Roued-Cunliffe, Henriette, and Andrea Copeland. 2017. "Introduction: What is participatory heritage?" In *Participatory Heritage.* Edited by Henriette Roued-Cunliffe, and Andrea Copeland, xv–xxi. London: Facet Publishing.

Samuels, Kathryn Lafrenz. 2015. "Introduction: Heritage as Persuasion." In *Heritage Keywords: Rhetoric and Redescription in Cultural Heritage.* Edited by Kathryn Lafrenz Samuels, and Trinidad Rico, 3–28. Boulder: University Press of Colorado.

Schwier, Ryan T. 2016 "Indiana Legal Archive." Website developed 2014-2016. http://www.indianalegalarchive.com/

Ševčenko, Liz. 2023. *Public History for a Post-Truth Era: Fighting Denial through Memory Movements*. New York: Routledge.

Simon, Nina. 2010. *The Participatory Museum*. Santa Cruz, California: Museum 2.0.

Toxic Heritage. "A Collaborative Research Project." Website managed by Elizabeth Kryder-Reid https://toxicheritage.com/

Vaughan, T. L., P. A. Stewart, S. Davis, et al. 1997. "Work in Dry Cleaning and the Incidence of Cancer of the Oral Cavity, Larynx, and Oesophagus." *Occupational and Environmental Medicine* 54: 692–695.

Zielhuis, Gerhard A., Ronald Gijsen, and Joost W. J. van der Gulden. 1989. "Menstrual disorders among dry-cleaning workers." *Scandinavian Journal of Work, Environment & Health* 15(3): 238.

Zimring, Carl A. 2015. *Clean and White: A History of Environmental Racism in the United States*. New York: New York University Press.

CASE STUDY 5

WHEN CLEANING UP THE BATTLEFIELDS FROM WHEN TIMES OF WAR HAVE POLLUTED SOILS IN TIMES OF PEACE: A CASE STUDY OF A SILENT BUT VISIBLE TOXIC LEGACY FROM THE GREAT WAR

Daniel Hubé and Tobias Bausinger

Introduction

Lasting physical and chemical environmental disturbances induced by wars are an invention of the XIXth century as a consequence of the increase of the power of new kinds of arms and explosives. During World War One (hereafter WWI), ammunition and energetic material were used at an unprecedented scale with an estimated 0.9 to 1 billion explosive shells fired on the Western Front, as well as hundreds of millions of grenades and trench artillery munitions (Linnenkohl 1996). The unexpected and sudden end of the war in November 1918 on the western front led to an accumulation of excess, unused, and obsolete munitions which were collected and stored in restricted dumps. These chemical and conventional munitions were composed of chemical elements and compounds with potential toxic effects on health, particularly as they rapidly deteriorated with time and weather. What had been essential weapons for achieving victory quickly became hazardous, undesirable war remains. There was an urgent need to safely neutralize this war waste and to recover the valuable material contained in it for commercial purposes. The materials included projectiles, explosives, propellants, and chemical warfare agents were composed of toxic elements (arsenic, lead, zinc, copper, cadmium, etc.) and chemical synthetic substances (like nitroaromatic compounds), and their danger was not only the threat of explosion but also contamination to soils and groundwater. The history of this hazardous war waste reveals the divergence of the valorized war memory promoted in the museums and battlefields of the Great War and the amnesia regarding its toxic heritage of environmental contamination.

DOI: 10.4324/9781003365259-25

Post-Conflict War Waste

The history of the demolition of WWI munitions and the resulting severe soil contamination and point sources of pollution are covered in detail (Hubé 2016, 2017; Bausinger et al., 2005, 2008). After the Armistice, the Allies had to manage their own surplus of salvaged ammunition, including both unexploded fired rounds and unused ammunition. Further, munitions were discovered in soils of the former battlefields during the restoration of the devastated grounds, or simply lying on the surface but hidden by mud. Another priority was the clearing of farmland from dangerous munitions when civilians returned and to avoid irresponsible and dangerous looting. In brief, the demolition and recycling process included dismantling, neutralizing, and detonating the weapons. Where that was not possible they were disposed of in underground and underwater deposits, including rivers, lakes, and off the coast. The process required managing not only an unprecedented quantity of artillery from the first industrial war, but also the new specialization of artillery including chemical, explosive, and incendiary shells. These dumps of the end of war were different from those of 1914 because of a larger diversity, origins of munitions in it, and the presence of a significant proportion of chemical shells. The projectiles that were stored outdoors quickly deteriorated. With corrosion, distinctive paint marks disappeared and the seals in the chemical rounds weakened, increasing the risk of leakage. The demolition process was dangerous for workers and residents of nearby villages. For example, archival records document that in 1927 "worker Eugène Retours was poisoned by phosgene gas when emptying a 150-mm shell" (L'Humanité 1927).

The Case of the Forest of Spincourt

One site in particular, "Place à Gaz" in the Spincourt Forest, exemplifies the history of the post-war weapons demolition and the complexities of toxic war heritage. Following the Armistice, French, British, American, and armies, helped by Prisoners of War (POW) and civilian labourers conducted salvage, transportation, sorting, and neutralization of ammunition operations until the end of 1919. Vast quantities of munitions, and especially old ex-German chemical ordnance, were accumulated in the region of Spincourt, c. 20 km northeast of Verdun (Meuse), The transportation and the storage of old-German ammunition was made possible by a dense and well-preserved railroad network installed by the German during WWI to serve the artillery batteries having fired on Verdun battlefield. The site in the Spincourt Forest was used by a private company under the name F.N. Pickett & Fils, one of the largest delaboration companies on the western front during the interwar period, to burn in open trenches an estimated 200,000 German chemical weapons (Forget 1928). While poorly documented historically, the "Place à Gaz" is one of the most visible because of the severe top-soil contamination and blackened ground, hindering the growth of vegetation and creating a clearing of about 1,000 m^2 (Figure 1).

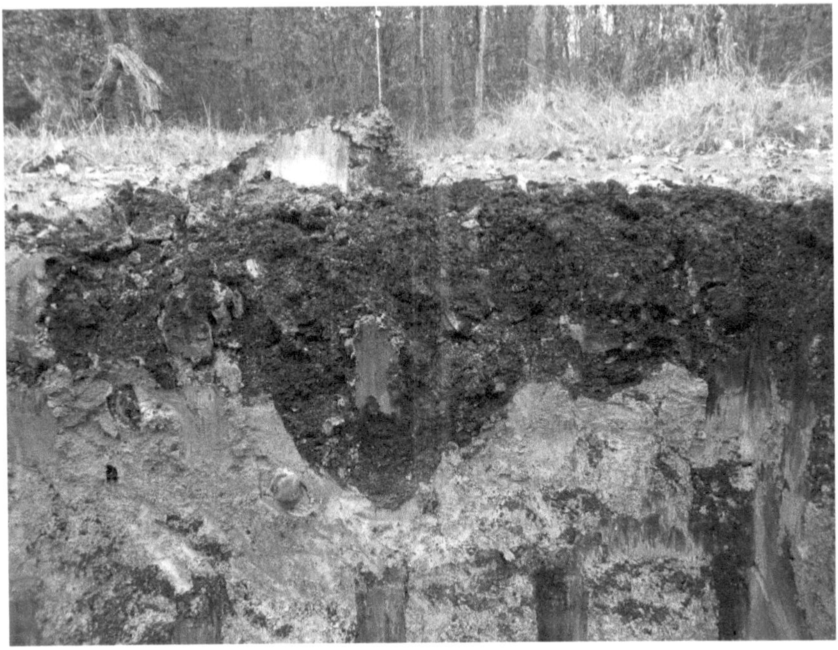

FIGURE 1 Place à Gaz. Cross-section of a trench shows the residues of Blackened Open-burning (OB) chemical shells overlying compact clay. (D Hubé, October 2014).

Since 2003 it has also been the subject of scientific testing and has seen extensive media coverage (Le Figaro 2007). Studies documented extremely high concentrations of dioxins, arsenic, and zinc in soils, residues, and nearby sediments. Based on those results, it has been recommended that the site be capped with a permanent and impermeable cover.

Conclusion. The Silent Legacy of the Great War

Despite its proximity to Verdun, there is no mention of the weapons destruction or its environmental impact at the battlefield interpretive center or public interpretation, lacunae echoed in other WWI heritage sites. This lack of collective memory may be explained by several factors. Firstly, at the end of the Great War, ammunition destruction had never before been experienced on an industrial scale, and there was no specific administration to manage and regulate such new industrial activities, particularly given the disruption of the French government at the end of the war. Furthermore, the archive resources are dispersed, rare, incomplete, largely relegated to administrative correspondence (Clout 1996), some of which were destroyed during WWII. Another explanation is that these operations, dealing with material remains of a passing war, objects of death and destruction, have been

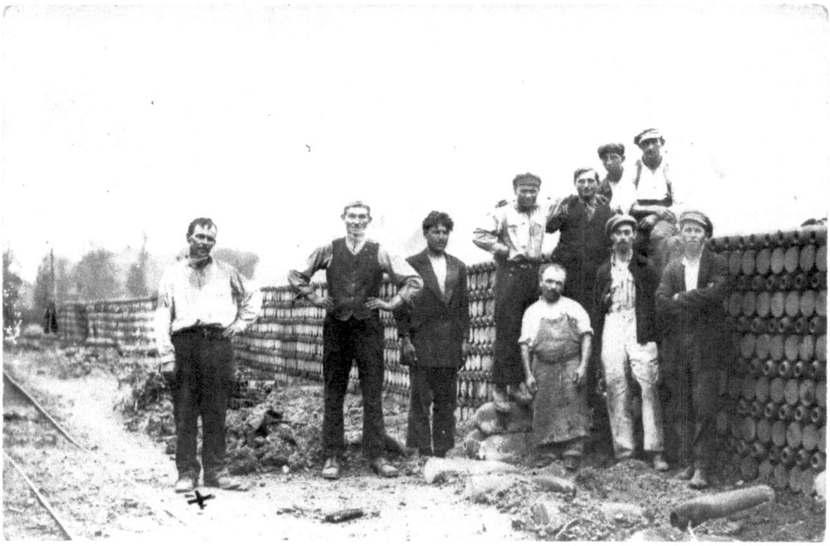

FIGURE 2 The cleaning of shell burning trench at Pickett's C-Factory in Trélon (Nord), c. 1922. The staff is standing on empty burned shells and mines (private collection).

suppressed or denied by the local population who just wanted to return to the normality of peacetime. Furthermore, numerous workers in these destruction sites were from the area, but there were also temporary workers hired from the east of Europe, from Spain, Portugal, and North Africa (Figure 2).

Lastly, there is a tendency for the residents who live on or close to former battlefields to trivialize contamination related to old ammunition and to forget any associated risks. For example, historian and journalist Olivier Saint-Hilaire and the author interviewed an inhabitant, Mr Y., 93 years old, of the village of Domléger in the Somme. He remembered only the most spectacular events, such as the accidental detonations or burns which caused injuries and the violent death of some workers, and not the environmental impact (Saint-Hilaire & Hubé unpublished). Similarly, an inhabitant of the village of Vaudoncourt, near Spincourt, reported that when he was young, he played in "deep craters with all colors" and had collected green-glass phials in the field "we used as inkpot at school". He added, after explaining where the phials came from, "I better understand why we suffer so often from headaches at school". When visiting three burning grounds of the former "C-Factory" at Trélon Liessies, the author met local historians who reported that in local memory the bare soils of the clearing were due to the fire intentionally caused by the Germans prior to their retreat in 1918. The three clearings were well-known to the inhabitants but what had caused the bare ground was less clear.

World War I museums, memorials, cemeteries, and historic sites abound across northern France. In contrast, there is no monument in France to remember the sacrifice of those who had been injured or have left their life to disarm, rendering peaceful the soils of former battlefields. Similarly, there is little public interpretation of the war's environmental impact or its ongoing contamination of soil and water. The greatest industrial war gave birth to an unseen industrial post-war which vanished from collective memory. Its chemical traces in the contemporary landscape exceed French and international toxicological standards. There are no natural processes that will attenuate these persistent risks; only remediation which, to date, has not yet started. Long after the end of the military conflict, pollution at Place à Gaz continues the slow violence of the war that produced it.

Bibliography

Bausinger, T., and J. Preuß. 2005. Environmental remnants of the first World War: soil contamination of a burning ground for arsenical ammunition. *Bulletin of Environmental Contamination and Toxicology* 74: 1045–1052.

Bausinger, T., E. Bonnaire, and J. Preuß. 2008. Exposure assessment of a burning ground for chemical ammunition on the Great War battlefields of Verdun. *Science of the Total Environment* 382: 259–271.

Clout, H. D. 1996. *After the Ruins. Restoring the Countryside of Northern France After the Great War*. University of Eweter Press.

Forget, J. 1928. La reconstitution forestière de la zone rouge dans la Meuse. *Bulletin de la Société des Lettres, Sciences et Arts de Bar-le-Duc et du Musée de Géographie*, 121–131.

Hubé, D. 2016. *Sur les traces d'un secret enfouis. Enquête sur l'héritage toxique de la Grande Guerre*. Michalon éditions. ISBN: 978-2-84186-829-2

Hubé, D. 2017. Industrial-scale destruction of old chemical ammunition near Verdun: a forgotten chapter of the Great War, *First World War Studies*, 8(2-3): 205–234, 10.1080/19475020.2017.1393347

Le Figaro. 2007/06/21 / Zoom Europa, Arte, 2007/11/07. *La destruction d'armes chimiques de la guerre de 14 a laissé'des traces*.

L'Humanité (Paris). 1927/10/09. Bibliothèque nationale de France.

Linnenkohl, H. 1996. *Vom Einzelschuss zur Feuerwalze*. Bonn: Bernard.

14

TOXIC CITY: INDUSTRIAL RESIDUES, THE BODY AND COMMUNITY ACTIVISM AS HERITAGE PRACTICE IN GLASGOW

Arthur McIvor

Glasgow was a city characterised by high levels of ill-health and premature mortality in working-class neighbourhoods of the city, with significantly higher rates of toxic contamination and related cancers amongst those who lived in communities adjacent to industrial works, chemical factories and mines. These health inequalities have become known in public health circles as "the Glasgow Effect". This undervaluing of working-class lives has been described aptly by Karen Bell as "environmental classism" (Bell 2020, 1–26). This chapter aims to contribute to the ongoing conversation around toxic heritage with an exploration of community activism and the legacies of industrialization in the Scottish city that hosted the COP26 conference in 2021. It focuses on campaigns around environmental justice associated with the chemical industry. There is engagement with some of the work of deindustrialization studies, environmental history and heritage scholars, and consideration of the assertion by Scottish environmental activist Kevin Dunion that "careless passivity and grumbling resignation" characterised Scotland and "the north" [of England] (Dunion 2003, 230). A key focus here, however, will be to explore community activism as heritage practice, evaluating how activism engages the toxic heritage of industry; how the industrial past has infused the present. We see this in community activism in creative modes of resistance – through advocacy and campaigning, but also in the deployment of memories and storytelling. I also want to reflect here on Laurajane Smith's idea of the "authorised heritage discourse" (AHD) in the context of Glasgow, where we see the virtual erasure of the toxic environment subject from the museums, heritage centres and memorialisation in the city (Smith 2006). The essay draws upon a range of sources, including some oral testimonies, which constitute a vital but contested strand of the intangible heritage of the city, frequently challenging, in essence, the AHD.

DOI: 10.4324/9781003365259-26

Glasgow: Industrial Legacies, Toxicity and the Limits of Regulation

Scotland's industrialisation in the nineteenth century was centred on the port city of Glasgow and industrial conurbation of wider Clydeside, and the scattered largely isolated coal mining communities in its hinterland of Lanarkshire and Ayrshire. Industries in this area included iron and steel making, shipbuilding, heavy engineering, textiles, clothing, plastics, electronics and a wide range of chemical manufacturing. Employment in transport and construction was significant too as the Victorian city bulged and expanded, attracting migrant labour from far afield. In 1821 the city population stood at 147,000; a century or so later population peaked at over 1 million, dropping back to 636,000 today (2020). Economic insecurity, grim working and slum living conditions germinated a powerful labour movement by the early twentieth century, with the city developing a reputation (much like Vienna) in the first half of the twentieth century for powerful and belligerent trade unions and a vibrant and eclectic socialist politics. "Red Clydeside" has been the topic of much debate and controversy within the contested historiography of the city (Figure 14.1).

Working-class tenements were closely clustered around the industrial workplaces (largely concentrated in the north, south and east of the city) and it was these working-class communities that bore the disproportionate brunt of toxicant

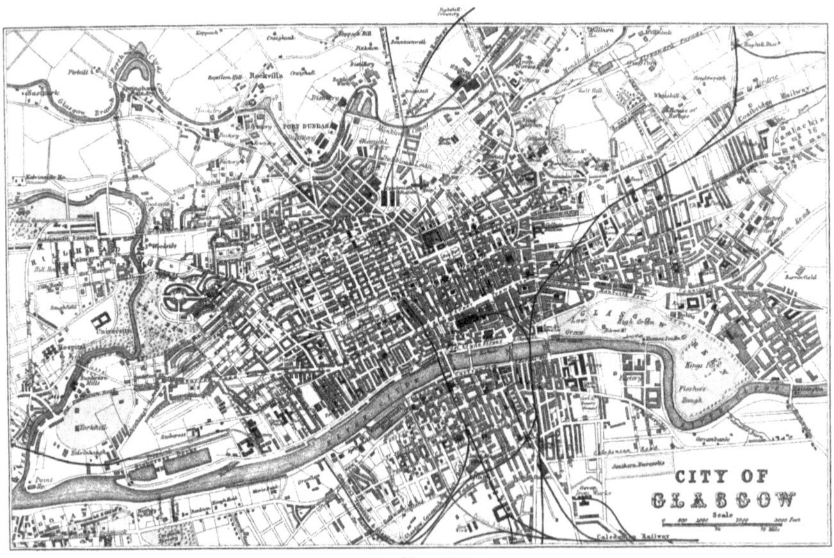

FIGURE 14.1 Map of Glasgow, 1878, depicting the industrial areas, docks and packed working-class tenements to the north, east and south of the city centre, and the middle-class neighbourhoods in the west). Courtesy of Mikhail Ryazanov (derivative work), via Wikimedia Commons. Accessed: https://commons.wikimedia.org/wiki/File:Glasgow_map_1878.png.

contamination, smoke, smog, noise pollution and the odour of industry. Its fumes, particulate, organic and chemical by-products were smelt, ingested and inhaled into the bodies of citizens. Brought up in a company house in the 1940s and 1950s in the environs of the massive Govan Ironworks in Glasgow (where her father worked) Marion Neil recalled something of this milieu:

> I can always remember everything was … even at home in the house and the garden, everything used to get a tremendous amount of soot and, you know, it was very, very dirty outside … Lots of smoke, yes. That caused lots of problems as well, you know … it came over towards the house of course and even with my mum hanging washing out there was a problem, you know. There was difficulty keeping things clean, you know.[1]

Marion added: "Oh there was lots of noise, yes, it was very noisy. You got used to that of course".

Industrial urban smells could also be pungent and persisting, especially around the neighbourhood of cattle and fish markets, abattoirs, bone meal factories, breweries and chemical works. Amongst the worse was the infamous "stinky ocean", a festering pond of sulphurous liquid chemical waste up to 80 feet deep dumped over 150+ years since the 1790s from Tennants St Rollox Chemical Works (which covered a 100 acres site), surrounded by solid chemical waste and "Jack's Mountain" – a massive chemical waste bing (Jack 2021). This blighted the Sighthill neighbourhood in north Glasgow with its rotten eggs (hydrogen sulphide) stink until the 1960s.[2] People living in the district (mostly consisting of working-class tenements) became inured to this assault on their senses and kids played across the polluted area. One recalled: "The stinky ocean smell was just about everywhere around there. Wit a braw smell … could be bloody hooching after a bit of wet weather and then the sun came oot, It really whiffed up then'.[3] And another ex-resident commented: 'Aye I remember playin' on Jacks mountain and the stinky ocean – used to get a tankin' from my Maw for coming home covered in grey 'clabber' from it!" Another resident described it as "real third world conditions right in the heart of Glasgow … we played there regularly along with most of the children from the Garngad, Townhead and Springburn areas".[4] Rancid odours also emanated from the canals (such as the Monkland Canal) and the rivers (such as the Kelvin) that fed into the Clyde, due to the chemical discharge from industrial works (Meighan 2008; *Glasgow Keelie*, May 1991). Paper mills, glassworks and brickworks were amongst the worst offenders.

Industrial "residue" lurked in people's bodies, especially their respiratory system (Sweeney 2021). Like other industrial cities, for much of the twentieth-century Glasgow experienced massive smoke pollution and suffered from recurrent smogs, where the dust, grit, soot and sulphur-dioxide-laden smoke from industrial chimneys and domestic fires were drawn downwards by atmospheric conditions. On average in the 1950s over 200 tons of "solid deposit" fell on every square mile of the city every year (Report of the Medical Officer of Health, hereafter RMOH 1958, 288–9). David Walker grew up in a heavily industrialised and deprived area of Glasgow, and recalled: "my own memory is of having bronchitis every year of my youth in Linthouse, Govan until we moved to a smokeless zone area in Mosspark in 1967. Linthouse has loads of tenements and everyone had a coal fire".[5] In the two weeks following the worst smog days in 1958 bronchitis deaths rose to three times the level of the same period in 1957 (RMOH 1958, 11–12). Thomas Ashford, the Senior Smoke Inspector for the city, reported a "growing intolerance of the public to atmospheric pollution" and "many complaints" (RMOH 1958, 287). These included belching emissions from railway shunting yards and from ships using the Clyde and berthed at the docks and wharves. The recurring smogs revealed the ineffectiveness of the Clean Air Act 1956, which introduced controls on emissions and smoke-free zones (Marsh 1963, 387). As with occupational health and safety, there remained a considerable gap between regulation and actual practice, with polluters continuing to breach the legal code with impunity.

Glasgow has the unenviable reputation of being one of Europe's most unhealthy cities, with high levels of multiple deprivations. And issues with poor air quality persisted. In a recent report by the World Health Organisation in 2017 Glasgow was identified as having the highest levels of air pollution of 51 UK towns and cities (including London).[6] Two recent studies have also highlighted long-standing legacies of industry and associated environmental concerns. Building on research by Balls in 1995–7, a recent study (2020) has identified the continuing high levels of historic pollutants in river sediment studies: "The Clyde Estuary, Scotland, is an optimal example site to determine environmental 'stress'", the researchers argue, "historically being one of the most contaminated estuarine environments in the UK" (Rodgers et al. 2020, 1058). And U.S. environmental geographer Juliana Maantay pointed to the "after life", or residues from industrialisation, identifying 927 separate "brownfield" vacant and derelict land (VDL) sites across Glasgow in 2011, mostly former factory sites (including the Govan Ironworks site), many of which were polluted from decades of industrial contamination (Maantay 2013). The vulnerability of communities in the present to this toxic legacy of the past is evident: 60% of the population of Glasgow were living within 500 metres of a VDL site, the highest proportion of people living in close proximity to such potentially polluted land of any local authority in Scotland. How this played out in one such neighbourhood – the Shawfield district of Rutherglen in the southeast of the city – is explored in the next section.

"Chemical Chernobyl": Environmental Injustice, Activism and the Glasgow Chemical Industry

Glasgow was the centre of a thriving heavy industrial chemical industry, including iconic plants such as the Tennants plant in Rollox, north Glasgow and J & J White's works in Rutherglen (established 1820). These chemical plants produced a wide range of products servicing industrialisation and consumer needs, such as bleach and soda powder (used extensively in the textiles industry) and chromates (used for ceramics, paints, plastics, corrosion prevention, chrome plating, pharmaceuticals and explosives). Work conditions could be terrible though. The socialist activist Keir Hardie commented scathingly on a range of occupational health and safety issues at Whites plant in the 1890s, including skin lesions from the chrome ('chrome holes'), liver and digestive disorders and respiratory problems from inhaling toxic fumes (including complete erosion of the nasal septum), describing the workers as 'white slaves' (Walker 2005; Walker 2007; Hunter 1959, 135). Ironically one of the company owners responsible for such historic bodily damage was James White II who was awarded a Lordship (becoming Lord Overtoun) and has a statue erected (in Cathedral Square, High St) in his honour for civic and philanthropic activities (Figure 14.2). The company continued to expand and by the 1930s was employing 900 workers, responsible for 70% of chromate production in the UK. It ceased production in Rutherglen in 1967 (Figure 14.3).

There are few remaining traces of what it was like to work in the White's chemical works. However, a rare interview conducted by David Walker for the Scottish Oral History Centre with ex-worker Richard Fitzpatrick provides some sense of prevailing work conditions and the toxic exposures workers faced.[7] Richard described how he worked as a labourer with his father and brother in the plant in the 1940s and 1950s. All of them lost their nasal septum – earning them the local nickname of "White's whistlers". Richard also suffered from skin lesions, experiencing chrome and soda corrosion holes on his arms and feet. He evoked the hard, hot, sweaty physical graft and the assault on the senses working in the antiquated plant, reflecting on the dust-laden atmosphere, the inadequate personal protection equipment ("muzzles") and the inefficient exhaust ventilation:

> There were always dust flying about ye know 'specially there [in the crystal department] again we're back tae the chromic acid ye know … there was about eight or nine furnaces but it was always dusty …, ye couldnae avoid that.

His awareness of the danger points and common-sense fatalism in regard to accidents and ill-health seeps through the testimony. When asked if he could have done any more to protect himself he retorted (Figures 14.4 and 14.5):

> I don't think I could have done anything bar put about ten muzzles [face covering] on [laughter], you know what a mean. If you were working with the stuff you covered yerself up but when you were walking about ye couldnae walk about wi a muzzle on all the time … . Just one of these things.

Richard Fitzpatrick's remarkable testimony from memory provides one story or discourse practice around toxic legacies. He shows through this embodied narrative

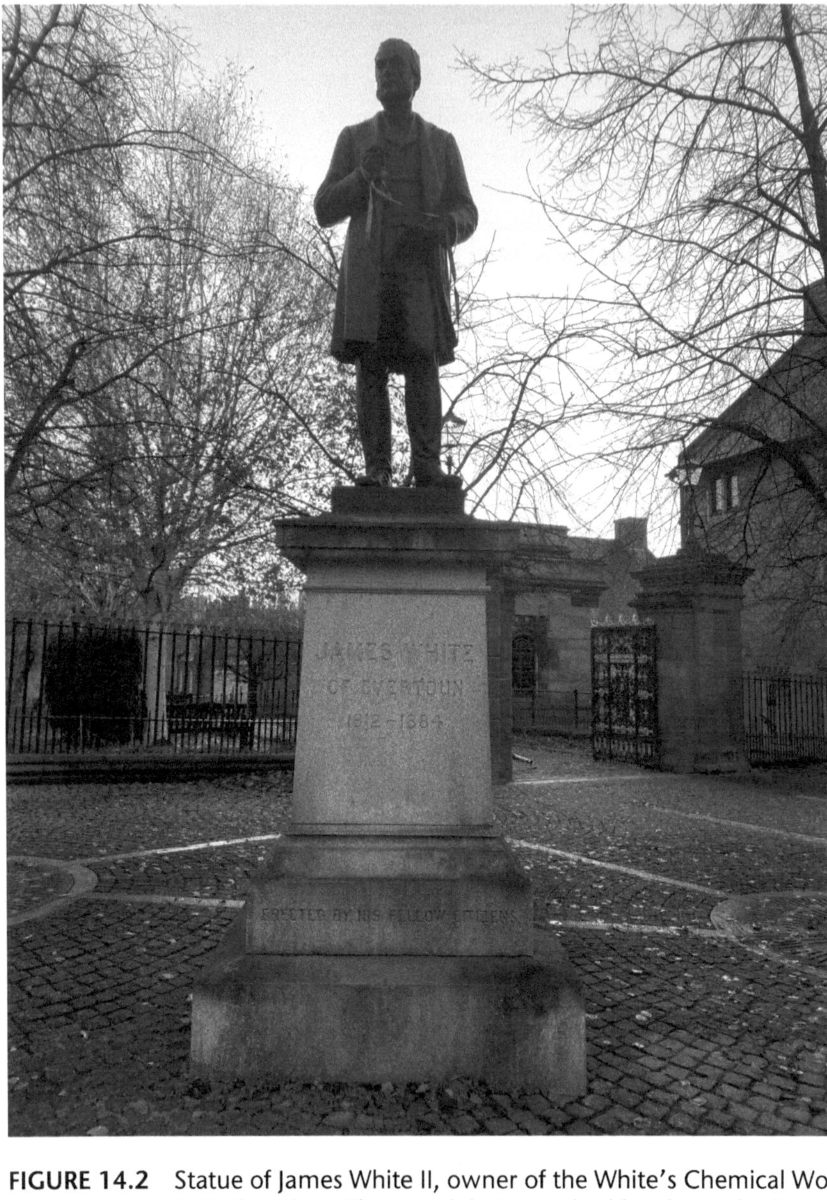

FIGURE 14.2 Statue of James White II, owner of the White's Chemical Works in Rutherglen, Glasgow (photographed by the author).

and physical scars how the residue from the plant persisted within former workers' bodies long after closure, evident in tissue destruction and breathing impairment.

The White's Rutherglen chemical plant site covered 30 acres (12 hectares), with 30–40% of this devoted to waste disposal into pits dug on site. Once capacity was reached on site, excess waste was disposed of around various other locations in East

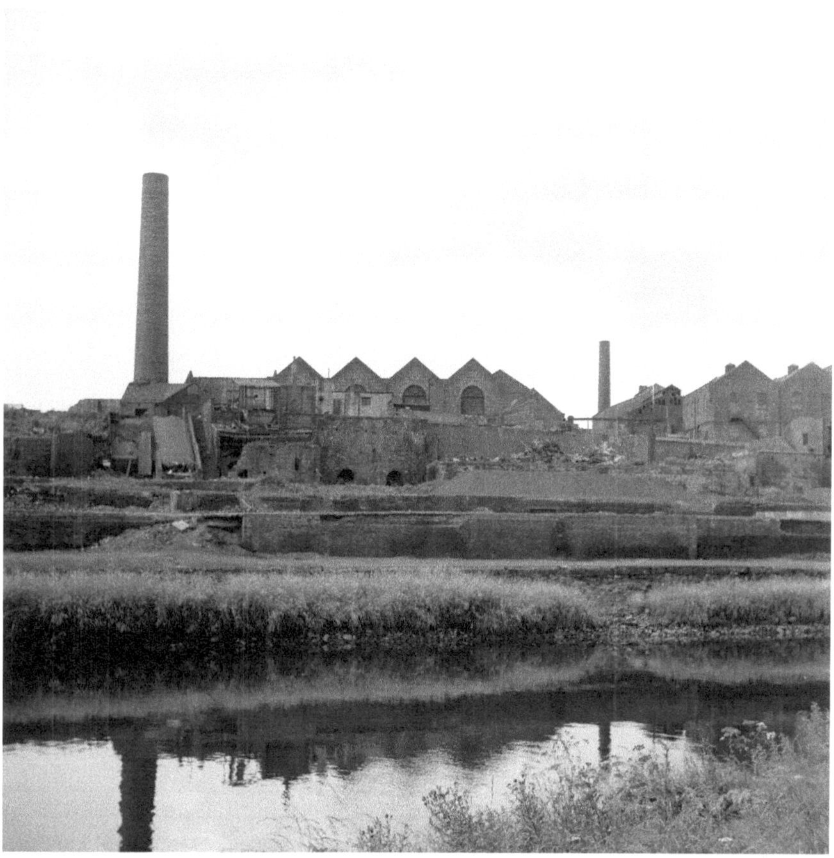

FIGURE 14.3 White's Chemical Works, Shawfield, Glasgow, 1967, showing an already derelict section of the works, with the river Clyde in the foreground. (From CANMORE and courtesy of Historic Environment Scotland and J.R. Hume, photographer).

Glasgow in and around Rutherglen, including old pits, mines and quarries (a legal practice at the time). Some 2.5 m tonnes of such waste from the plant was dumped outside the boundary of the chemicals works itself.[8] Awareness of toxicant contamination from these sites appears to have first surfaced in 1990 when a land survey for building development on nearby playing fields discovered massively high levels of chromium ore processing residue. Chromium (Cr) was found to be leaching into the water table and flowing into local streams (called 'burns' in Scotland) and through to the river Clyde, contributing to the dire pollution of that waterway and the destruction of animal, fish and other organic life (*The Herald*, 21 March 2019). Recently, sediment studies of the river Clyde show chromium to be one of the most prevalent of the base metals detected, with continuing unknown impacts on organic and microbiological life (Rodgers et al. 2020). The carcinogenic hexavalent

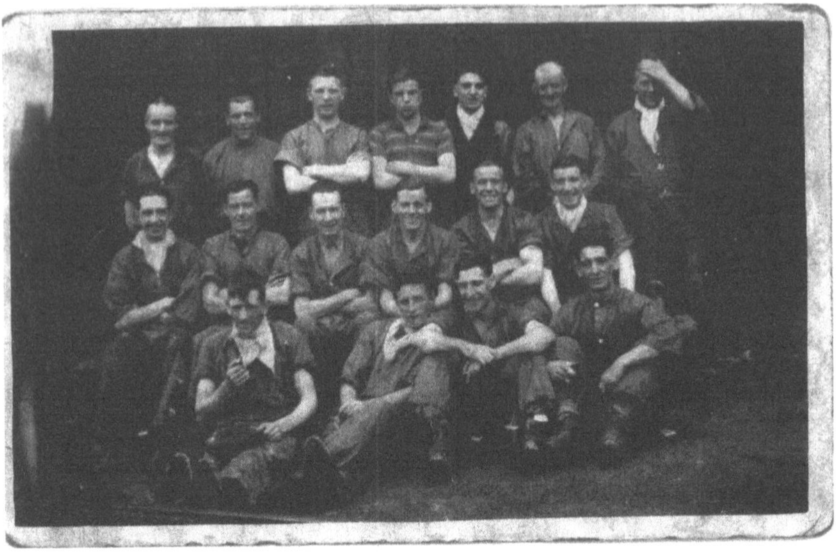

FIGURE 14.4 White's Chemical workers, c1940s. Richard Fitzpatrick is fifth from the left, middle row. Courtesy of Richard Fitzpatrick and David Walker (interviewer).

chromium (Cr VI) contamination is the same as that featured in the movie *Erin Brockovich* (2000), based on a true story about local residents' cancer prevalence in Hinckley, California. Locals in Rutherglen recall the streams – such as the Polmadie Burn – running bright green and "luminous yellow" with chrome.

The Polmadie Burn and known chromium waste dump sites (including the Glencairn football ground, Toryglen playing fields and the Dukes Road playing fields) were fenced off with "Danger: Keep Out" and "Keep Out Contaminated Land" signs erected. Rutherglen resident Tommy McAvoy, who had a brother and grandfather who worked in White's, recalled in 1995:

> I was born and brought up just a couple of hundred yards from that factory and used to play in the streams and the burns adjacent to it, despite the fact that the burn ran all sorts of colours as a result of the chemicals dumped in it. One did not realise the dangers at the time; it was only in later years that one realised the environmental mess that the place was in.
>
> (Hansard 1995)

McAvoy (born 1943) is recalling a childhood in the 1940s and 1950s when the plant was operational. His comments in 1995 were made when McAvoy was the Labour Member of Parliament (MP) for Rutherglen (1987–2010) and the leading senior politician campaigning on the health risks and for remediation of the site at the time. Local resident Alison Tait also recalled playing as a youngster amidst the

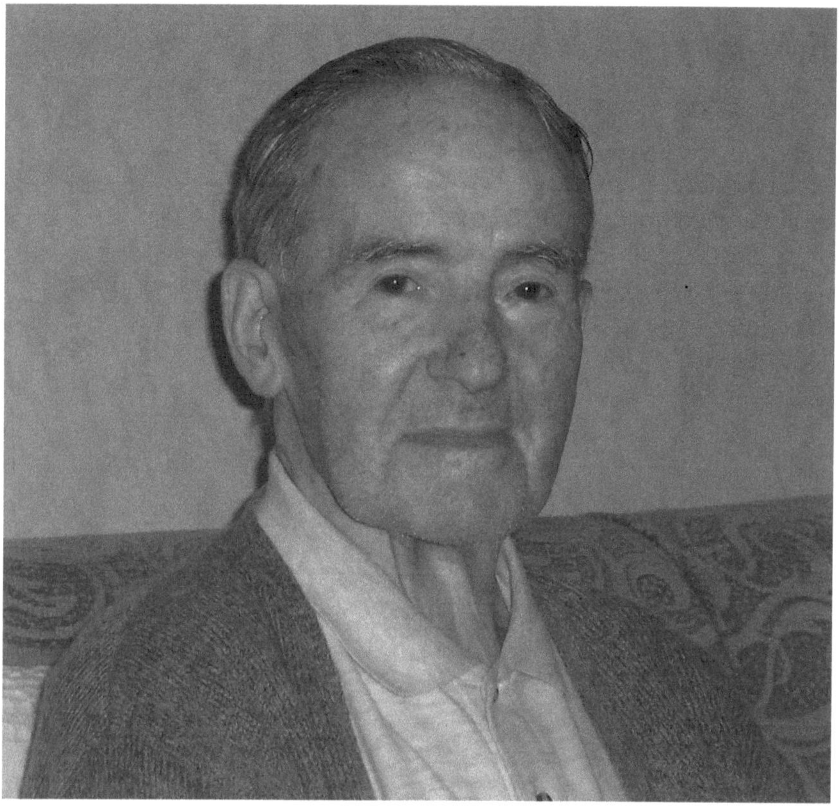

FIGURE 14.5 Richard Fitzpatrick at the time of his interview with David Walker 2004. Courtesy of Richard Fitzpatrick and David Walker (interviewer).

polluted area. She told how her father who worked in the White's plant died of cancer aged 51 in 1966 (which she suspects was work-related) and how local people were acutely aware of the issues:

> When I was older I heard my aunts talking of the number of male neighbours from Farme Cross, who had worked at White's and died prematurely. These ordinary women pointed the finger at White's even then but there was no evidence.
>
> (*The Herald*, 6 March 2019)

Rose Mary Brown was one of those. She recalled:

> My father Harry Bradley worked in Whites until it closed in 1967. He died in 1973 at the age of 59. When he went for radiotherapy at Belvedere [hospital] it was like an old or really more truthfully, a young boys reunion of Whites workers.[9]

Two 'official' epidemiological enquiries into high leukaemia rates in the area, however, found no proven correlation with the CrVI contamination, or heightened risk. Nonetheless, as McAvoy commented:

> Greater Glasgow Health Board concedes that these sites are health risks if the chromium waste is disturbed; the dust can be breathed in, which can cause health problems. Problems can be caused if the chromium is touched. If young children play on the sites and then put their hands in their mouth, swallowing the chromium can also cause them health risks.

"Families are quite right to be concerned" said MP McAvoy (in 1995), "and to pressurise and harass me to try to get something done". Families like the Fitzpatricks, Browns, Taits and the McAvoys were affected cumulatively by the lifelong risks of both working in the chemical industry and by the environmental exposure that emanated from it.

Grassroots community activism coalesced in a local action group being formed: Cambuslang, Carmyle and Rutherglen Against Pollution (CCRAP). It organised a local meeting at the Rutherglen Burgh Hall in April 1992, dubbed "Chrome Day" where the full details of the pollution in this "chemical Chernobyl" (as one witness described it) were revealed to the public (*Glasgow Keelie*, July 2022). Scottish environmentalist Kevin Dunion recounted that local activists located ex-company lorry drivers who gave information about where they dumped the toxic waste. The community was not invited to engage with the prior and ongoing scientific studies, so activists produced their own report, *How Safe is Safe?* Such "citizen science", tracing "exposure pathways" and "toxic trespass" into bodies, and challenging in the process more "orthodox" science and knowledge, replicated what was happening within community environmental activism elsewhere (Brown, de la Rossa and Cordner, 2020). "By the end of our research" a CCRAP activist later noted, "we listed 65 sites throughout the areas from Cambuslang to Dumbarton".[10] As the sites were investigated for contamination it was found one site (Duke's Road Playing Fields) had over 40 times the threshold (normal) chromium VI level and another site (the banking at Rutherglen Glencairn Football Club) had over 150 times the threshold level. Dunion argued: "without the efforts of the local activists and their links with the community, the extent of contamination was unlikely ever to have been discovered" (Dunion, 113). The radical paper *Glasgow Keelie* (1990–93) reported the activities of CCRAP and noted in May 1992 that there were "clusters of cancers, leukaemias, abortions and respiratory complaints in the area". A trade union official Pat Graham argued: "the pollution in the Cambuslang district is so intense and widespread, it should be classified as a disaster area and evacuation, not development, should be considered as the only safe alternative" (*Glasgow Keelie*, May 1992). Celtic Football Club abandoned plans to relocate to the area partly as a consequence of the furore resulting from CCRAP's campaigning (after having failed to extract an apology from CCRAP).

CrVI is both poisonous and carcinogenic. The 2000 study by Glasgow Corporation Health Board based on *self-reported health* found 25% of those residents in the affected area *believed* chromium levels to be harmful to their well-being and health. These lay beliefs embedded within a large minority of the population speak to distrust of "official" medico-scientific knowledge and the significance of what the researchers called "perception and possible anxiety". Dunion argued:

> The health of the people may still have been affected by the exposure, not in the manner in which occupational exposure afflicts workers but in the commonly experienced diminution of well-being and increase in stress … The people of Rutherglen are the unknown complainants of environmental injustice.
>
> (Dunion, 117)

Who was to be responsible for the costs of remediation was a primary sticking point in the 1990s. J & J Whites and successor chemical companies were all in liquidation and chasing the company insurers also proved fruitless. Legally there was no recourse. This is a recurring theme as polluters rarely pay fully for their actions.[11] MP Tommy McAvoy then lobbied the Scottish Office for funding (effectively from taxpayers pockets). This was refused, ostensibly because the site did not meet what was then the criteria for funding environmental remediation – which was that the site should have a demonstrable economic end-use. Environmental and health criteria were not accepted, illustrating blatant injustice, the power dynamics of the day and almost total disregard for the concerns of local residents, themselves living in an area of multiple deprivations, with heightened levels of vulnerability to chronic ill-health. McAvoy commented on the Scottish Office's intransigence in the face of what he called an 'environmental disaster' as 'like bashing my head against a brick wall' (Hansard 1995).

Just how much environmental/ecological/human damage was caused by two centuries of toxic pollution from White's chemical plant is unknown. After the issue was exposed in the early 1990s, it was still over a decade before serious work began on remediating the site, by a consortium called Clyde Gateway, funded by Glasgow City Council and South Lanarkshire Council. This was linked to the economic redevelopment of the site and the idea of a cleaned-up revitalised Clyde river being central to Glasgow's aspirations as a tourist attraction (see the chapter on Swansea in this volume for a discussion of similar issues). Health and environmental concerns appear to have again been less important than perceived economic benefits. Over £30 million was spent in the first phase of three planned remediation phases, with an estimated cost (in 2019) of over £54 million reckoned to be needed to complete the job of cleaning up the full site. In 2019, the present local MP Ged Killen was trying to exert pressure on the UK government to step in to fund the shortfall (*The Herald*, 6 March and 21 March 2019). At this point, the Scottish Environmental Protection Agency had stepped in to initiate monthly monitoring checks on the Polmadie Burn and the Oatlands Community Council had joined the campaign. The issue had also attracted the attention of the UK monitoring agency environment *analyst UK,* who produced at least four commentaries on the White's chromium contamination in 2019.[12] The fundamental problem, environmental health researcher and campaigner Andrew Waterson argued, was the control of Scotland by the UK government and their lack of will to address the issue – hence nationalism became entwined with the protection of citizen's bodies: "each year more people die from exposures to these dangerous substances … causing cancers, respiratory and other diseases than from road traffic fatalities, murders and suicides combined" (*The National*, 20 April 2019). Waterson pointed to much more robust regulatory regimes for historic toxicant control and toxic use in Ontario, Canada and Massachusetts, USA. In Scotland, Brexit and COVID-19 have subsequently intervened, so the problems of exposure to dangerous chemicals, including chromium VI contamination, in Scotland's most toxic city endure. The Polmadie Burn (where not culverted) continues to be observed running "luminous yellow", causing much anxiety for local residents (*Glasgow Times*, 26 April 2021).

Concluding Thoughts: Narrating, Curating and Memorialising a Toxic City

Community activism on the chemical industry's toxic heritage drew upon and found inspiration in a long tradition of irreverent radical protest in the city associated with "Red Clydeside". This was fuelled by a strong sense of community, equality and fairness linked to ideas around the moral economy, and leftist politics that found expression in strong support for socialism and the Labour Party in the city, and, as deindustrialisation deepened, support for a left-leaning populist Scottish National Party. The activist-led campaign against the toxic heritage of chemical pollution in the city mirrored other eruptions of community activism on environmental issues. Notably, a powerful grassroots anti-asbestos movement emerged in the 1980s – Clydeside Action on Asbestos (CAA) – to advocate for victims, and this morphed over time into an influential environmental justice movement. Working class environmentalism, epitomised in voluntary citizen-led groups like CAA and the chemical pollution campaigners (CCRAP) in Rutherglen, and widely supported politically and in civic society from the 1980s, was important in shaping a vibrant and influential environmental justice movement. Here resistance and community activism, fuelled by memories and storytelling, was a primary heritage practice.

Storytelling and "banter" has a long tradition in Scotland, and in Glasgow toxic heritage legacies are recollected vividly in memory practices. In these stories, dark humour often mediates the real risks to health being navigated every day. Toxic pasts are curated, archived and in turn recreated today, offering lessons and inspiration, feeding and energising community activism.[13] Such citizen stories are witnesses to environmental harm and a vital source enabling, as Hilary Orange put it, "reanimation" of the industrial past (Orange 2015). The city is also home to pertinent documentary archive collections, including the papers of the Occupational and

Environmental Diseases' Association (OEDA), set up by Nancy Tait, the wife of a mesothelioma victim, and other environmental justice activists, such as Alan Dalton and Harry Flynn (at the University of Strathclyde Archives and Special Collections). Nonetheless, what is remarkable is the extent to which toxic industrial legacies, from asbestos to chemicals, fail to feature significantly in Glasgow and West of Scotland's many museums and heritage centres – much like the history of "Red Clydeside" is sidelined (McIvor 2019). With some exceptions, instead of funding reinterpretations based on environmental history and the intangible heritage embedded in people's memories, the City leaders have focused attention on art, culture and the display of material objects in their museums, such as the Museum of Religious Life, the famous Kelvingrove Museum, the Burrell Collection and the Riverside (transport) Museum. Nor have the untold number of lives lost to the occupational and environmental legacies of industrialisation been publically memorialised in the city. The statues of industrialists who were also major polluters, however, remain standing (such as the Lord Overtoun statue in Cathedral Square). The nearest to such memorialisation lies 8 miles away in Clydebank, where the remarkable international asbestos memorial was erected in 2015 (Figure 14.6). This impressive memorial (by the artist Jephson Robb) to "all those known and unknown" who have died from asbestos-related disease was commissioned by the local victim's organisation, Clydebank Asbestos Group, and opened in 2015.

What is evident in the city then is a virtual erasure of the environmental story – of toxic industrial Glasgow. Reminders of pollution and toxic residues are not good

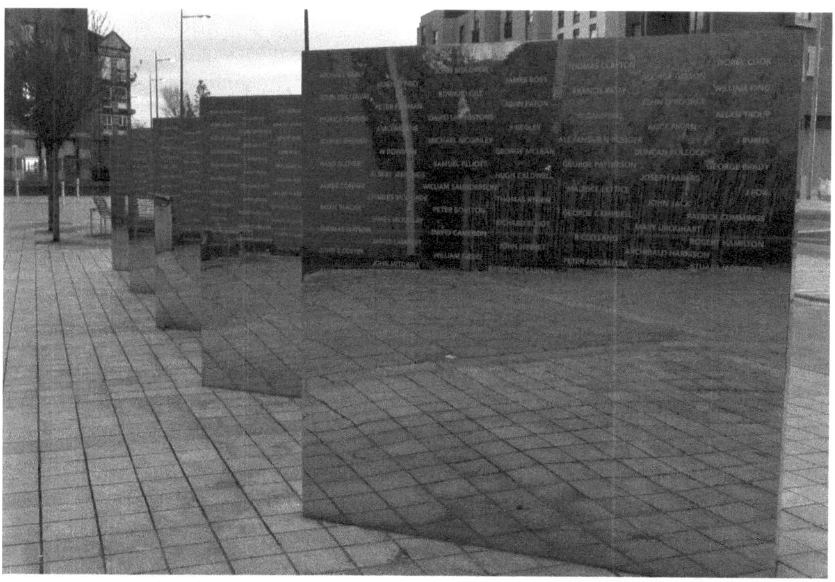

FIGURE 14.6 International asbestos memorial, Clydebank (photographed by the author).

for tourism; not resonant with the kind of image the city wishes to promote. What Laurajane Smith (2006) has referred to as the "authorised heritage discourse" (AHD) dominates the heritage space and public memorialisation, with a focus on the contribution of Glasgow to science, technology and the industrialisation process – celebrating inventors and businessmen, ships, steam engines and locomotives – and, to a lesser extent, religious and cultural developments. The environmental disaster of Glasgow's industrialisation has largely been airbrushed out – though it is still evident in the black staining of some tenement blocks in neighbourhoods that missed the gentrification cleaning frenzy of the 1980s and 1990s. The ecomuseum movement of France has found no place as yet in Glasgow and the wider Clydeside urban conurbation. And the physical heritage of industry has also largely gone. Only rarely were industrial edifices saved from the bulldozer, preserved and re-purposed in the city (Fairfields shipyard offices and the iconic Templetons Carpet factory are two such exceptions).

Glasgow, then, was blighted by environmental damage and harm caused by industrialisation but the response was not "passive". In the face of "recalcitrant" polluters, erasure, invisibility and what Steven High has called "class cleansing" and "forced forgetting" there was considerable resistance (High 2021, 173). In relation to asbestos and toxic chemicals, environmental activism was significant. However, for a long time economic imperatives, jobs, wages, housing and occupational and public health issues superseded environmental ones in city and workplace politics and community activism. Only relatively recently has environmentalism moved to centre stage politically (from the 1980s). Whilst there was an identifiable transition from occupational health and safety to wider issues of environmental harm and injustice, this transition to advocacy for sustainable development appears to have come later to working-class communities and the labour movement in Glasgow (Sellars 1997). Class played a part here. Glasgow was an archetypal proletarian industrial city. As Bell has argued, working-class activists could find themselves alienated from a predominantly middle-class environmental movement (Bell 2020, 163–192). Environmental justice organisations, like Friends of the Earth Scotland and the Scottish Campaign for Nuclear Disarmament, had a predominantly middle-class membership.[14]

The industrial past continues to linger on in the present in polluted soil, rivers, dirty buildings and in the bodies and memories of Glasgow's people. This dark shadow is an object lesson in what Linkon argues is the persisting "half-life" of deindustrialization (Linkon 2013) and Dudley defines as the 'trauma' of deindustrialization where "our well-being does not matter to the systems of power upon which we depend" (Dudley 2021, 202). The intangible heritage of people's bodies and memories urgently requires to be harvested to reconstruct these interactions between environment, health and well-being and stories of power, injustice and community activism. And these aspects of the city's past, however unsavoury, deserve to be featured more in public history.

As Sweeney has persuasively contended in her cultural heritage work on asbestos, heritage practice needs to think more about "dispersed relics" in the landscape and the intangible: "to attend to the ways in which the industrial past has permeated the everyday present" (Sweeney 2021, 50). Much more research is needed to comprehend how people were affected by toxic industrial environments, what might be described as the texture of feeling – what they felt about this, how they responded, and the work of trade unions and community activists on the environment (for a great example see MacKinnon 2020). The emerging field of environmental labour studies will contribute to this, and the green shoots of such enterprise are evident in Scotland (see Räthzel, Stevis and Uzzell 2021). The gendering of agency and activism also deserves more attention. And we need to deepen the conversation and develop our understanding of the discourses around industrial legacies and residues, environmental harm, health and injustice and how narratives were shaped within the power dynamics of city politics and heritage institutions around the environment, deindustrialization and ruination in cities like Glasgow.

Acknowledgements

I am grateful to Dr David Walker for his wonderful interview with Richard Fitzpatrick in 2004, for additional leads, information and photographs (14.4 and 14.5), as well as to the editors for many helpful comments on this paper in the drafting stage.

Notes

1 Marion Neil interviewed by David Walker, 28 November 2008 (SOHC Archive). The Govan Ironworks closed in 1958.
2 See 'Contaminated land around Glasgow and Lanarkshire', posts from 2002-2009. http://www.hiddenglasgow.com/forums/viewtopic.php?f=25&t=6215&start=15
3 http://discuss.glasgowguide.co.uk/index.php?showtopic=18264. This post 3 March 2010.
4 http://discuss.glasgowguide.co.uk/Port-Dundas-and39stinky-Oceanand39-and39jackand39s-Mountain-t12403.html Residents posts, 20 November 2007 and 14 December 2020.
5 Personal email, David Walker to Arthur McIvor, 3 May 2022.
6 BBC News, 31 October 2017. Accessed at: https://www.bbc.co.uk/news/uk-scotland-glasgow-west-41816722
7 Richard Fitzpatrick interviewed by David Walker, Friday 13 August 2004 (Scottish Oral History Centre Archive, University of Strathclyde).
8 https://www.groundsure.com/resources/toxic-glasgow-burn-under-investigation-what-can-historical-maps-reveal/
9 Cited at https://rutherglenheritage.wixsite.com/website-46/whites-chemical-company.
10 Cited at https://steeltoun.wordpress.com/tag/m74/
11 Book co-editor Sarah May has made the point that the temporality of this, where toxins outlast legal entities responsible for them, is important - and interesting in relation to nuclear waste which is the only industry with responsibility for management beyond the lifespan of the company.
12 See for example https://environment-analyst.com/uk/74926/row-hots-up-over-rutherglen-chromium-pollution
13 Made accessible, for example, through the Scottish Oral History Centre Archive accessed through the University of Strathclyde Archives and Special Collections at https://www.strath.ac.uk/humanities/schoolofhumanities/history/scottishoralhistorycentre/
14 See the Friends of the Earth Scotland website. https://foe.scot/about-us/our-history/

Bibliography

Bell, Karen. 2020. *Working-Class Environmentalism*. London: Palgrave Macmillan.

Brown, Phil, Vanessa de la Rosa and Alissa Cordner. 2020. 'Toxic Trespass', in Tom Davies and Alice Mah (Eds.), *Toxic Truths: Environmental Justice and Citizen Science in a Post-Truth Age*. Manchester: Manchester University Press.

Dudley, Kathryn Marie. 2021. 'Precarity's Affects: The Trauma of Deindustrialization', in Michele Fazio, Christie Launius and Tim Strangleman (Eds.), *Routledge International Handbook of Working Class Studies*. London: Routledge.

Dunion, Kevin. 2003. *Troublemakers: The Struggle for Environmental Justice in Scotland*. Edinburgh: Edinburgh University Press.

The Glasgow Keelie, May 1991, May 1992, July 1992 accessed at: http://www.workerscity.org/keelie/

Glasgow Times, 26 April 2021. https://www.glasgowtimes.co.uk/news/19260608.sepa-called-investigate-toxic-glasgow-burn/

Hansard, 264, 24 October 1995. https://hansard.parliament.uk/Commons/1995-10-24/debates/f49a8a2a-b70c-48c7-8359-48f633b7b391/Environment(Scotland)?highlight=mcavoy#contribution-3e478409-18fc-49d8-b96b-66b4318c8aac

The Herald, 6 March 2019: https://www.heraldscotland.com/news/17440856.polmadie-burn-glasgow-sealed-off-dangerous-levels-cancer-causing-chromium/

The Herald, 21 March 2019: https://www.heraldscotland.com/news/17515581.keir-hardie-broke-story-clydes-green-abomination/.

High, Steven. 2021. 'Deindustrialization and its Consequences', in Michele Fazio, Christie Launius and Tim Strangleman (Eds.), *The Routledge International Handbook of Working Class Studies*.

Hunter, Donald. 1959. *Health in Industry*. Harmondsworth: Penguin.

Jack, Ian. 'The Stinky Ocean', Granta 154, February 2021 accessed at https://granta.com/the-stinky-ocean/

Linkon, Sherry Lee. 2013 'Narrating Past and Future: Deindustrialized Landscapes as Resources', *International Labor and Working Class History* 84: 38–54.

Maantay, Juliana A. 2013. 'The Collapse of Place: Derelict Land, Deprivation, and Health Inequality in Glasgow, Scotland', *Cities and the Environment* 6: 1. Available open access at: https://digitalcommons.lmu.edu/cate/vol6/iss1/10

Mackinnon, Lachlan. 2020. *Closing Sysco: Industrial Decline in Atlantic's Steel City*. Toronto: University of Toronto Press.

Marsh, Arnold. 1963. 'The December Smog. A First Survey', *Journal of the Air Pollution Control Association*, 13(8): 384–387, https://doi.org/10.1080/00022470.1963.10468197

McIvor, Arthur. 2019. 'Where is Red Clydeside? Industrial Heritage, Working-Class Culture and Memory in the Glasgow Region', in Stefan Berger (Ed.), *Constructing Industrial Pasts: Heritage, Historical Culture and Identity in Regions Undergoing Structural Transformation*. New York: Berghahn Press.

Meighan, Michael. 2008. *Glasgow Smells*. Stroud: The History Press.

The National, 20 April 2019. https://www.thenational.scot/news/17586631.academic-calls-fresh strategy-chemical-hazards-workplaces/

Orange, Hilary. (ed.) 2015. *Reanimating Industrial Spaces: Conducting Memory Work in Post-Industrial Societies*. Walnut Creek, CA: Left Coast Press.

Räthzel, Nora, Dimitris Stevis, David Uzzell (Eds.). 2021. *The Palgrave Handbook of Environmental Labour Studies*. Cham, Switzerland: Palgrave Macmillan.

Report of the Medical Officer of Health (RMOH), City of Glasgow. 1958. Accessed at: https://wellcomecollection.org/works/r4wqheu7/items?canvas=297

Rodgers, Kiri, Iain McLellan, Tatyana Peshkur, Roderick Williams, Rebecca Tonner, Charles W. Knapp, Fiona L. Henriquez, Andrew S. Hursthouse. 2020. 'The legacy of industrial pollution in estuarine sediments: spatial and temporal variability implications for ecosystem stress', *Environmental and Geochemical Health* 42: 1057–1068; Accessed at https://link.springer.com/content/pdf/10.1007/s10653-019-00316-4.pdf

Sellars, Chris. 1997. *Hazards of the Job: From Industrial Disease to Environmental Health Science.* Chapel Hill: University of North Carolina Press.

Smith, Laurajane. 2006. *Uses of Heritage.* London: Routledge.

Sweeney, Josephine. 2021. *Approaching Asbestos as Heritage.* Masters thesis in Cultural Heritage Studies, University College London.

Walker, David. 2005. 'Working in it, through it and among it all day: Chrome dust at J & J White of Rutherglen, 1893-1967', *Scottish Labour History* 40: 50–69.

Walker, David. 2007. *Occupational Health and Safety in the British Chemical Industry, 1914–1974.* PhD thesis, University of Strathclyde.

CASE STUDY 6

RUBBER AS (TOXIC) HERITAGE: AMAZONIAN KNOWLEDGE AND THE RUBBER INDUSTRY

Tiago Silva Alves Muniz

Introduction to natural rubber

Natural rubber latex (hereafter NRL) is harvested from latex produced by rubber trees. Latex is an important tool for the plants' defences against herbivores, and the laticifers occur in many plants across different botanical families. Latex can be colourless, yellow, orange, red, and, most commonly, white. Sometimes it is a milky product, such as the latex from the botanical families Euphorbiaceae and Moraceae. Chemically, natural rubber latex consists of a complex mixture of polyisoprene, lipids, phospholipids, and proteins. As rubber production developed, a variety of chemicals were added to vulcanize rubber which makes it more rigid and durable (Table 1). Many of these materials present significant health hazards. Besides vulcanizing, chemicals were added as a "biological protector," and ammonia was standardized as a "universal protector" at least since 1853 (Blackley 1997).

Rubber heritage in the Amazon

In the Brazilian Amazon, the only registered heritage related to rubber is the House of Chico Mendes (Xapuri, Acre) which was recognized due to its historical and social relevance as the home of the famous rubber tapper, environmentalist, and unionist. The stately home represents remembrance of a glorious and utopic past, while also connoting the sense of loss and the sense that those who continue to live and work in the area have been left behind. The toxic heritage of rubber is also represented at Fordlândia, the industrial property of the Ford Motor Company, which is currently being considered for designation. These properties are the visible remnants of the rubber industrial heritage, but they also encompass the often marginalized intangible heritage of Indigenous knowledge and the often invisible toxic consequences of working in the rubber industry.

DOI: 10.4324/9781003365259-27

TABLE 1 Chronology of vulcanizing methods' discoveries. Adapted from: Hills (1971)

Date	*Inventor*	*Vulcanization agents*
1839	Goodyear	Sulfur
1842	Hancock	Sulfur
1846	Parkes	Sulfur chloride
1847	Burke	Antimony pentasulphide
1913	Klopstock	Halogens of Se and Te
1915	Ostromislensky	Polynitrobenzenes
1915	Ostromislensky	Benzoyl peroxide
1918	Peachey	SO_2 H_2S
1918	Boggs	Selenium
1921	Buizov	Diazoaminobenzene and derivatives
1921	Romani	Disulfides of tetra-alkylthiurams
1925	Le Blanc and Kroger	Sulfur thiocyanates
1931	Fisher	Quinone halogens
1932	Edland	Tellurium
1933	Fisher	Phenols or amines + oxidizing agents
1934	Midgley, Henne, and Shepard	Organic metal compounds
1936	Fisher	Quinone amines
1939	Rubber-Stichting	Active phenol-formaldehyde resins
1940	Dufraisse and Compagnon	Vulcanization by active resin formation

It is yet uncertain which or how many methods were used to vulcanize rubber in the Brazilian Amazon during the rubber boom (1850–1920), but the most known vulcanization system consisted of heating and smoking rubber until it coagulated. At the time there was little concern about the harmful effects of the process and workers had little protective equipment. Instead, they were exposed to the dangers of working with hot materials and to breathing the polymers emitted during the smoking process (Figure 1).

The short and long-term health impacts are difficult to document, particularly since life expectancy in Brazil in 1900 was just 33 years (Pontes 2009). In contrast, Indigenous rubber tappers used different types of coagulants, such as the rubber tree's own flower, lemon juice, tucupi (manioc bitter juice), and resins. These ingredients are now considered innovative techniques (Teixeira 2009), but the local knowledge, which would have been less harmful to workers, was not adopted by the industrial producers of NRL. The intangible heritage of Indigenous knowledge of working and shaping materials in many ways with forest products continues to be applied, but rubber tappers who once created non-hegemonic methods to vulcanize rubber now benefit from machinery that shapes NRL into rubber sheets and they use their own recipes to colour them and produce rubber bio-jewelries (Fig. 2).

Rubber is only one example of exploitation and coloniality worldwide (Muniz 2020). In the Amazon, Indigenous peoples were "recolonized" to tap rubber and,

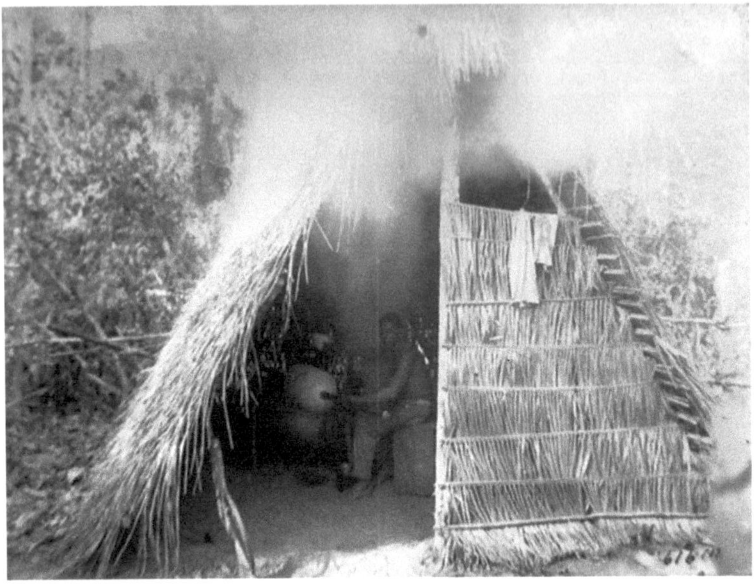

FIGURE 1 Latex smoking process and rubber balls solidification; highlights on unsanitary work with the aspiration of toxic gases. Photo rendered with MyHeritage app.

Source: Ule (1908–1909).

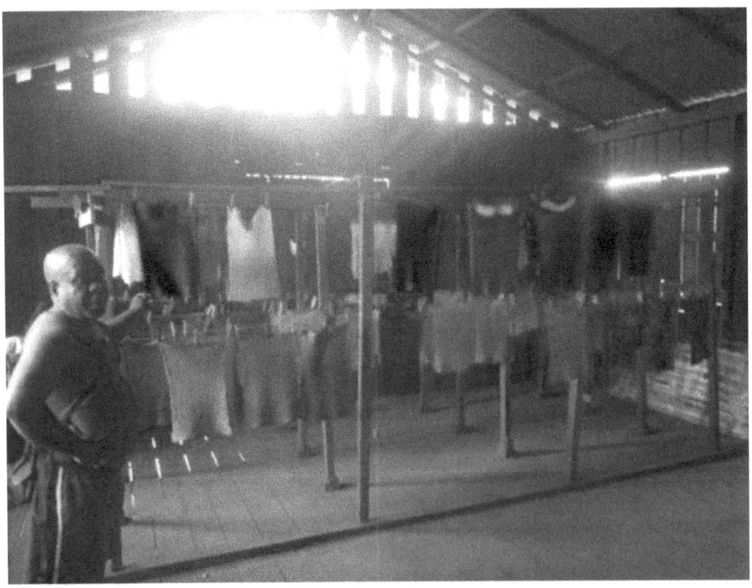

FIGURE 2 Chemist showing natural rubber sheets drying after being coloured. Photo: Tiago Muniz.

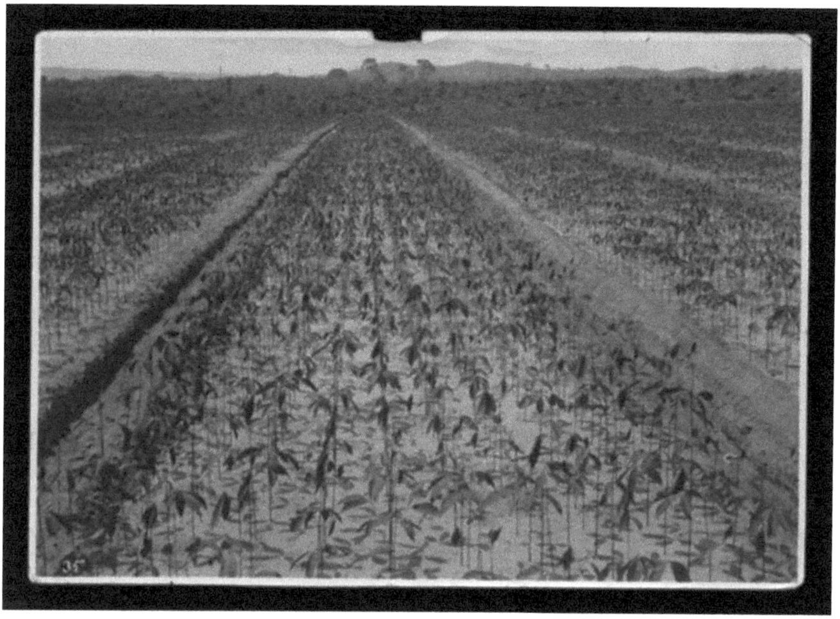

FIGURE 3 *Hevea brasiliensis* plantation at Fordlândia (Pará State, Brazil). Highlight: the small distance between the rubber seedlings plantation, which possibly made the growth of the trees unfeasible and contributed to the proliferation of the fungus (leaf-sickness).

Source: The Henry Ford Foundation.

while their labour was exploited, their knowledge system about tapping and growing rubber was not recognized. The consequences of this erasure can be seen in Fordlândia (Fig. 3).

When Henry Ford decided to establish rubber plantations in Brazil and build planned cities in his so-dreamed Amazonian Metropolis, plants were placed too close to each other spreading the growth of the leaf-blight fungus and in addition to tropical diseases., Ford's plans failed and the plantations were abandoned. Today Fordlândia is commonly described as a 'ghost town' (Reed 2016). Rarely, however, do the accounts mention the harms which haunt it – the fact that Amazonian rubber toxicity rests upon not only chemical residues in the environment but also on colonial extractivism and the legacy of exhausted soils and peoples.

References

Blackley, David Charles. 1997. *Polymer Latices Volume 2: Types of Latices*. 2nd ed. London: Chapman & Hall.

Hills, D. A. 1971. *Heat Transfer and Vulcanisation of Rubber*. Elsevier Publishing Company.

Muniz, Tiago Silva Alves. 2020. Towards an archaeology of rubber: por uma arqueologia da borracha. *Brasiliana: Journal for Brazilian Studies*, 9(2), 233–251.

Pontes, Ricardo J. S., Ramos Junior, A. N., Kerr, Ligia R. S., and Bosi, Maria L. M. 2009. *Transição Demográfica e Epidemiológica*. In: MEDRONHO, R.A. et al. Epidemiologia. 2.ed., 123-151. São Paulo: Atheneu.

Reed, Drew. 2016. Lost cities #10: Fordlandia – the failure of Henry Ford's utopian city in the Amazon. *The Guardian*. 19 August, https://www.theguardian.com/cities/2016/aug/19/lost-cities-10-fordlandia-failure-henry-ford-amazon

Teixeira, Carlos Correa. 2009 *Servidão humana na selva: o aviamento e o barracão nos seringais da Amazônia*. Manaus: Valer Editora.

CASE STUDY 7

THREE MEMORY FRAMEWORKS ON CHERNOBYL

Matteo Benussi

In the wake of the 1986's Chernobyl accident, different actors and constituencies in Ukraine articulated multiple memory frameworks about the disaster. Each being endowed with distinct manifestations as heritage sites, spatial as well as immaterial, these frameworks reveal how Chernobyl's toxic legacy plays out at different levels, from the transnational to the local. Two of these narratives – the "radioactive hazard" and "national martyrdom" stories – emerged as dominant, while the "lost paradise" story remained local and marginal.

The "radioactive hazard" framework centers upon Chernobyl's pathogenic effects. The most cosmopolitan of the three, this memory narrative has found enduring resonance not only in metropolitan Ukraine but also among anxious international audiences. Through an array of Chernobyl-themed documentaries, works of fiction, videogames, etc., several of which enjoy international circulation, cosmopolitan – and mostly urban – actors remember and construe the 1986 disaster as the first major man-made global public health crisis, the ominous implications of which forebode the possibility of humankind's technogenic extinction (Heise 2006; Kit 2012; Fuller 2016). Remembrance here includes a warning: a planetary *memento mori* (Beck 1987; Žižek 1991). Central to this narrative are the tropes of illness and its sensationalistic twin: mutation. The hazard-centric Chernobyl memory regime incorporates elements from science (the epidemiological measurement of the accident's consequences) as well as science fiction (Chernobyl as a source of monsters), activist journalism (factual reports about survivors and/or ecological damage) alongside mythopoetic licence (Hollywoodesque sci-fi appropriations of the disaster). In terms of concrete heritage sites, the "radioactive hazard" framework finds its material and visual embodiment in the iconic ghost town of Pripyat, now the epicentre of an international disaster-tourism industry, as well as the countless pop-culture representations of the Chernobyl Exclusion Zone, often depicted as a post-apocalyptic hellscape.

DOI: 10.4324/9781003365259-28

Operating at the nation-state level, what I call the "national martyrdom" narrative frames the 1986 disaster as the climax of the USSR's misrule of Ukraine. Under this framework, a multitude of national storytellers (civil society actors, independent media outlets, official representatives, etc.) have mobilized the public health crisis in support of a moral-political argument about historical responsibility, with the Soviet regime's ineptitude and Moscow's disregard for Ukrainians occupying a central position therein (Petryna 1995; Phillips 2004; Zhukova 2018). This narrative casts the toxic legacy of the radionuclides unleashed by the Chernobyl fire as a materialization of the toxic legacy of Soviet rule. Memorialization efforts highlight Ukrainian heroism – personified by the firefighters who "saved the world" – and collective victimhood. Within this framework, Chernobyl is one of the battle sites of the "memory wars" raging across post-Soviet Eastern Europe: for example, the Ukrainian Wikipedia entry defines the disaster as the "Chernobyl *catastrophe*" using emotionally and morally connoted language, while the equivalent Russian-language page resorts to the neutral terminology of "Chernobyl *accident*." Not by coincidence, Russia's occupation of the Chernobyl Exclusion Zone and the Zaporizhzhia atomic plant during the 2022 invasion rekindled powerful associations with the 1986 disaster, touching the "atomic nerve" of independent Ukraine's national consciousness. The national martyrdom framework is anchored to an array of *lieux de mémoire* that perpetuate national storytelling about the catastrophe. Memorials and statues dedicated to "Chernobyl heroes" and related themes can be found in Chernobyl, Slavutych (a purpose-built town designed to house Chernobyl workers after 1986), and Kyiv. Commemoration of Chernobyl as a quintessentially Ukrainian tragedy also takes place through eco-patriotic demonstrations, official commemorations, state media, and so forth (Dawson 1996; Nonjon 2020).

It is worth observing that these two narratives may interweave: the celebrated 2019 UK-US miniseries on Chernobyl – a powerful, if immaterial, transnational *lieu de memoire* – combines scientific exactitude, inclusive of graphic depictions of acute radiation syndrome symptoms, with a stereotyped depiction of Soviet society rife with Cold War tropes. An important physical heritage site such as the Chernobyl Museum in Kyiv also appears to draw on both memory frameworks, as it documents the ecological and epidemiological ramifications of the 1986 accident while also emphatically depicting the country's heroism and suffering.

The third framework, which is almost entirely absent from national and international conversations about the Chernobyl disaster, reflects the intimate experiences of the rural communities inhabiting the territories around the power plant, in the region of Ukrainian Polesie. After the explosion, the local population was hastily relocated from the most contaminated areas. As a result of this large-scale population transfer, necessary but marred by poor logistics and communications that compounded the trauma of uprooting, a vast chunk of Polesie was turned into a no-go zone studded with decaying ghost towns. Doing ethnographic research on representations and memories of Chernobyl at the ground zero, in Chernobyl-scarred Polesie (2008, 2012–2013, 2016), I soon realized that when many local survivors talk about the disaster, they tend to refer first and foremost to the evacuation (*pereselenie*) that forever changed the Polesian landscape. Chernobyl's "catastrophe of places" (*katastrofa mists'*) affected both evacuees from the Exclusion Zone and the communities located in the "Outer" Zone – a territory, inhabited to this date, which borders the depopulated area. Evacuees and Outer Zone residents share a common sense of place based on a sense of belonging to the same Polesian "native land" (*ridnyi krai*).

These demographics jointly partake in processes of memory-making, with Polesian cultural activists based in the diaspora or the Outer Zone playing a key role in memorialization practices. Local commemorative activities, overwhelmingly grassroots, often involve the self-production and circulation of poetry, folk music, and artworks bemoaning the disappearance of an emotional geography that Polesians often nostalgically described as their "lost paradise" (*vtrachenyi rai*). For example, many Polesian painters and photographers produce memory-infused landscapes combining realistic elements, childhood recollections, evocations of idyllic pre-catastrophe village life, and mournful intimations of loss. Some of the finest examples of regional poetry, carrying such titles as *Native River*, *The Land of Childhood*, or *My First Address*, bring back to life, frequently in painful detail, concrete sites – villages, courtyards, street corners – that have disappeared from the official map but remain very much present in the affective topography of Chernobyl survivors.

A panoply of local commemorative sites supported, sponsored, and animated by survivor communities spatially manifest the "lost paradise" memory narrative. The town of Krasiatychi, close to the Exclusion Zone, hosts a memorial complex with a cenotaph commemorating the dozens of abandoned settlements nearby (Figure 1):

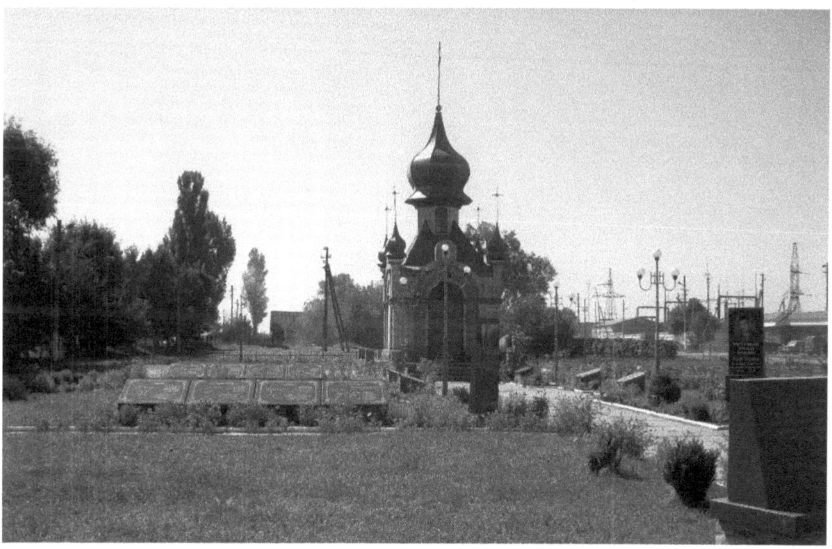

FIGURE 1 The cenotaph of lost villages near Krasiatychi. Photo by author.

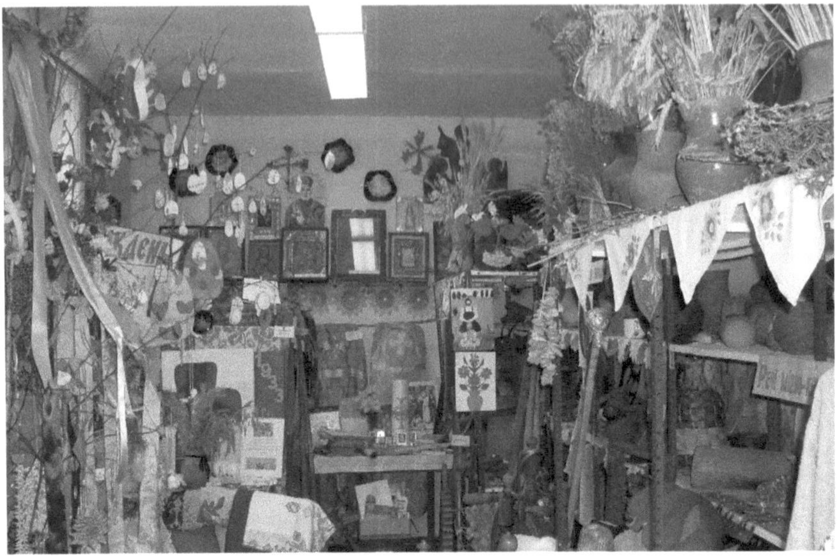

FIGURE 2 A local museum in Krasiatychi. Photo by author.

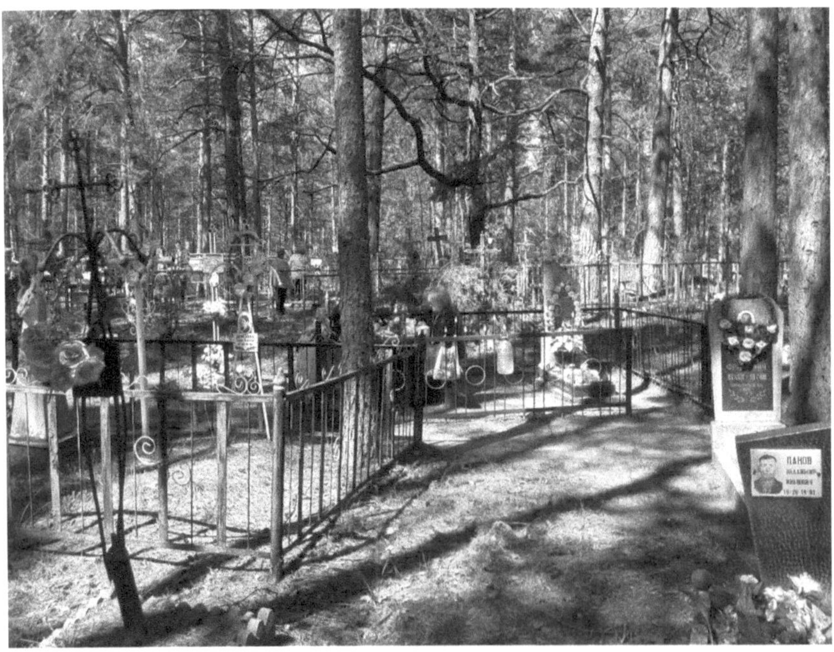

FIGURE 3 The cemetery near the ghost town of Vilcha. Photo by author.

the markers, similar to gravestones, that surround a little onion-roofed chapel are engraved with the ghost villages' toponyms and maps. Several small historical-ethnographic museums exist in Outer Zone settlements (Figure 2), and in 2004 the Polesian diaspora opened a larger "Museum of Memory" in the city of Pereiaslav, home to a large evacuee population. Perhaps the most striking grassroots *lieux de mémoire* associated with Chernobyl, however, are the impeccably well-kept cemeteries that stud the Exclusion Zone (Figure 3). These sites predate the disaster, as do the ritual practices carried out therein, but they took on particular relevance in the wake of the disaster. As the ruins of deserted settlements become engulfed in vegetation, thousands of ordinary Polesian survivors, year after year, keep returning to their ancestral burial grounds to carry out maintenance and perform commemorative rituals, thereby turning a landscape of loss into one of remembrance and resilience (Benussi 2022). While the Exclusion Zone cemeteries are absent from mainstream heritage regimes in Chernobyl, nowhere else is the Polesian population's commitment to their lost paradise embodied so powerfully.

References

Beck, Ulrich. 1987. "The Anthropological Shock: Chernobyl and the Contours of the Risk Society." *Berkeley Journal of Sociology* 32: 153–165.

Benussi, Matteo. 2022. "The Vital Afterlife of Chernobyl Cemeteries." *Allegra Lab.* https://allegralaboratory.net/the-vital-afterlife-of-chernobyl-cemeteries/

Dawson, Jane. 1996. *Eco-Nationalism: Anti-Nuclear Activism and National Identity in Russia, Lithuania and Ukraine.* Durham: Duke University Press.

Fuller, Daniel. 2016. "S.T.A.L.K.E.R. Shadow of Chernobyl and Memory of Utopia: How a Ukrainian Computer Game Transfigured Folkloric Processes of Remembering." *EEFB East European Film Bulletin* 26 https://eefb.org/retrospectives/how-a-ukrainian-computer-game-transfigured-folkloric-processes-of-remembering/

Kit, Zorianna. 2012. "Horror film 'Chernobyl Diaries' draws some protests." *Reuters* (25/05/2012) https://www.reuters.com/article/entertainment-us-chernobyldiaries-idUSBRE84O0Y420120525.

Nonjon, Adrien. 2020. "The new path of Ukrainian eco-nationalism." *Forum for Ukrainian Studies* (09/04/2020), https://ukrainian-studies.ca/2020/04/09/the-new-path-of-ukrainian-eco-nationalism/

Petryna, Adriana. 1995. "Sarcophagus: Chernobyl in Historical Light." *Cultural Anthropology* 10 (2): 196–220.

Phillips, Sarah. 2004. "Chernobyl's Sixth Sense: The Symbolism of an Ever-Present Awareness." *Anthropology & Humanism* 29 (2): 159–185. 10.1525/ahu.2004.29.2.159

Zhukova, Ekatherina. 2018. "Chernobyl, Responsibility and National Identity: Positioning Europe and Russia in the Media of Belarus and Ukraine (1992-2014)." *Europe and Asia Studies* 70 (7): 1055–1082. 10.1080/09668136.2018.1499876

Žižek, Slavoj. 1991. *Looking Awry: An Introduction to Jacques Lacan Through Popular Culture.* Cambridge, MA: MIT Press.

15

THE TOXIC ANTHRACITE = TOXIC HERITAGE

Paul A. Shackel

Introduction

Northeastern Pennsylvania, also referred to in the U.S. as the anthracite region, is a landscape of contradictions. The region is situated in the northern portion of the Allegheny mountains and is home to waterfalls and numerous lakes. It is also one of the most disturbed terrains in the world. Billions of tons of debris can be found in the form of coal waste piles, mine dumps, and abandoned strip mines scattered throughout a region of about 484 square miles. Two centuries of mining coal have exposed heavy metals that continue to pollute the waterways. Streams and rivers are dead. Historical documents and several generations of oral histories describe how these communities were exposed to poisonous air quality conditions and drank contaminated well water. Daily, coal workers endured the perils associated with underground mining, thinking they might not survive until the end of the day. The coal workers were often underpaid and always on the verge of starvation. Today, the anthracite region is a place of poverty, sickness, and addiction, and these devastating conditions are prime for a critique of unchecked capitalism. For some, the region is now a destination for dark heritage tourism. For others, the place is a destination to remember and commemorate the struggles of a familiar past (Shackel 2018).

The toxic heritage of northeastern Pennsylvania is tied chiefly to the anthracite mining industry. It is a history that focuses on men's work and making do during difficult times. This toxic heritage of destruction and devastation is quite visible in the landscape. At the same time, some forms of toxic heritage have their most significant impacts underground and are difficult to see. For instance, for over a century, coal companies used deep mining methods, which involved a system of shafts and slopes, some as deep as several hundred feet. These deep mines were filled with water and had to be pumped out continuously. Therefore, when mining came

DOI: 10.4324/9781003365259-29

to a halt, the deep mines filled with water. The water dissolves heavy metals like copper, lead, and mercury. With a lower-than-normal pH, the water eventually flows through these abandoned mines and empties into rivers and streams, a process known as acid mine drainage (AMD) (Marsh 1987). This acid water is also capable of dissolving the substructure of bridges.

While the mining industry is nearly extinct in the region, the communities continue to be impacted by the deteriorated landscape. While capital has moved on to exploit other regions and communities, northeastern Pennsylvania endures the perpetual effects of a created toxic landscape. Despite these detrimental impacts of the coal mining industry, communities search for a heroic past. In some cases, communities celebrate the power of industry. In other cases, they honor the individual struggle to overcome overwhelming odds to survive, and in some cases, to confront exploitation. Nostalgia plays a vital role in the memorialization of this toxic landscape.

Background

Anthracite coal was first discovered in northeastern Pennsylvania, U.S., in the 1760s, although it took several decades before technologies developed to market it for home heating and industrial purposes (Miller and Sharpless 1985, 15). By the 1820s, East Coast industries began replacing water power with steam power, creating a greater demand for anthracite coal. In addition, the iron industry developed rapidly with increased coal extraction, and iron became cheaper and more accessible to industries (Keil and Keil 2015, 7; MacGaffy 2013, 4). As a result, many regional historians credit the advent and acceleration of anthracite mining for the development of the U.S. industrial revolution (Tarone 2004).

The early nineteenth-century anthracite workforce came from Germany, England, Scotland, and Wales. By the 1840s and 1850s, many new workers came from Ireland, escaping famine and political oppression. In the 1880s, many coal operators began recruiting workers from eastern and southern Europe. The coal operators believed that ethnically dividing the labor movement by recruiting workers who spoke many different languages would make it difficult for the coal workers to organize (Barendse 1981, 7–8, 24–28; Brooks 1898; Greene 1968; Miller and Sharpless 1998, 170–173; Roberts 1970 [1904]).

Peter Roberts (1970 [1904]), a sociologist studying the impact of the coal industry on the new immigrant community, commented on the impact of the anthracite coal industry on the region. He explained how several generations of coal extraction had severely impacted the environment. He reported that "Trunks of trees stand in the valley, veritable ghosts of stately pines which no more know spring-time and summer." In many places acres of culm heaps, which are the refuse of a century of mining, "stand as black monsters defiling our fairest valleys" (Roberts 1970 [1904], 155) (Figure 15.1).

Roberts (1970 [1904]) also explained how mining impacted the local streams and rivers, which some mining communities relied on for drinking water. "A great change has come over this charming landscape … But in nothing is the change so

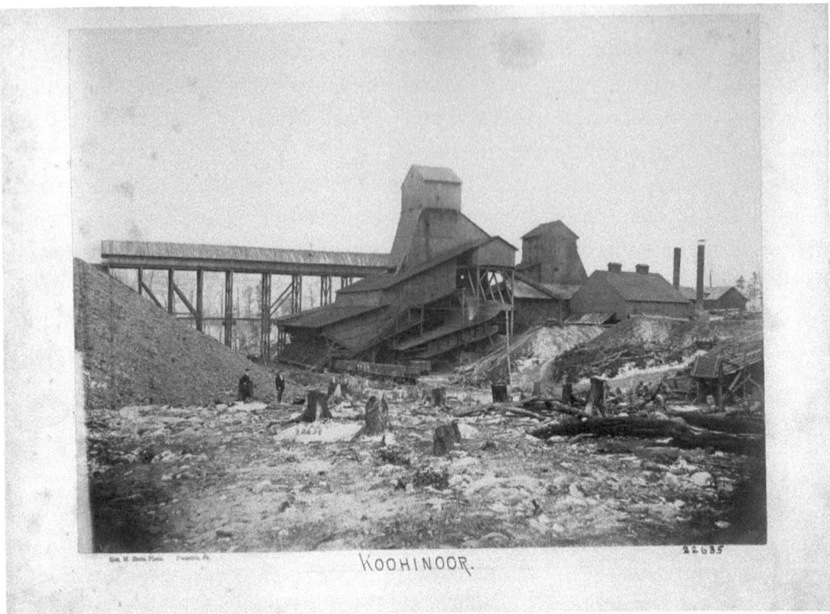

FIGURE 15.1 Kohinoor Colliery [Koohinor] Colliery at Shenandoah City, Pennsylvania, c. 1884. From the mid-nineteenth century these barren landscapes with breakers surrounded with coal ash were common scenes throughout the anthracite region (Bretz Collection, University of Maryland, Baltimore County, Accession Number 73-02-018).

marked as in the character of these mountain streams … Now the rain and snow have no natural reservoirs … Every storm means a flood … The mining industry perfects the work of destruction" (Roberts 1970 [1904], 6). Roberts (1970 [1904], 155) later explains how the waterways were being polluted with coal dust, "The contamination of our streams, the black creeks full of water laden with coal-dust, the dismal acres where the refuse from washeries has long been turned – these make a dreary environment."

A few decades earlier, in 1886, in a case related to the contamination of the drinking water of an individual property owner, the Pennsylvania Supreme Court, in a 5–2 decision, ruled against water quality in favor of the coal company. The court claimed that the property suffered a "mere personal inconvenience," which "must yield to the necessities of a great public industry" (Conlogue 2013, 36). The public good of the industry outweighed the interest of the individual and the good of the working communities (Shackel 2019, 2020). The destruction of the landscape and the lack of regard for communities' general health and welfare continued into the twentieth century. The coal companies were not held responsible for the general welfare of the communities they operated.

Coal operators had little regard for the life and health of their workers. In response to the 1902 Anthracite coal strike, one sympathetic observer, Rev. John McDowell, wrote about the working conditions of the miners. He explained, "The air is impure, the mine damp; often men are compelled to work standing in water to their knees, and water dripping on them from above. Impaired sight, total blindness, and bodies maimed for life are some of the results of this most dangerous work" (McDowell 1902:9–10). He also described the living conditions within the patch towns. "The atmosphere, because of nearness to the breaker, is laden with dust; the water is tainted with Sulphur; and the sun has to fight its way through dense clouds of black smoke which blackens everything in sight" (McDowell 1902:10). In addition, he noted, "The squalor and dirt found in the patches surpasses anything known in the slums of our cities. Coal dirt is everywhere and in everything. The food is gritty with it. The clothing grates with it. It rubs itself into the souls of those who live in it" (McDowell 1902:10).

Boys went to work in the coal breakers as early as eight years of age, starting their long-term exposure to breathing coal dust. One former coal miner explained that he migrated to the U.S. in 1909 at age 13. He initially went to school and learned English; at age 15, he began picking slate for 10 hours a day for six days a week. He described, "They dumped 2,000 cars in the breaker before they released the water (to tamper down the coal dust), and then the dust was all over the building. When I came home, I looked like a n-----" (Victor Bartus interview 1973). He eventually became a miner, and in 1973 he claimed he had black lung disease. Another contemporary worker explained, "I was a laborer at the time and the miner that I was with. We were alongside of each other, and we couldn't see each other for the dust that was between us. There was too much dust in the place" (Anthony Naugin interview 1973). In general, these dust-filled underground chambers meant that most of the coal workers suffered from black lung disease, limiting their physical abilities and impacting their household income.

In northeastern Pennsylvania, the waste coal piles are called "culm heaps" or "culm banks." Since the earliest miners in the region migrated from England and Wales (as well as Scotland and Germany), the term culm is probably derived from the Middle English "colme" and/or the Welsh term cwlm" ("knot or tie"), referring to a type of coal found in balls or knots in some parts of Wales. Both refer to the fine anthracite dust too small to ship to consumers. In other regions, these waste piles are referred to as "slag heaps," "gob piles," or "boney piles." The culm banks consist of small bits of coal, shale, and other rock materials. For example, in 1898, Henry Edward Rood, writing for the *Century* magazine, described a scene in the coal patch town of Lattimer:

> The level land between the culm and the hillside whereon the foreigners live is devoid of vegetation. Grasses and wild flowers once were luxuriant there, but for many decades, rains have been washing from the huge pile some of the deadly black particles that smother plants, even trees, as we realize by noting the gaunt,

> leafless, lifeless trunks scattered here and there, with naked grayish limbs uplifted as if crying to Heaven for help. Were it not for the green hillsides and the kaleidoscopic sky, this would be indeed a somber picture. To the immigrant just arrived from Italy the colliery town must seem a realization of desolation itself.
>
> (Rood 1898: 809)

However, they are more than stagnant markers on the landscape. Because coal and heavy metals are part of the culm bank, rainwater leaches through these culm banks, reacting with the coal and other traces of heavy metals, and drains into the groundwater and streams (Ewall 2017). As a result, communities relying on wells and streams for potable water often suffered, unknowing the cause of their maladies. Also, because culm banks have a low amount of coal, they can catch on fire and burn for years, releasing toxic chemicals into the atmosphere.

Poor sanitary and environmental conditions and the lack of a proper diet also meant that the mortality rate for children tended to be much higher than the average American. Diseases such as measles and typhoid ran rampant in these communities. In some mining communities, as many as 40% of the children died before they made it to adolescence (Miller and Sharpless 1998:195: Roberts 1970 [1904]). All these conditions existed while the coal and railroad companies made significant profits, with little regard for the workers, their families, or the communities in which they lived. The workers were interchangeable and replaceable.

By 1917, at the height of the industry, anthracite coal extraction increased to about 100 million tons a year, and the industry employed 180,000 workers. During WWI, the coal industry had a difficult time meeting demand. However, after the war, oil and natural gas gained a more significant market share, and the anthracite industry began its rapid decline (Rose 1981, 77). By 1922, the production of coal had dropped by nearly 40%. While there was a slight uptick in coal extraction during World War II, the industry further declined in the 1950s. The Knox Mine disaster occurred on January 22, 1959, resulting in twelve miners' deaths when the Susquehanna River broke through the roof of a mine. This tragedy was the death knell of coal extraction in the region. With the dramatic closing of the mines, many of the region's women became the sole economic provider for families as they found jobs in the garment industry. Many men were unemployed or "gone to New Jersey" for work (Wolensky 2020; Shackel 2023).

Nostalgia in a Toxic Environment

Generally translated from Greek, nostalgia means "a longing to return home." Nostalgia becomes useful during times of stress and uncertainty, especially when the present does not conform to one's expectations. It can help create a sense of comfort in a time of traumatic stress. Nostalgia can cushion the shock of change. Individuals call upon nostalgia during times of distress, loneliness, and anomie to help provide identity continuity (Rudacille 2015). The use of nostalgia and claiming a past is a way to veil the disharmony and the angst of contending with the deindustrialization of the region and dealing with new cultural values in a relatively new post-industrial environment (Shackel 2016). Smith and Campbell (2017, 612–627) argue that nostalgia can "assert a sense of communal belonging and sense of place in the context of rapid deindustrialization and social change." Mah (2010) explains, "The concept of the ruin implies sad beauty, majesty, glorious memory, tragedy, loss, and historical import .... [R]uins reflect pastness, romance and nostalgia, while at the same time representing risk, commodification and neglect." In the case of the anthracite region, there is a strong relationship between the decline of the coal industry and the growth of nostalgia.

During the demise of the anthracite region, coal companies abandoned the region, leaving behind a ruined landscape. Penn State geographers Deasy and Griess (1961) came to the area in 1960 to assess the landscape and suggested revitalizing the economy by developing a sense of heritage. They explain that some people might be attracted to the distressed anthracite landscape. They claim that it is not uncommon for people to go out of their way to visit the slums of Paris and New York. While some might find the decimated anthracite landscape repulsive, they believed it could not be ignored. The anthracite region is one of the world's largest concentrations of disturbed lands. The area is littered with coal waste piles and abandoned strip mines. Deasy and Griess compare the amount of earthmoving to the "engineering feats of the Suez and Panama canals and the pyramids of Egypt" (Deasy and Griess 1961, 3). Within this context, they describe the landscape as "man-made Bad Lands."

Now that anthracite mining is nearly non-existent in northeastern Pennsylvania, the intrusive culm bank on the landscape serves as a reminder of the industrial past. Many culm banks throughout the region are over 100 feet high and can be viewed from significant distances. Goin and Raymond (2004, 39) note that some former miners and descendants of the miners consider these features, as memorials to the labor performed by the coal workers. – They are monuments to the many anonymous workers who toiled in this uncompromising industry. "Culm banks in Wyoming Valley are the remaining markers of a once-powerful industry that gave meaning to the place and to the lives of those who lived in it" (Goin and Raymond 2004, 39). To others, these features are now part of the vernacular landscape (Shackel 2020) (Figure 15.2).

Thomas Kiety Blomain's poem provides a scene of old miners and their relationship to the landscape (quoted in Goin and Raymond 2004, 40–41):

I've seen the old men from the slag valley
wearing their worn gray caps
inhaling Lucky Strike and anthracite
congregating at the aluminum-sided bars
that line the dim street with familiar signs.
I've seen them malinger
outside vast cathedral doors

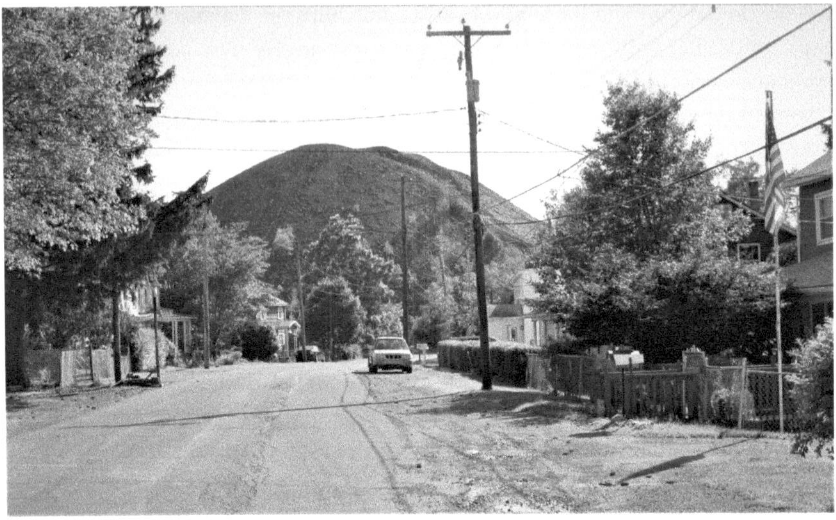

FIGURE 15.2 The culm bank at Beaver Meadow, Pennsylvania stands about 100 feet tall. The town developed adjacent to this culm bank. These culm banks are common throughout the anthracite region's landscape (Courtesy, Paul A. Shackel).

fidgety as children
in parochial Easter suits,
their catechism
on earth's dusty shelf
a bound history
of culm mountains.

The poem provides a faint perspective of the past, referring to the landscape of slag valley (a coal by-product) and inhaling cigarette smoke and anthracite dust. The cathedral doors open to the earth's dusty shelf, which is the dust and grime-filled environment. In the end, the toxic culm mountain looms large and dominates the toxic landscapes.

Since the coal industry's decline, monuments and statues commemorating the workers have sprung up in many town parks and public spaces. Anthracite mining nostalgia increased in the 1960s with the widespread economic devastation in the region. For instance, in 1961, the Ashland Community Enterprises developed a mine tour and a half-mile train ride to view the results of an abandoned strip mine, an imposing 250-foot high carved solid rock wall. Griess and Deasy (1964, 215) exclaimed that this site is "one of the great engineering wonders of the world." In some coal mine tours, former miners will interpret their mining experiences. Similar to what Dicks (2008) found in her ethnography of mine tours in Wales, these new tour guides present work as a form of dignity and, at the same time, acknowledge how they were subjugated and defeated as capital moved on. Forms of nostalgia helped prop the value of working-class citizens that helped to ignite the industrial revolution.

The drive for a nostalgic past accelerated with the 1969 filming by Paramount Pictures of the movie *The Molly Maguires* in the coal patch town of Eckley. The film portrays coal mining life in northeastern Pennsylvania during the reign of the Molly Maguires. After filming and because Eckley was well preserved, the Anthracite Historic Site and Museum Corporation, affiliated with the Hazleton Chamber of Commerce, purchased the town in 1971 and then donated the village to the Commonwealth of Pennsylvania (Warfel 1993, 6). In 1975, Eckley Miners' Village became one of several museum sites in the anthracite region developed to interpret the history of the anthracite coal industry. Eckley Miners' Village reinforces a nostalgia for past working-class life, interpreting the life and working conditions people once endured under the rule of the coal barons.

In Mahanoy City, a memorial was erected in 2010 to commemorate the Molly Maguires. The Molly Maguires became active in the anthracite region since at least the Panic of 1873 and probably earlier. Consisting of miners of Irish descent, the Mollies confronted the coal operators for their unjust labor practices, and they were eventually accused of murder, arson, and other crimes. The Pinkerton National Detective Agency infiltrated the Molly Maguires, which led to the conviction of 20 men for murder, arson, kidnapping, and other crimes. They were executed by hanging in 1877 and 1878. Some histories characterize the Mollies as working-class laborers fighting their exploitations, while others condemn the organization for its violence. A memorial is a blindfolded man about to be hanged. It is located on the town's main street, although it sits behind a block wall that is not visible from the road (Hand 2015; Shackel 2019, 2020).

The Anthracite Heritage Museum in Scranton, Pennsylvania, documents the impact of the region's dominant industry. The museum explores the industry's impact on life and community during its growth and eventual demise. Adjacent to the museum is the Lackawanna Coal Mine Tour. It is one of three mine tours in the region (in addition to one in Lansford and the other in Ashland).

Another effort to support anthracite heritage is an oratorio for choir and chamber ensemble by Julia Wolfe titled *Anthracite Fields*. It premiered in Philadelphia in 2014 and was awarded the 2015 Pulitzer Prize for Music (Shackel 2019).

Living in a toxic environment has led to the creation of different forms of heritage. The anthracite region is in a visible state of ruin. The environment, landscape, and communities have been severely impacted. Within this devastated landscape, communities are striving to remember a past. Some forms of remembering glorify capital, while others condemn the region's destruction of people and communities.

Conclusion

There is often a question about how to view and interpret these ruined landscapes. Interestingly, Rudd and Davis (1998) show how a Utah corporation promoted copper mine tourism intending to address the community's concern about the industry's impact on the environment. The corporation interprets its industry as "a gift of nature" developed only through active mining, although they do not address the activity as polluting the region and impacting the landscape. In northeastern Pennsylvania, the mining industry has left a toxic and ruined environment. However, the region's people have found ways to commemorate a heroic industrial past. Amid the ruined landscape, there are monuments, memorials, and museums.

In many cases, the glorification of capital overrides the industry's devastating impact on people and the environment. Looking beyond the memorialization of capital, a toxic heritage is found throughout the landscape and is a reminder of the coal industry's heedless destruction of people and communities (Figure 15.3).

The anthracite region has the most disturbed landscape in Pennsylvania. Currently, there are eight known underground mine fires in the region. In addition, there are about sixty-four square miles of abandoned mine lands and five hundred and four miles of impaired streams in the anthracite region (Clark 2011). There are black mounds and hills of coal waste, known as culm banks, some reaching 125 feet high. Water running through abandoned mines empty into streams and rivers. This water has a very low pH, and some water is orange in color,

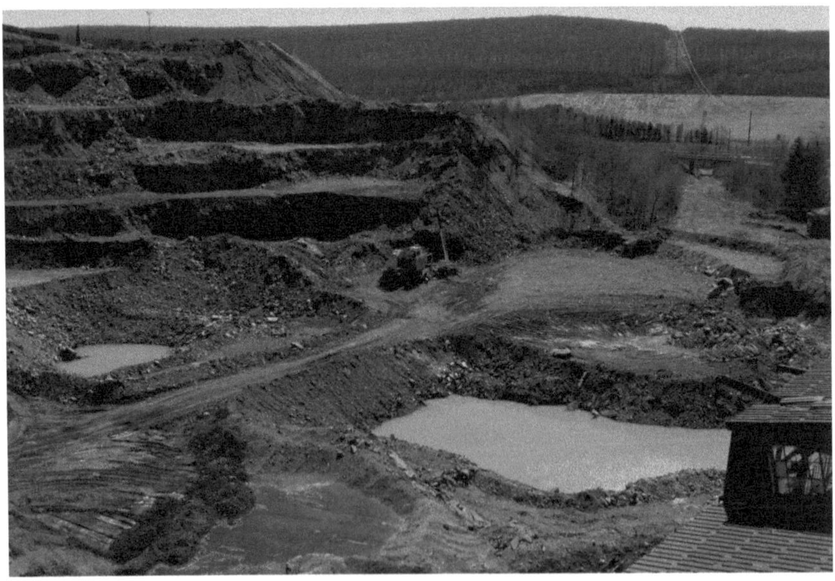

FIGURE 15.3 Acid mine drainage near Mahanoy City, Pennsylvania. This orange water has high levels of iron hydroxide, a common feature found in waters impacted by mine water runoff. While this pool is found in a strip mine, the impact of acid mine drainage can be found throughout the anthracite region (Courtesy Gabby Zawacki, 2016).

a product of acid mine drainage (Figure 3). As early as 1924, the *New York Times* (May 30, 1924) reported that the abandoned mines in northeastern Pennsylvania were filled with water, mixing and reacting with coal. Communities had basements flooded with water. In order to address this situation, boreholes were drilled in abandoned mines. This work helped drain the mine water away from the property and into surrounding rivers and streams, thereby creating another environmental disaster, acid mine drainage (AMD) into major bodies of water (also see Conlogue 2013, 30). The Jeddo Tunnel, completed in 1894, was the largest mine drainage tunnel in the world. The tunnel still drains about 40,000 gallons of water per minute, with an average pH of 4.3. An average of 2,900 pounds of aluminum, 1,350 pounds of manganese, and 860 pounds of iron flow from the tunnel daily (*Coal Age* 1914, 391; Mendinsky and Dempsey 2004; Shackel 2017).

Hardwood forests in the region were once filled with diverse forms of wildlife are now replaced with mountainous culm banks. It takes several generations before the acid soils can support some primary growth, such as briar bushes, scrub vegetation, and eventually birch trees. There are several land reclamation projects in process. The focus is to bring the land back to its natural contours. After filling the open-pit mines, a layer of topsoil is added and then seeded to begin the reclamation process. A local cable talk show often interviews coal industry representatives. At times, they will state that when they finish mining, and the land is reclaimed, the landscape will be better than its original condition. Berger (2019: 48) found a similar situation in the industrial Ruhr valley. While the industrial site polluted the air and poisoned the lands, industrialists are claiming the emergence of an "industrial nature," which is the re-emergence of vegetation on former industrial sites. The scarred landscape is healing, and industrialists claim that there is more plant and animal life diversity than before.

It is important to recognize the people who continue to live in these landscapes since the toxic landscape still has an impact on their daily lives today (Stoler 2013, 9). Today, the region's population skews older than the rest of the state, with a large majority of the residents being descendants of the mining community. The region is not well. Heart disease deaths are 20% higher than the state's average. In addition, the cancer death rates are about 24% higher, and the rate of deaths by respiratory disease is more than 30% higher than in the rest of the state (Huang 2019). As one author notes, "the coal region's industrial economy left behind potential health hazards" (Huang 2019). Nevertheless, the community strives to find ways to remember a selective past.

References

Barendse, Michael A. 1981. *Social Expectations and Perception: The Case of the Slavic Anthracite Workers*. The Pennsylvania State University Studies, no. 47. University Park, PA: Pennsylvania State University Press.

Berger, Stefan. 2019. "Industrial Heritage and the Ambiguities of Nostalgia for an Industrial Past in the Ruhr Valley, Germany." *Labor* 16(1): 37–64.

Brooks, John Graham. 1898. "Notes." *The Yale Review: A Quarterly Journal for the Scientific Discussion of Economic, Political and Social Questions* 6: 306–311.

Coal Age. 1914. *Coal Age: Devoted to Coal Mining and Coal Manufacturing* 6 (10).

Clark, Tom. 2011. "Anthracite Region Mine Drainage Remediation Strategy." *Susquehanna River Basin Commission*, December: 1–3.

Conlogue, Bill. 2013. *Here and There: Reading Pennsylvania's Working Landscape*. University Park, PA: The Pennsylvania State University Press.

Deasy, George F., and Phyllis R. Griess. 1961. "Tourism for the Anthracite Region: An Alternative for Unemployment." *Mineral Industries* 30(7): 1–8.

Dicks, Bella. 2008. "Performing the Hidden Injuries of Class in Coal-Mining Heritage." *Sociology* 42 (3): 436–452.

Ewall, Mike. 2017. "Waste Coal." Energyjustice.net. http://www.energyjustice.net/coal/wastecoal

Goin, Peter, and C. Elizabeth Raymond. 2004. *Changing Mines in America*. Santa Fe, NM and Staunton, VA: The Center for American Places.

Greene, Victor R. 1968. *The Slavic Community on Strike: Immigrant Labor in Pennsylvania Anthracite*. Notre Dame, IN: University of Notre Dame Press.

Griess, Phyllis R., and George F. Deasy. 1964. "Economic Impact of a Pennsylvania Tourist Facility." *Land Economics* 40(2): 213–220.

Hand, Mark. 2015. "West Virginia Miners Play Second Fiddle to the Molly Maguires." *CounterPunch*. September 29. http://www.counterpunch.org/2015/09/29/west-virginia-miners-play-second-fiddle-to-the-molly-maguires/

Huang, Binghui. 2019. "Pennsylvania Coal Region's Industry Burned Out. What Remains are Pockets of Poverty where Sick People Get Sicker." *The Morning Call*. January 27. https://www.mcall.com/news/watchdog/mc-nws-health-coal-country-project-20181216-htmlstory.html

Keil, Thomas, and Jacqueline M. Keil. 2015. *Anthracite's Demise and the Post-Coal Economy of Northeastern Pennsylvania*. Bethlehem, PA: Lehigh University Press.

MacGaffey, Janet. 2013. *Coal Dust on Your Feet: The Rise, Decline, and Restoration of an Anthracite Mining Town*. Lewisburg, PA: Bucknell University Press.

Mah, A. 2010. "Memory, Uncertainty and Industrial Ruination: Walker Riverside, Newcastle upon Tyne." *International Journals of Urban and Regional Research* 34 (2): 398–413. 10.1111/j.1468-2427.2010.00898.x

Marsh, Ben. 1987. "Continuity and Decline in the Anthracite Towns of Pennsylvania." *Annals of the Association of American Geographers*. 77(3): 337–352. http://www.jstor.org/stable/2563271

McDowell, Rev. John. 1902. "The Strike of the Anthracite Miners: Its Causes, Justice, and Significance." *The Pilgrim*, July: 9-10.

Mendinsky, Justin J., and Brian A. Dempsey. 2004. "Effects of AMD Pollutant Loading On Streams in the Hazleton PA Area." *Proceedings America Society of Mining and Reclamation*. http://www.asmr.us/Portals/0/Documents/Conference-Proceedings/2004/1289-Mendinsky.pdf

Miller, Donald L., and Richard E. Sharpless. 1985. *The Kingdom of Coal: Work, Enterprise, and Ethnic Communities in the Mine Fields*. Philadelphia, PA: University of Pennsylvania Press.

New York Times. 1924. "Veteran Buried by Mine Cave." May 30.

Roberts, Peter. 1904 1970. *Anthracite Coal Communities*. New York: Arno Press and the New York Times.

Rood, H. E. 1898. "A Pennsylvania Colliery Village: A Polyglot Community." *Century*, April 0055(6): 809–821.

Rose, Dan. 1981. *Energy Transition and the Local Community: A Theory of Society Applied to Hazleton, Pennsylvania.* Philadelphia, PA: University of Pennsylvania Press.

Rudacille, Deborah. 2015. "Steel Mills, Coal Mines, and Blast Furnaces Have Killed and Poisoned Thousands of Americans – So Why Mourn their Passing?" *Aeon*. https://aeon.co/essays/how-toxic-is-smokestack-nostalgia

Rudd, M., and J. Davis. 1998. "Industrial Heritage Tourism at the Bingham Canyon Copper Mine." *Journal of Travel Research*. 36:85–89.

Shackel, Paul A. 2016. "The Meaning of Place in the Anthracite Region of Northeastern Pennsylvania." *International Journal of Heritage Studies* 22(3): 200–213. 10.1080/13527258.2015.1114009

Shackel, Paul A. 2017. "Anthracite Heritage: Landscape, Memory and the Environment." *Open Rivers: Rethinking Water, Place and Community* 7(Summer). http://editions.lib.umn.edu/openrivers/article/anthracite-heritage-landscape-memory-and-the-environment/

Shackel, Paul A. 2018. "Transgenerational Impact of Structural Violence: Epigenetics and the Legacy of Anthracite Coal." *International Journal of Historical Archaeology* 22(4): 865–882. 10.1007/s10761-017-0451-0

Shackel, Paul A. 2019. "Structural Violence and the Industrial Landscape." *International Journal of Heritage Studies* 25(7): 750–762. 10.1080/13527258.2018.1517374

Shackel, Paul A. 2020. *An Archaeology of Unchecked Capitalism: From the American Rust Belt to the Developing World.* NY: Berghahn Books.

Shackel, Paul A. 2023. *The Ruined Anthracite: Historical Trauma in Coal Mining Labor Communities.* Champaign, IL: University of Illinois Press.

Smith, Laurajane, and Gary Campbell. 2017. "'Nostalgia for the Future': Memory, Nostalgia and the Politics of Class." *International Journal of Heritage Studies* 23(7): 612–627.

Stoler, A. L. 2013. "Introduction: "The Rot Remains": From Ruins to Ruination." In *Imperial Debris: On Ruins and Ruination*, edited by A. L. Stoler, 1–35. Durham: Duke University Press.

Tarone, L. A. 2004. *We Were Here Once: Successes, Mistakes, & Calamities in Hazleton Area History*. Hazleton, PA: Citizen Publishing, Company, Inc.

Turner, Victor, and Edith Turner. 1978. *Image and Pilgrimage in Christian Culture.* New York: Columbia University Press.

Warfel, Stephen G. 1993. *A Patch of Land Owned by the Company*. Harrisburg, PA: Pennsylvania Historical and Museum Commission.

Warne, Frank Julian. 1904. *The Slav Invasion and the Mine Workers*. Philadelphia, PA: J.B. Lippincott.

Wolensky, Robert, ed. 2020. *Sewn in Coal: An Oral History of the Ladies' Garment Industry in Northeastern Pennsylvania, 1945-1995.* University Park, PA: The Pennsylvania State University Press.

SECTION 5

INTRODUCTION: APPROACHES AND INTERVENTIONS

Elizabeth Kryder-Reid and Sarah May

This section highlights the ways in which individuals, organisations, and industries operate in response to the creation and consequences of toxicity. It asks the question, 'What is to be done?' The other sections of this volume explore how we frame, understand, engage with, and explain toxic heritage. This section focuses on action and on how heritage is deployed by people as a response to toxic materials to achieve specific goals. These actions create futures – often with an emphasis on remediation, amelioration of harm, education, and raising awareness – with the hope of creating change.

These contributions highlight examples of how the strategies and tools of heritage - making, including exhibitions, art interventions, toxic tours, museum activism, and citizen science projects, have tried to galvanise action. Some raise new questions. Sevcenko explores how reparations for the harms of toxic heritage might take a form similar to reparations for other historic harms. Baptista details how mobilising toxic tours has been a strategy for community advocacy and resistance in one of the most densely polluted cities in the USA. Filippelli shares an example of how citizen science can be a form of participatory heritage-making as people investigate their own exposomes and make connections with the broader patterns of community contamination. Holtorf investigates nuclear waste curation at a storage facility. McKenzie shares the work of Climate Museum UK as an example of the activism of cultural organisations as change agents. Valderrama presents the participatory design and activist practices of a group of architects for social and environmental justice on the Saladillo stream in Rosario, Argentina. Joyce examines how artistic practices can respond to and articulate risks that are unfathomable and explores how they support communities living with the risks of nuclear waste.

DOI: 10.4324/9781003365259-30

In these examples, toxic heritage inspires actions by highlighting the impacts and meanings of toxic materials and by reactivating the past as a source of power in the present. By offering a counterpoint to places where heritage practices have silenced the environmental burdens and their consequences, these authors demonstrate how communities can advocate for change by deploying the histories of toxic pasts.

16

ENVIRONMENTAL JUSTICE TOURS: TRANSFORMATIVE NARRATIVES OF STRUGGLE, SOLIDARITY, AND ACTIVISM

Ana Isabel Baptista

Introduction

Toxic tours or environmental justice (EJ) tours have a long history in the environmental justice movement (EJM). More than a form of tourism, EJ tours represent an important movement strategy. The practice of sharing stories, memories, and experiences of place has always been an integral part of many social justice movements, particularly place-based struggles. The political and spatial conditions of marginalization make the act of sharing firsthand narratives and physical locations a powerful form of resistance. This chapter explores EJ tours as a tactic of the US EJM to (1) build solidarity and organize residents, (2) reconstitute and amplify community-centered narratives about EJ, and (3) advance grassroots-led solutions and inspire action to demand accountability. EJ tours' history and purpose are examined from the perspective of activists who, over several decades, worked with the Ironbound Community Corporation (ICC), a community-based EJ organization in Newark, New Jersey.

EJ tour reflections in this chapter are grounded in my experiences as a protagonist for EJ tours in the Ironbound where I grew up, lived, and worked for many years, and the insights from other EJ tour guides at ICC such as Nancy Zak, who has worked at ICC for over four decades, and Maria Lopez-Nunez, who currently leads ICC's EJ work. We supported each other in learning how to conduct EJ tours and in leading EJ tours in different eras of the EJM within the organization. I have also had the privilege of being a participant in EJ tours in communities around the US through my work as an EJ activist and scholar.

Phaedra Pezzullo (2009) examines traditional forms of tourism and their combination or adaptation to EJ cases. Using toxic tours from around the country, Pezzullo discusses the power of tours for amplifying the cultural, political, and democratic

DOI: 10.4324/9781003365259-31

practices of EJ communities. The form and function of tours are transfigured by EJ activists seeking to reconnect and reclaim the places that modern, global society has sought to degrade and invisibilize. Pezzullo builds her arguments around the notion that toxic tours are political acts of resistance in the way that they reconstitute memory and build communities (Pezzullo 2009). These themes are not unfamiliar to the EJ scholar Giovanna DiChiro (2003) who highlights the important role that toxic tours play among activists trying to draw attention to the realities of polluted environments. DiChiro describes the ways in which activists leading EJ tours use the "act of seeing with one's own eyes" as a means for trans-communal cooperation, an action which she describes as "a method that entails face-to-face contact and mutual trust" (2003, 221). Thus, the act of showing and witnessing is an essential part of making connections between affected communities, allies, and anyone seeking EJ.

Tours are also part of the fabric of toxic heritage studies because activists and scholars interrogate the creation of places and people that have been deemed "toxic," delving into the histories, traditions, and practices that have shaped the people and places in sacrifice zones so severely impacted by pollution. Toxic heritage studies examine these communities as places of memory and as active terrains for mobilizing action: "Toxic heritage is often the site of resistance, resilience, and social action, as communities mobilize to demand mitigation, embrace citizen science, advocate for environmental causes, and document the histories of homelands and neighborhoods" (Toxic Heritage n.d.). In many ways, EJ tours uncover the "slow violence" of environmental injustice buried over time and hidden from most of society (Nixon 2011). Donna Houston describes EJ storytelling as, "Environmental justice storytelling provides a framework for understanding how multiple realities of environmental injury come together in ways that are not always readily discernible through policy or scientific practice" (2013, 419). The practice of EJ storytelling is embedded in tours and provides insights into the lived experiences of residents at the fenceline and frontlines of polluted environments. EJ tours go beyond sharing toxic injuries, recasting EJ communities as rich, culturally important places that are abundant in treasures amid the toxics. This recasting, via the act of storytelling on tours, can be transformative but it is not without risks.

While the literature examining toxic tours often comes from the perspective of outsiders who see the power and potential of tours, this gaze misses the painful experiences that tours elicit. Tours highlight historical and current conditions of violence and suffering, which often engender feelings of despair and surface trauma, conflicts, and unsettled histories. Additionally, tours place affected residents in direct contact with perpetrators of harm or in vulnerable positions with institutions of power, which expose activists to injurious comments. Moreover, in many places around the world, the act of exposing toxic conditions puts activists' lives at risk. Thus, it is important to note that EJ tours are not removed acts of sharing statistics or highlighting destinations on a map. Rather, tours are acts of resistance that take a physical and mental toll on the people living the everyday reality of environmental injustice.

History of Ironbound Community Corporation (ICC) and Ironbound, Newark, NJ

The Ironbound Community Corporation (ICC) was founded in 1969 in the wake of the Newark rebellions. It was formed as a community-based non-profit that arose out of resident outcry for affordable childcare. The organization began focusing on EJ issues when affiliated residents and parents organized to fight pollution in the community. In 1981, the organization founded a related group called the Ironbound Committee Against Toxic Waste (ICATW), which focused on the rising cases of industrial accidents and proposals for noxious facilities in the neighborhood.

The Ironbound neighborhood has a long history of toxic facilities. The community is located on the eastern edge of Newark, New Jersey, and is bounded on three sides by rail lines and the Passaic River. The name "Ironbound" itself is a source of diverse oral histories. Some have suggested it derives from the presence of rail lines encircling the community and others suggest it relates to the presence of foundries that once surrounded the area. Whatever the origin, the name recalls spaces that have been intimately linked to industrialization. The Ironbound was built atop what was once a teeming meadowland ecosystem with an abundance that sustained the Munsee Lunaape people that are the living and sovereign Indigenous People of the area. Over time these marshes were infilled with garbage and industries as urbanization expanded throughout the region (Newark Public Library 1996). In the 20th century, the Ironbound was also redlined as the least desirable type of neighborhood (D or 4th-grade) by the Home Owners' Loan Corporation (Hagstrom Company Inc. 1939). The city also experienced periods of intense disinvestment, white flight, and deindustrialization. Today, the area, which encompasses four square miles and is home to 60,000 people, hosts the second-largest seaport in the US, the state's largest garbage incinerator and sewage treatment facility, a regional airport, rail yards, and hundreds of polluting industries (City of Newark, NJ 2021). Historically the Ironbound was home to successive waves of ethnic European immigrants. Today the community is very diverse, and the population is comprised mostly of Latinx, Black, and low-wealth immigrant residents (US Census Bureau n.d.).

History & Evolution of EJ Tours at ICC: Building Solidarity and Organizing

Nancy Zak has worked at ICC for over four decades and was likely one of the first people to lead informal EJ tours of the Ironbound along with her partner, Arnold Cohen. In the 1980s, they led visitors to the former Diamond Shamrock herbicide plant situated along the Passaic River. The plant produced millions of pounds of Agent Orange between 1952 and 1968 for the Vietnam war (Bruno 2010) (Figure 16.1).

Agent Orange is a highly carcinogenic compound called dioxin, which can lead to cancer, and reproductive, developmental, and immune system problems (US EPA 2022). By the early 1980s, the facility was abandoned and former factory workers and Vietnam veterans launched class-action lawsuits against corporate

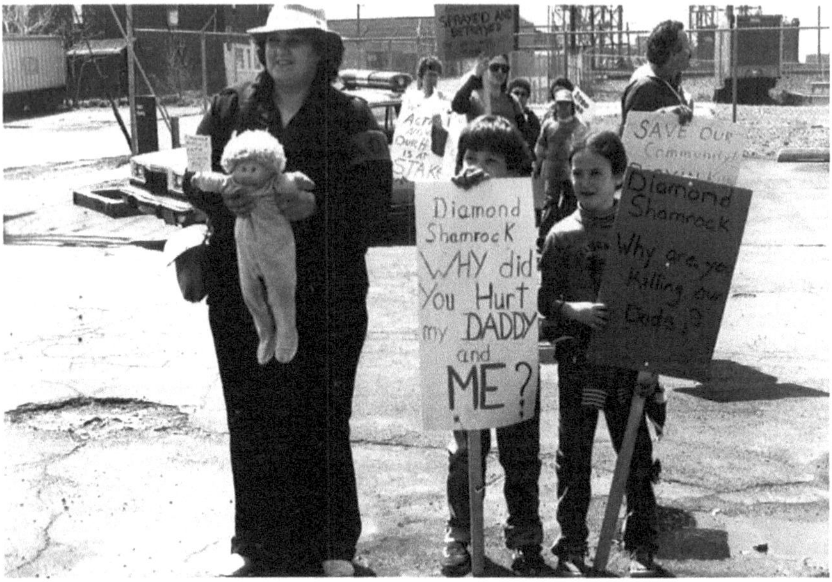

FIGURE 16.1 A photo of Ironbound residents protesting Diamond Shamrock. June 1983, Ironbound Community Corporation (https://picturingjustice.tumblr.com/tagged/dioxin).

owners for long-term health damages suffered over many years of exposure (Parisi 1983, 2). These lawsuits highlighted the malfeasance of corporations that knowingly exposed people to toxic chemicals and the complicity of government agencies that failed to intervene and protect residents living near the plant.

During this time, ICC organized residents to demand government and corporate actors remediate the site and examine the health impacts of dioxin exposure. Zak invited reporters and allies to the neighborhood around the plant, which allowed residents and workers to share how their lives were affected by Agent Orange. She notes that these visits included planned actions: "When we went out to the site, people would start to tell us about when they worked or played near the site and all the health problems that happened to them and it was very jarring to hear how the dioxin really had a terrible effect on ordinary people" (Personal communication, July 20, 2022). While Zak did not have access to detailed technical data about the extent of the site's pollution, she had access to many personal accounts from residents. One of these accounts, which continues to be recounted on EJ tours decades later goes something like this,

> A young woman in her thirties shared that she lived most of her life a block from the Diamond Shamrock plant. Many of her family members worked at the plant and she remembers them coming home covered in a fine orange dust that was ubiquitous around the plant. She recounted children like herself playing on the

> dirt piles near the plant. Years later she suffered multiple miscarriages and was diagnosed with a rare form of reproductive cancer that afflicted her and others in the area. In 1983, the plant was listed as a Superfund site and the governor issued an emergency declaration for the areas around the plant (Bruno 2010). Zak describes the shock of residents as they sat on their front porches seeing men dressed in moon suits who were vacuuming the streets and placing the contaminated dirt in drums for removal.

The first time I heard Zak recounting this story, it painted a jarring picture of the juxtaposition of residents going about their daily lives while government officials donned hazmat suits to protect themselves from something residents were exposed to for many decades (see Figure 16.2). This story continues to be shared on EJ tours today. We do not know the full impact of a lifetime of exposure to these toxins and the story does not definitively state whether the dioxin contamination caused the young woman's health problems, but it points to the devastating impacts that linger over those who are exposed to poison.

This story illustrates the power of memory in conveying the harm of pollution to individuals, families, and communities. Amelia Fiske (2018) points to the ways in which the embodied knowledge of toxins, held by those organizing against corporate polluters, are powerful markers of EJ struggles. She also points to the

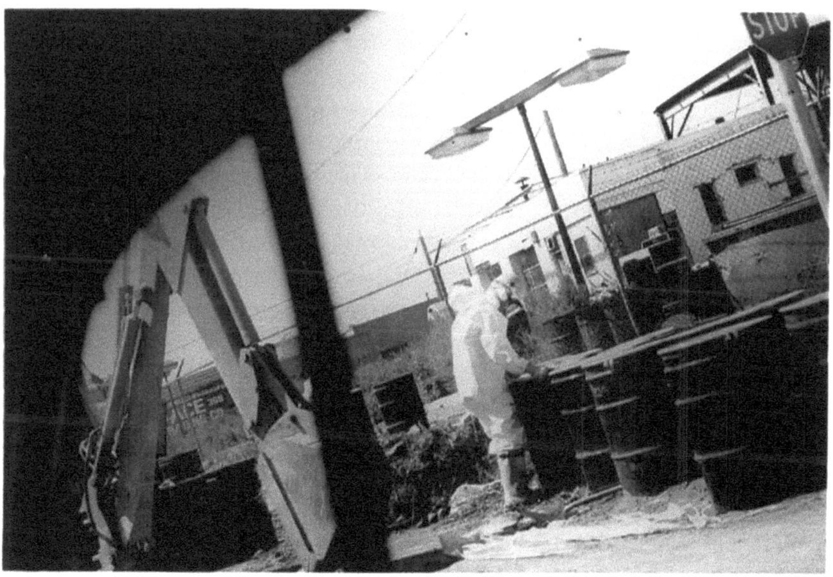

FIGURE 16.2 A photo of men from the New Jersey Department of Environmental Protection in hazmat suits. Ironbound Community Corporation (https://picturingjustice.tumblr.com/tagged/dioxin).

material harm that activists are exposed to in the process of storytelling: "Simply put, there are burdens implied through the engagement of bodily knowledges in the politics of denunciation that require further consideration beyond their potential to disrupt and contest powerful structures – whether in activist work, citizen science initiatives, political campaigns, or as the subject of ethnography" (Fiske 2018, 404).

Toxic legacies are often buried both figuratively and literally. Unearthing them through speech can combat government and corporate obfuscation of the negative impacts of toxins. This oral tradition materializes the bodily harm that scientific risk analysis mostly denies. For instance, Melissa Checker notes, "'But I know it's true', is the oft-repeated refrain of EJ communities across the country … Although these people know that they and their neighbors suffer from uncommon health problems, they have been unable to secure scientific proof that the chemicals are the cause" (2007, 113). Furthermore, Checker discusses the limitations of scientific risk assessments for capturing the impacts facing EJ communities, resulting in environmental decision-making that fails to account for their lived experiences. As a result, EJ communities have responded with their own forms of popular epidemiology and citizen science to reclaim their embodied experiences (Brown, De la Rosa, and Cordner 2020).

The Diamond Shamrock story also connects to storytelling as a tool for political education and collective political framing for impacted people. EJ tours that highlight sources of industrial production and fallout, like Diamond Shamrock, link EJ communities trans-locally, to locations of contestation and facilitate a critique of capitalism and extractivism (Chagnon et al. 2022). Doug McAdam (1982) suggests that the ability of movements to respond to injustices partly rests on a process of consciousness-raising in which oppressed people become aware of the conditions of injustice. McAdam calls this process "cognitive liberation," which is also referenced by Temper et al. (2018) to describe how EJM actors productively use conflict to contest hegemonic power structures and push for social change. In this regard, EJ tours serve as a pathway for cognitive liberation for those living within EJ communities. They also serve as means by which to contest negative stereotypes of polluted communities, reshaping those narratives into sources of power.

While EJ struggles may be viewed as strictly focused on a fair distribution of benefits and harms, the politics of the EJM reach well beyond mainstream reformist environmentalism. For instance, David Harvey (1996) discusses the radical politics of the grassroots EJM, which exposes the contradictions of the neo-liberal state and capitalism. He notes the power of the movement to use moral reasoning and symbolic meaning-making to contest expert-driven discourses deployed by the state and corporate actors that undercut EJ activists' claims (Harvey 1996). In this regard, tours highlight the material contradictions inherent in capitalism and make critical, discursive connections to similarly situated communities, pushing back against global extractivism in its many forms.

Evolution of EJ Tours at ICC: Audiences and Movement Goals

ICC's EJ tours have evolved over more than four decades from informal, often site-specific visits by small groups to larger, organized multi-stop events. Today, Maria Lopez-Nunez fields a multitude of requests for tours with a range of stakeholders due to a growing interest in EJ. These requests necessitate a significant amount of time and resources. When asked about how she understands the role and prioritization of tour requests, Lopez-Nunez reflects,

> To be honest, these tours can be a huge drain, but they can also be politically important. If they're done in a way that drives our demands and puts community voices in the driver's seat, then they can showcase community wins and power. I also want people to feel uncomfortable on these tours, to get a little taste of what accountability looks like if we take seriously how implicated we all are in the creation of sacrifice zones. If you walk away from the tour feeling sorry for us then we haven't done our jobs, but if you walk away feeling like communities know best what the solutions are, then maybe it was worth the time.
>
> (Personal interview, September 12, 2022).

When EJ groups curate tours, they include neighborhood maps and descriptive handouts to introduce participants to the history and context of the host organization and the community. It is also a form of counter-mapping, describing the locations of both past experiences or memories of environmental injustices and the active, contested spaces of current fights or victories. The narratives about tour stops differ dramatically from official state or corporate statements about the tour stops. The map itself reflects the changing dynamics of activist campaigns and areas of focus over time.

Figure 16.3 illustrates a map of an EJ tour I helped guide with residents in 2005. The tour included stops at a local park that residents saved from demolition, Chemical Row where many industrial accidents occurred, and an active campaign to clean up the garbage incinerator. It also showcased the community's vision for Riverbank Park, a proposed future park on an abandoned industrial waterfront along the Passaic River. While Riverfront Park was just a dream expressed on the 2005 EJ tour, today EJ tour participants can visit it in its completed form, which is a testament to the power of community visioning and organizing. The importance of mapping counter-narratives and a shared sense of history and place is critical for communities that have been made to feel invisible or disposable in the places they call home. In Rodney Harrison's article on counter-mapping in Australia, he describes the purpose of counter-mapping as a way to "undermine power relations and challenge the dominant political and social geographies of power" (2011, 7). He goes on to describe the value that such counter-mapping can have, "This process has the potential to give voice to politically marginal and subaltern understandings of the past, empowering them by drawing attention to them in the

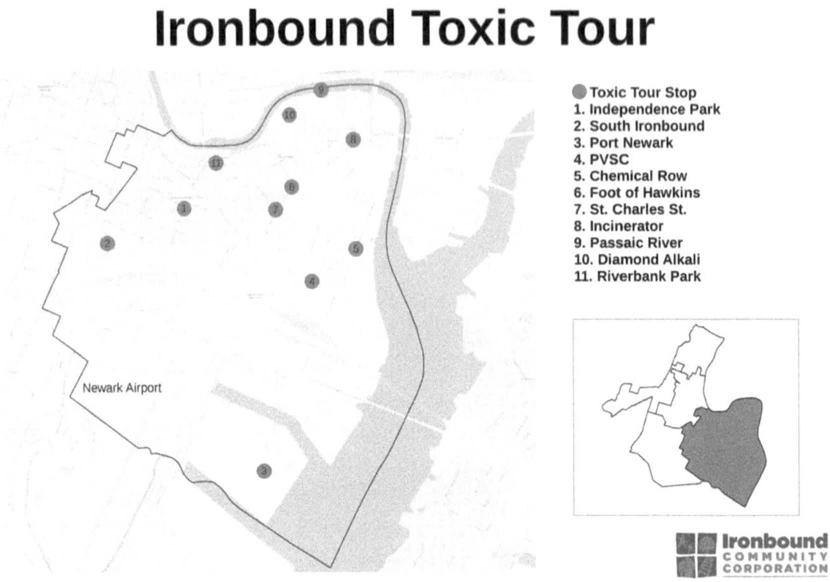

FIGURE 16.3 Map of ICC Ironbound EJ Tour, circa 2005. Prepared by Ana Isabel Baptista.

present … Such intimate, everyday attachments of people to place are at the heart of contemporary approaches to heritage" (Harrison 2011, 10).

We Speak for Ourselves: Reclaiming Spaces of Resistance and Reconstituting Possibilities

One of the most powerful characteristics of the EJM is the principle of "we speak for ourselves." This very simple statement carries a powerful message about who has a legitimate role in conveying the lived experience of environmental injustice. In *From the Ground Up*, Luke Cole and Sheila Foster describe how "Environmental justice activists usually have an immediate and material stake in solving the environmental problems they confront" (2001, 33). The material experience of being part of a community that is directly impacted by pollution and having an identity or multiple identities associated with a history of oppression, marginalization, or dispossession, determines how you speak in an EJM context. Those with lived experiences who are the closest to the problems are the ones whose voices should be most heard and uplifted. EJ tours, as movement strategies, make the position of the EJ tour leader central to considering whose story is being told and what perspectives and solutions are centered. If you take an EJ tour that is not grounded in or directly accountable to those most impacted by pollution, then you are on a tour, but it is likely not an EJ tour. The curation of tours by activists and residents who are directly affected by pollution is one of the principal elements that connect the tour to an act of heritage-making or place-making. Without material connections to places and links to community organizing strategies, tours are devoid of political framing and context that make tours part of the EJM.

EJ tours can also bring into focus the interdependence and culpability of the people and places outside EJ communities. In Laura Pulido's article exploring the socio-spatial dynamics of racism, she notes, "… pollution concentrations are inevitably the product of relationships between distinct places, including industrial zones, affluent suburbs, working-class suburbs, and downtown areas, all of which are racialized" (2000, 533). EJ tours with groups that reside outside of EJ communities allow the unpacking of sedimented racism and how this has, directly and indirectly, benefited white people. On ICC's tours, guides pause to point out the garbage incinerator, power plants, sewage treatment, and warehouses that serve the entire region, but whose pollution impacts are felt very locally. Places like the Ironbound cannot exist without the existence of white, wealthy suburbs that have excluded industrial development and people of Color. These white suburbs benefitted from the legacy of racist housing policies while pushing noxious industries into redlined communities (Rothstein 2017). On tours, Maria Lopez-Nunez is careful to point out the complicity of those living outside sacrifice zones. She reminds participants, "Look around, these are the places and people that have been sacrificed to allow you to live a comfortable existence. We have been told we are not valued; our lives don't matter as much. As long as the lights still come on and the garbage gets dumped somewhere else, people take for granted that these places exist" (EJ tour participant observation, October 15, 2022).

While curating EJ tours provides opportunities to surface hidden histories, tours also allow stigma to be transformed into a source of strength. In Carolyn Merchant's critical history of western conservation movements, she points to white supremacy's stigmatization of non-white land: "The control of the wild represented the kind of state that Western societies could export throughout the world to colonized "Other" lands … The "Others" were the colonized indigenous people, immigrants, and people of color who were outside the controlled, managed garden" (Merchant 2003, 389). Sacrifice zones are associated with people whose identities have also been "othered" and exploited along with the land. Tours are opportunities to challenge these stigmatizing tropes of communities as "wastelands" and reclaim them as sources of power and inspiration (Di Palma 2014).

EJ tours also represent powerful accounts of what Alice Mah (2010) discusses as "living memories" of the complex relationships of people living in communities in a post-industrial transition. According to Mah, "many people who have lived through processes of industrial ruination focus on imminent regeneration rather than mourning or celebrating the industrial past" (2010, 398). Places like Newark have experienced hundreds of years of "ruination" or rather diverse formations of capitalist and settler-colonial projects of land development. EJ tours reveal this laden history while also revealing the contemporary processes of global capitalism that are actively being contested (Harvey 1982; Mah 2012). In the narration of EJ tours,

Lopez-Nunez reflects on both the struggles to resist the constant incursions from global commerce, or what she calls "fighting the bad," and the power of contested visions for what the post-industrial landscape could be in the future, or "building the new."

EJ tours can be powerful portals for reclaiming our homes and histories and projecting the futures we want. As a young person working for ICC, my first EJ tours left me with a sense of pride and appreciation for my community. I was able to share the richness of our culture, the beauty of social networks of care among diverse neighbors, the power of resistance and history of organizing and action, and the love of community that made the Ironbound such a formative part of my life. EJ tours as a movement tactic draw inspiration from Robin Kelley's *Freedom Dreams*, inviting us to imagine the radical possibilities for liberation that come out of intersecting social movements, a feature that is certainly also present in the EJM (Kelley 2002; n.d.).

ICC EJ tours feature important landmarks such as Riverfront Park and Down Bottom Farm (See Figure 16.4) as spaces that have been reclaimed and shaped by the people in the community. These spaces express community culture and identity, where people gather, celebrate, and organize. The spaces are physical reminders of a community that is organized and awake to the possibilities of environmental justice. The act of narrating the tour and bearing witness also creates opportunities to dream about a future with EJ. It invites the participants and guides to be in dialogue about how we engage in the process of liberation and the creation

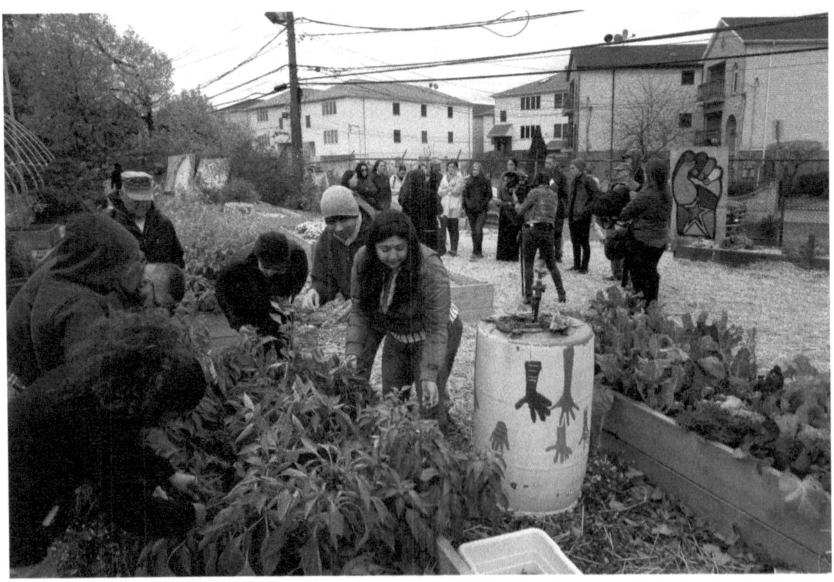

FIGURE 16.4 Down Bottom Farm, Ironbound Community Corp, EJ tour stop 2018, photo courtesy of Kristin Reiman.

of new spaces that are free from environmental degradation, racism, oppression, and capitalism's rapacious exploitation. EJ tours give us a chance to reflect on the future we can build together.

EJ Tours as Movement Tools for Making Demands for Accountability and Action

EJ tours serve as powerful movement tools for resistance and organizing. ICC's EJ tours have always been linked to direct actions, campaigns, and a set of demands, which are often seen as contentious, especially by those in power. Tours create opportunities to confront decision-makers for their complicity in the harms suffered by communities. EJ tours are also an unmediated way for EJ communities to demonstrate their power and make direct demands through unfiltered interactions that are not typically part of official policy-making processes.

While many EJ activists see through the "liberal illusion" of governments intervening in the business-as-usual of industries, they use EJ tours to challenge the technocratic environmental regulations that leave EJ communities unprotected. On many ICC EJ tours led by Lopez-Nunez, she reminds lawmakers and regulators that "you are not doing your jobs" and that the community must constantly step in to fill gaps. EJ tours also challenge traditional forms of public participation like public hearings because they present a more directly democratic, unfiltered, approach to participation and diminish the distance between residents and regulators by bringing them into direct contact with EJ communities. On EJ tours, organizers can set the agenda, direct the action, and place material realities on display that are difficult to capture in policy briefs or legal complaints.

EJ Tours with politicians and agency officials also allow EJ groups to direct their attention to specific sites and campaigns under active contestation. It is much harder for decision-makers to obfuscate or skirt questions and demands when residents confront them with the material realities of pollution. Maximizing the public accountability potential of tours requires a great deal of organizing, such as securing decision-maker participation and following up with them to ensure accountability. If an EJ group is perceived as a powerful voice with an organized base, decision-makers would be remiss to reject the invitation to a tour. On the other hand, they may also see EJ tours as performative in that they can provide photo opportunities with little accountability. Whether EJ tours are low stakes or high stakes in part depends on the power of the EJ groups to go beyond the tour and pressure decision-makers. This organizing frame is critical to ensuring that tours are not reduced to rituals devoid of power.

Conclusion – EJ Tours as Portals to Environmental Justice Futures

EJ tours are windows into the past and current realities of environmental injustice as well as the richness of the communities that call these places home. They are an important tool for EJ movement organizations to shape policy agendas, make demands, shift popular narratives, build relationships of solidarity, center frontline voices and solutions, and make tangible improvements in communities that have existed as sacrifice zones for far too long. EJ tours are also part of the fabric of toxic heritage and place-making, opening up portals for connecting subaltern histories, embodied experiences, and contemporary resistance to global extractivism among movement activists around the world. As a movement tool, tours have many benefits such as facilitating firsthand experiences that uncover complex forces that drive environmental injustice and reorienting people, both within and outside the community, to the possibilities for action and solidarity. The public, unfiltered perspective that an EJ tour delivers is also a powerful tool for making demands and holding decision-makers accountable. Despite all these benefits, EJ tours are not without risks, particularly for the EJ activists and movement groups that deploy them. Nevertheless, EJ tours will continue to be an important movement tactic. As the EJ movement evolves, EJ tours will continue to play a role in visioning justice, deepening political education, and building solidarity. Most importantly, EJ tours pass down rich oral traditions to new generations of EJ organizers and residents who will be the future stewards and storytellers of the EJ movement.

Bibliography

Brown, Phil, Vanessa De la Rosa, and Alissa Cordner. 2020. "Toxic Trespass: Science, Activism, and Policy Concerning Chemicals in Our Bodies." In *Toxic Truths: Environmental Justice and Citizen Science in a Post-Truth Age*, 34–58. Manchester: Manchester University Press.

Bruno, Mary. 2010. "How We Poisoned the Passaic." Grist. June 11, 2010. https://grist.org/article/2010-06-03-from-paradise-to-superfund-afloat-on-new-jerseys-passaic-river-p/

Chagnon, Christopher W., Francesco Durante, Barry K. Gills, Sophia E. Hagolani-Albov, Saana Hokkanen, Sohvi M. J. Kangasluoma, Heidi Konttinen, et al. 2022. "From Extractivism to Global Extractivism: The Evolution of an Organizing Concept." *The Journal of Peasant Studies* 49 (4): 760–792. 10.1080/03066150.2022.2069015.

City of Newark, NJ. 2021. "Newark, NJ, 2021 Environmental Resources Inventory." ArcGIS Hub. 2021. https://hub.arcgis.com/documents/NewGIN::newark-nj-2021-environmental-resources-inventory/about

Di Chiro, Giovanna. 2003. "6. Beyond Ecoliberal "Common Futures": Environmental Justice, Toxic Touring, and a Transcommunal Politics of Place." In *Race, Nature, and the Politics of Difference*. Durham: Duke University Press.

Di Palma, Vittoria. 2014. *Wasteland: A History*. New Haven: Yale University Press.

Fiske, Amelia. 2018. "Dirty Hands: The Toxic Politics of Denunciation." *Social Studies of Science* 48 (3): 389–413. 10.1177/0306312718781505

Hagstrom Company Inc. 1939. "Essex County, New Jersey HOLC Map, 1939." University of Richmond Digital Scholarship Lab. https://s3.amazonaws.com/holc/tiles/NJ/EssexCounty/1939/holc-scan.jpg

Harrison, Rodney. 2011. "'Counter-Mapping' Heritage, Communities and Places in Australia and the UK, Rodney Harrison." In *Local Heritage, Global Context: Cultural Perspectives on Sense of Place*, 79–98. Heritage, Culture and Identity. Farnham, Surrey; Burlington, VT: Ashgate.

Harvey, David. 1982. *The Limits to Capital*. Oxford: Blackwell.
Harvey, David. 1996. *Justice, Nature, and the Geography of Difference*. Cambridge, Mass: Blackwell Publishers.
Kelley, Robin D. G. 2002. *Freedom Dreams: The Black Radical Imagination*. Boston, Mass: Beacon Press.
Kelley, Robin D. G. n.d. "Twenty Years of Freedom Dreams." Boston Review. Accessed October 10, 2022. https://www.bostonreview.net/articles/twenty-years-of-freedom-dreams/
Mah, Alice. 2010. "Memory, Uncertainty and Industrial Ruination: Walker Riverside, Newcastle upon Tyne: Memory, Uncertainty and Industrial Ruination." *International Journal of Urban and Regional Research* 34 (2): 398–413. 10.1111/j.1468-2427.2010.00898.x
Mah, Alice. 2012. *Industrial Ruination, Community, and Place: Landscapes and Legacies of Urban Decline*. Toronto: University of Toronto Press.
McAdam, Doug. 1982. *Political Process and the Development of Black Insurgency, 1930–1970*. Chicago: University of Chicago Press.
Merchant, Carolyn. 2003. "Shades of Darkness: Race and Environmental History." *Environmental History* 8 (3): 380–394. 10.2307/3986200
Newark Public Library. 1996. "The Ironbound's Early History. Part I: Immigrants, Industry Remake a Once-Bucolic Down Neck." Knowing Newark. September 12, 1996. https://knowingnewark.npl.org/the-ironbounds-early-history-part-i-immigrants-industry-remake-a-once-bucolic-down-neck/
Nixon, Rob. 2011. *Slow Violence and the Environmentalism of the Poor*. Cambridge, Mass: Harvard University Press.
Parisi, Albert J. 1983. "Newark Farmers' Market: Dioxin Imperils Future." *The New York Times*, June 12, 1983, sec. 11.
Pezzullo, Phaedra C. 2009. *Toxic Tourism: Rhetorics of Pollution, Travel, and Environmental Justice*. University of Alabama Press.
Rothstein, Richard. 2017. *The Color of Law: A Forgotten History of How Our Government Segregated America*. First edition. New York, London: Liveright Publishing Corporation, a division of W. W. Norton & Company.
Temper, Leah, Mariana Walter, Iokiñe Rodriguez, Ashish Kothari, and Ethemcan Turhan. 2018. "A Perspective on Radical Transformations to Sustainability: Resistances, Movements and Alternatives." *Sustainability Science* 13 (3): 747–764. 10.1007/s11625-018-0543-8
Toxic Heritage. n.d. "Where's the Harm? Investigating Our Toxic History." Toxic Heritage – Collaborative research. Accessed October 11, 2022. https://toxicheritage.com/
US Census Bureau. n.d. "QuickFacts: Newark City, New Jersey." QuickFacts. Accessed June 30, 2022. https://www.census.gov/quickfacts/newarkcitynewjersey
US EPA. 2022. "Learn about Dioxin." Overviews and Factsheets. June 1, 2022. https://www.epa.gov/dioxin/learn-about-dioxin

VISUAL ESSAY 3

GETTING THE LEAD OUT, ONE COMMUNITY AT A TIME

Gabriel Filippelli

Understanding and eliminating pollution exposure pathways to communities is part science, part awareness, and part activation. We have been developing and implementing a host of citizen science programs in partnership with communities, particularly revolving around eliminating lead poisoning. These projects have involved multiple novel tools and techniques, but first and foremost they have involved active community leadership and participation (Map My Environment), which have allowed us to help the community to map and potentially eradicate soil lead contamination. Figure 1 is an example of the types of visualisations developed, with pollution hotspots indicated by shifts to the red spectrum. It is critical to break down the long-held assumption, which arises from the educational and classist privilege that universities are the real holders of knowledge, to be applied to "help" communities. We need to reframe this partnership as one where the communities are not only holders of important local knowledge, but indeed are critical to developing and implementing community-based solutions. This reframing can be considered a heritage-making practice – shaping the stories neighbourhoods tell about themselves, understanding their histories in a new light, and bringing new voices and perspectives into the narrative.

Science needs new approaches, and willing partners, to help inform and begin to solve many of our current challenges. One challenge is simply clear communication of the science, with clear messaging provided to citizen scientists to understand the sources and risks of lead exposure in homes (Figure 2). One approach to meaningful communication and engagement adopted by some environmental health scientists from the birding community is "citizen science," where communities co-create research with university faculty and students to make everything more understandable and actionable, from educational material to participating in the research process to activating and disseminating research products that then inform practice and policy. As noted by Davis and Mah in their edited volume *Toxic Truths* (2020),

DOI: 10.4324/9781003365259-32

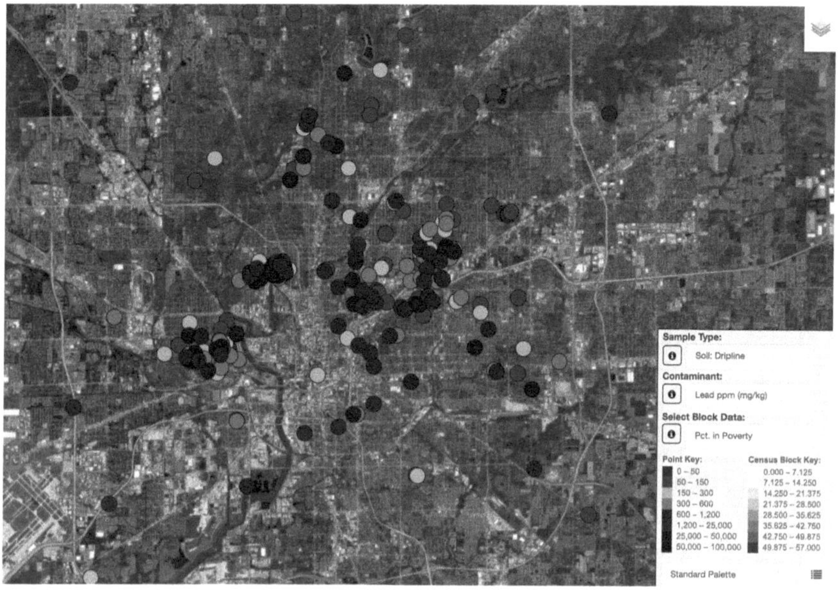

FIGURE 1 An aerial photograph of Indianapolis with shaded (B&W image) or red, orange, yellow, and green (color image) dots. A key in the lower right identifies them as lead pollution hotspots indicated by shifts to darker tones or to the red spectrum.

FIGURE 2 Flyer explaining childhood lead exposure with infographics and a QR code that links to the Anthropocene Household project.

"for decades, environmental justice activists have campaigned against the misuses of science, while at the same time engaging in community-led citizen science," and I argue that some of this misuse centres around exploitative research, where the citizen-scientists themselves were engaged largely as data collection agents for research purposes, and not as research partners themselves.

We have focused our community engagement and co-creation approaches to address one of the massive public health challenges confronting many cities – legacy lead contamination, which permanently impacts the brain development of children (Filippelli and Laidlaw 2010). In its current form, we have robust partnerships with local communities, including faith-based communities constructing testing kits to be distributed in their communities (Fig. 3). Ten years ago, we started off our efforts to understand the environmental legacies of lead contamination from a purely scientific frame – where is the lead, how are children being exposed, and how can we reduce that (Filippelli et al. 2018). We quickly found that communities have many of these same questions. We at the university are hampered in our efforts because we may lack access and participants to adequately address the key scientific questions (Hayhow et al. 2021). Meanwhile, communities are hampered by a lack of affordable access to analytical and technical resources to adequately constrain the community problem. By working together, could both problems be solved? The short answer is yes, but not without facing a bit of a learning curve.

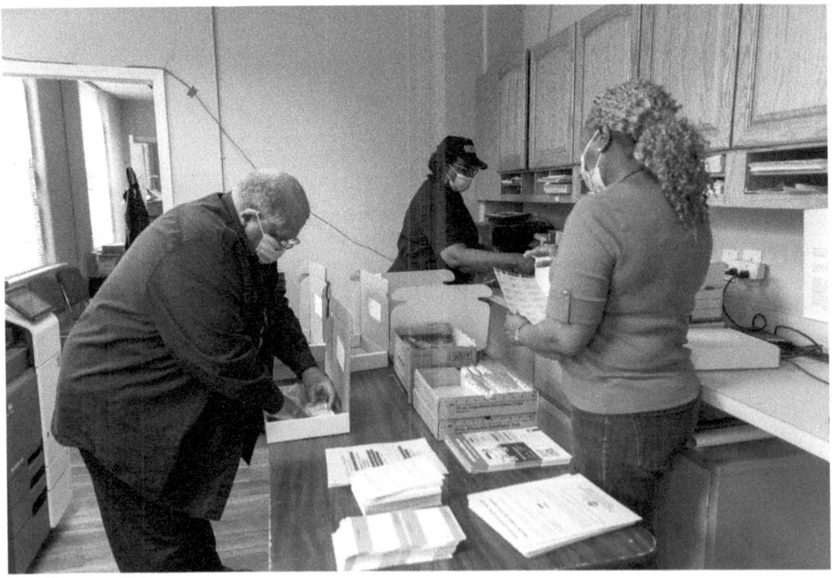

FIGURE 3 Volunteers from a faith-based community partner assemble testing lead kits to be distributed in their communities.

One key lesson learned from our own work on lead contamination is that university-community partnerships have to be like all healthy relationships – supportive, nurturing, trusting, and reciprocal. The needs of each partner are different. Universities produce knowledge and train students, and thus a partnership needs to result in products that are of value to them, such as publications and student theses. Neither of these products have inherent value for communities, who may instead value actionable data to improve the community. Thus, universities must approach these partnerships as not just opportunities to do research **on** communities, but rather **with** and **for** them.

When community partners are brought into the team from the conception phase, they bring with them their heritage of the place, and often new approaches and viewpoints that are relevant and effective within the community context. A naturalist term for including these perspectives is "traditional ecological knowledge," which pertains to urban communities, with their own ecosystems of organisations, leaders, and environments. It is the cultural and community heritage and history of the place that provides rich opportunities for exploration and engagement, including translating scientific practice into workable approaches for both assessing environmental issues and extrapolating the results of research practices into tools that are useful to the community. For example, community voices might bring more straightforward approaches to environmental sampling, such as doing away with the lengthy and often confusing sample labelling protocols used in many laboratories for a simpler and more effective scheme (Fig. 4).

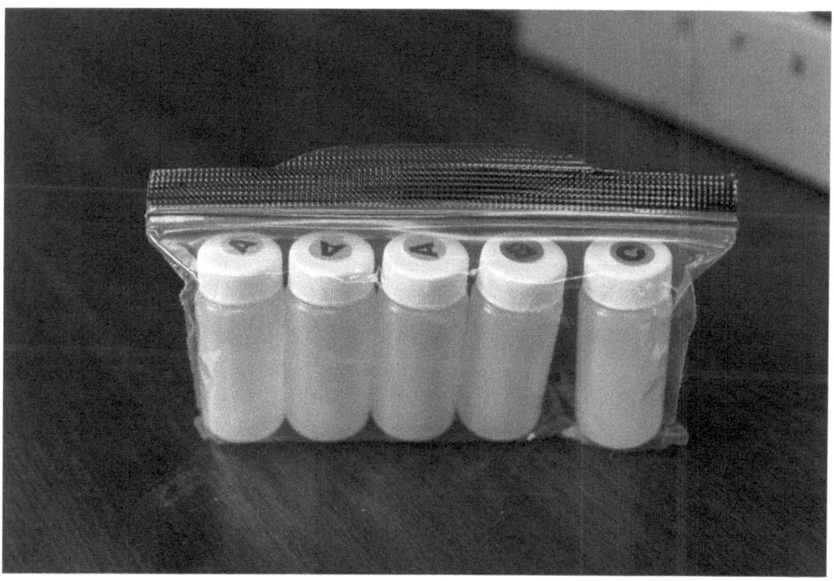

FIGURE 4 Plastic vials for lead testing to be distributed to community citizen scientists.

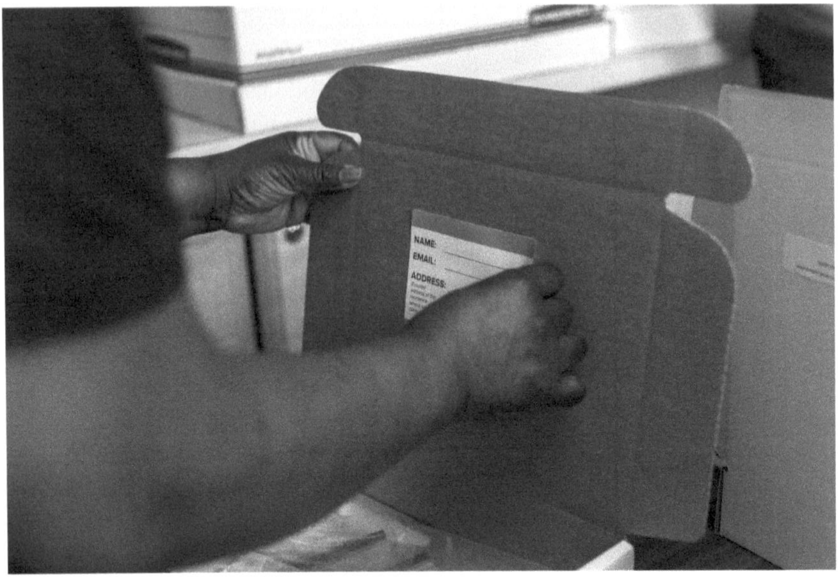

FIGURE 5 Volunteers assemble a lead testing kit box.

Working together with community partners has its own benefits, especially when it comes to accessing difficult sampling locations and mobilising a team of community volunteers to package and distribute sampling kits—one example is "kit construction parties" held as part of our home sampling efforts (Fig. 5). Community members have access to their own backyards, can take samples themselves, and can deliver those samples to central locations, such as churches, for collection by researchers.

Communities often lack the expertise and technologies to confront some of the environmental challenges that they face, leaving them with few options. Universities, on the other hand, have ample expertise and resources (Fig. 6) and are often literally blocks away from these same under-resourced neighbourhoods, but often narrowly view environmental problems as those that can be solved by scientific research alone. The very fact that legacy contaminants continue to overburden many communities is not a scientific problem per se, but rather one that requires the voices, talents, and actions of those communities – without those perspectives, no amount of science will solve the problem. Partnerships can effectively bridge that knowledge chasm, bringing tools and resources to bear to solve problems. Additionally, students can learn more about how to apply their training to real-world environmental problems.

Educating the next generation of young learners is critical to overcoming the intergenerational burdens of community contamination. Young people are additional research partners and citizen scientists, bringing their own unique worldviews to many of these issues. In an informal setting, youth are more open and free to express their ideas, and their emotions, about the environment surrounding them – especially

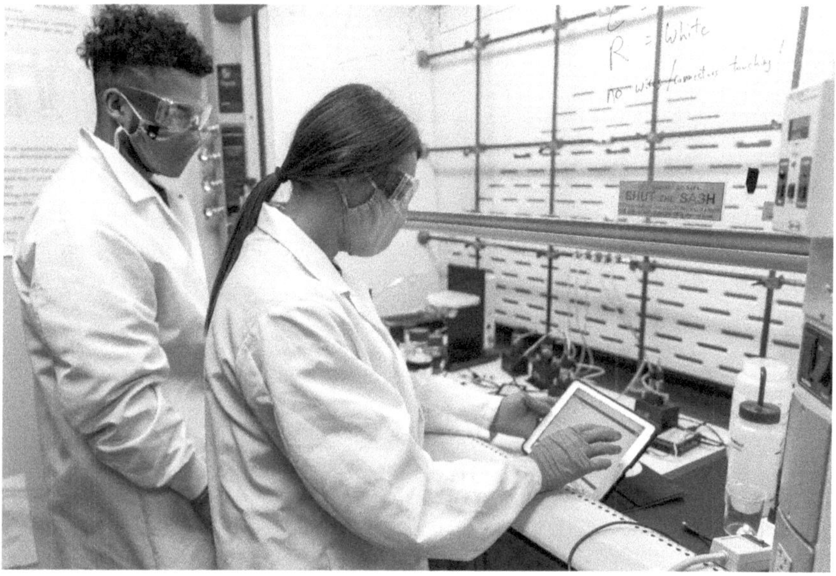

FIGURE 6 Lead testing samples analyzed in the lab.

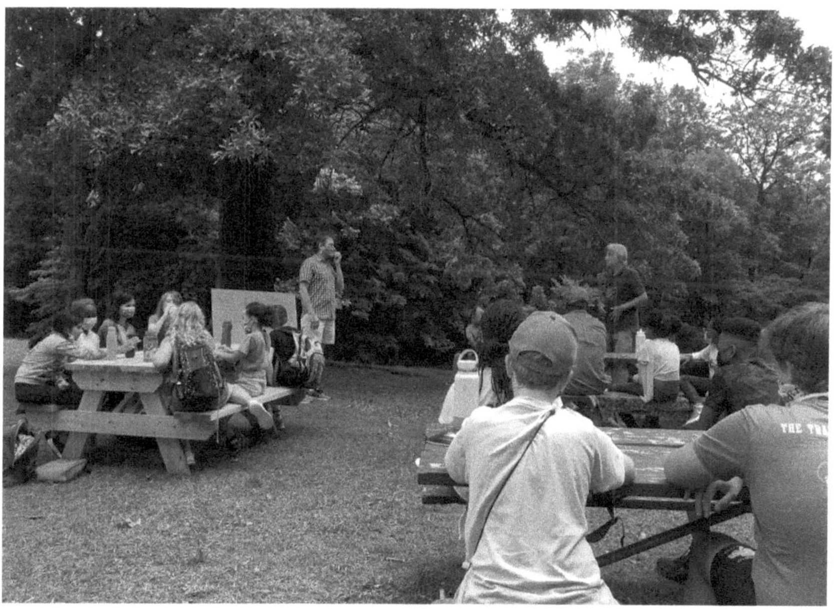

FIGURE 7 Project leaders (including the author) present information about lead contamination to community members.

if conducted in the actual environment of natural space as opposed to a classroom (Fig. 7). Furthermore, teachers sometimes grapple with implementing technical curricula that are relevant to some communities, and by providing opportunities for children and teachers to learn about and confront real problems in their neighbourhoods. One example is education about the history of racist policies such as redlining and zoning that contributed to the cycles of disinvestment and disproportionate impacts of pollution in the first place, thus engaging with the troubling heritage of contaminated communities. By engaging with their community heritage, emerging learners can find relevance, value, and agency in the concepts that they learn in school.

All images, except the soil lead map, are courtesy of Liz Kaye, Indiana University.

References

Davis, T. and Mah, A. (Eds.). 2020. *Toxic Truths: Environmental Justice and Citizen Science in A Post-truth Age.* Manchester University Press, ISBN 9781526137029.

Filippelli, G.M. and Laidlaw, M.A.S. 2010. The Elephant in the Playground: Confronting lead-contaminated soils as an important source of lead burdens to urban populations. *Perspectives in Biology and Medicine*, 53: 31–45.

Filippelli, G.M., Adamic, J., Nichols, D., Shukle, J., and Frix, E. 2018. Mapping the urban lead exposome: A detailed analysis of soil metal concentrations at the household scale using citizen science. *International Journal of Environmental Research and Public Health*, 15(7), 1531. 10.3390/ijerph15071531

Hayhow, C., Brabander, D., Jim, R., Lively, M., and Filippelli, G.M. 2021. Addressing the need for just GeoHealth engagement: Evolving models for actionable research that transform communities. *GeoHealth*. 10.1029/2021GH000496

CASE STUDY 8

CLIMATE MUSEUM UK: PRACTICES IN RESPONSE TO THE TRAUMASPHERE

Bridget McKenzie

Eco-psychologist Zhiwa Woodbury (2019) has proposed that Earth's systems can presently be understood as the Traumasphere. Over centuries, ecocidal activities have created traumatised, toxic localities and a global condition of biosphere trauma. In particular, burning fossil fuels has generated climate trauma, the biggest recent wounding of the biosphere, with worse impacts to come.

Our organisation Climate Museum UK (abbreviated to CMUK) aims to detoxify and regenerate culture by curating and gathering responses to the Earth crisis and designing trauma-sensitive 'activations'. (The term culture is meant in both senses: 'culture' as an expression of social paradigms and 'Culture' as a sector of arts and heritage.)

There are three aspects to CMUK's definition of a museum. Firstly, we are an **experimental museum,** rethinking institutions in an Earth crisis, working with partners to deepen responses and invite cultural audiences to see with 'eco-lenses'. Secondly, we are a **distributed museum**, a network of creative practitioners activating our local communities, helping people to express 'Earth emotions', to explore toxic histories behind the crisis, and to open imaginations to possible futures. A collective that collects, we do not extract and accumulate objects in one place; instead, each individual gathers and releases material that is personally owned by themselves or participants. Thirdly, we are a **digital museum**, creating a 'Possitopia Online', with stories of the impacts of the Earth crisis and the potential for regenerative change.

We play a provocative role in advocating an emergency response to professional communities tackling the challenge of engaging communities with these problematic issues, particularly in our founding role with Culture Declares Emergency,[1] a movement of cultural sector individuals and organisations telling truths, taking action and seeking justice. As network members of the Happy Museum Project,[2]

DOI: 10.4324/9781003365259-33

we have contributed to its manifesto calling upon museums to be agents of change in response to the multiple crises. We present regularly at a growing number of conferences, journal issues and research initiatives on questions of curating and collecting climate, such as our chapter in the publication, Reimagining Museums for Climate Action (Harrison & Sterling 2021).

We foreground **activation** as our purpose, distinct from traditional museums that exist to collect and exhibit objects. If all museums can be redefined through practices that we model as imaginative change-makers co-creating a distributed commons, we can help to detoxify the sector while somewhat detoxifying the planet (Figure 1).

As dispersed practitioners with diverse approaches, we are united by core principles, three of which are: aiming to be **Participatory,** involving people democratically; **Intersectional,** acknowledging intersecting factors of injustice; and **Planet-kind**, beyond doing no harm we aim to heal in practical ways.

Our most unique principle is to be **Possitopian** in future thinking. (McKenzie 2020). This expands the cone of possible futures, drawing on geophysical realities while also harnessing the powers of the imagination. We help people imagine future scenarios which are potentially more catastrophic or more abundant than

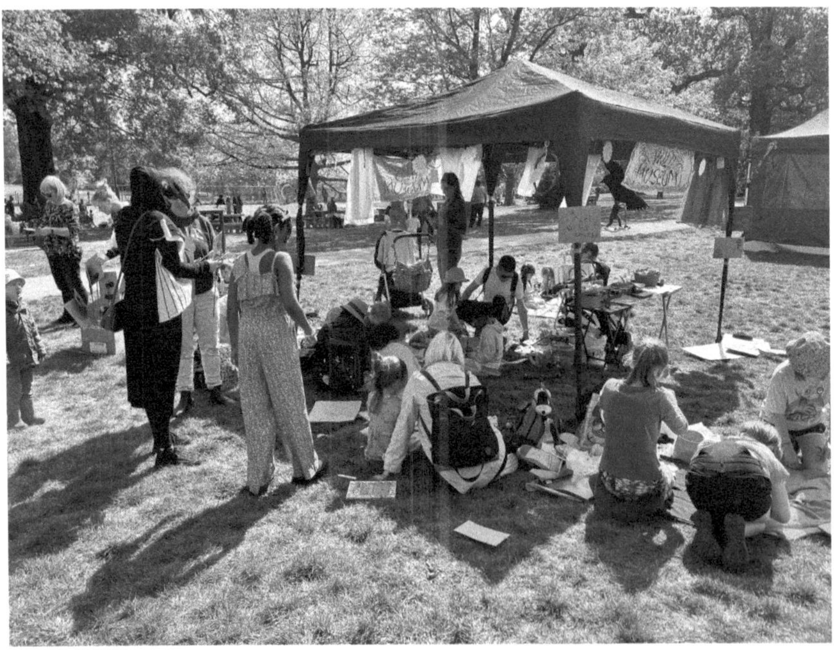

FIGURE 1 CMUK practitioners in the guise of animal curators, as The Wild Museum. Some 'activations' are outdoors helping us reach diverse audiences and linking climate to ecology. Photograph by the author.

they might normally think, weaving these together in reference to real contexts. Our Digital Museum plan is to activate people to create personal lives, communities and ecosystems that are rich with possibility. In a planned platform, our practitioners will showcase their offers and capture stories of resistance, resilience and regenerative change in their places.

Another principle is being **Holistic**. An ongoing collective conversation is how to help participants expand perspectives to see interconnected emergencies and their causes, making visible links between colonialism and extraction; ecocidal harm and climate breakdown; toxicity and illness; biodiversity loss and zoonotic diseases.

We enable this Holistic principle in a range of ways, for example, by working with partner museums to put an 'eco lens' on their collections and to expand perceptions of the Earth crisis. With Tullie House in Carlisle, through staff training and the recruitment of a young artist practitioner, Megan Bowyer, we are exploring links between collecting natural specimens, local histories of material extraction and climate change. As a CMUK associate, Megan has engaged local groups in creative responses to collections of fossils and natural history records, and has contributed her artworks to the Museum's exhibition 'Human change, not climate change'.[3]

Another way to enact our Holistic principle is by designing graphic tools which we use in workshops, toolkits and 'Print It Yourself displays'. Amy Scaife has commissioned illustrators with Global South perspectives to create posters for her mobile experience, **hubRen**. She rides her cargo tricycle to events around Waltham Forest and unfolds banners and posters to stimulate discussion about planetary boundaries and climate justice. She shares with us what she hears, most frequently a version of these words: *'I am scared, I am worried about climate change, I don't know what to do, what can I do?'* (Scaife 2022).

Another example of a graphic tool is 'Earth Crisis Blinkers', used in workshops for participants to expand the framings of the crisis (Figure 2). It builds on a diagram by Jan Konietzko (2022) 'Carbon Tunnel Vision', adding more detail in three zones: Impacts on Lives; Planetary Boundaries; and Solutions.

A final detoxifying principle is being **Compassionate**, increasing empathy and care for others in the human and more-than-human realms, and developing trauma-sensitive approaches. We carried out a Listening Project for A New Direction in 2021[4], which showed that people need considerately designed spaces that acknowledge the Earth crisis. In particular, young people are not empowered to speak, contrary to media portrayals of youth as natural eco-warriors. With eco-anxiety layered onto developmental sensitivities, they feel their only recourse is to speak because they are so limited in realms of action, and yet they do not feel educated enough to speak confidently against dominant suppressive discourses.

It is challenging for anyone, of any age or expertise, to grasp the scale and complexity of what is broken. As practitioners, we each hope to be agents for

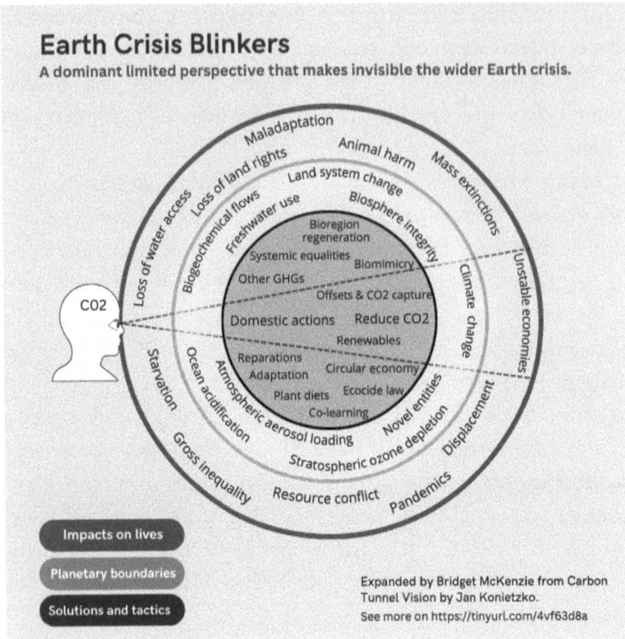

FIGURE 2 Diagram showing the concept of Earth Crisis Blinkers, used in CMUK training.

imaginative expansion, to support people to see systemic harms and the potential of regenerative change. We must balance aiding the expression of feelings with a more transpersonal uncovering of systemic issues. In every public interaction, we experiment in small and varied ways on how to manage this.

Notes

1 Culture Declares Emergency https://www.culturedeclares.org/
2 Happy Museum Manifesto https://happymuseumproject.org/our-happy-museum-manifesto/
3 Tullie House project 'Once Upon a Planet', 2022. https://www.tulliehouse.co.uk/events/once-upon-planet
4 https://www.anewdirection.org.uk/research/listening-projects

References

Harrison, Rodney & Sterling, Colin eds. 2021. *Reimagining Museums for Climate Action*, Museums for Climate Action, UCL, https://www.museumsforclimateaction.org/mobilise/book

Konietzko, Jan. 2022. *Moving Beyond Carbon Tunnel Vision*. Cognizant website. https://digitally.cognizant.com/moving-beyond-carbon-tunnel-vision-with-a-sustainability-data-strategy-codex7121

McKenzie, Bridget. 2020. *Explaining Possitopia.* CMUK website. https://climatemuseumuk.org/2020/10/15/explaining-possitopia/

Scaife, Amy. April 2022. *Joining the Dots in Waltham Forest.* CMUK website. https://climatemuseumuk.org/2022/04/29/joining-the-dots-in-waltham-forest/

Woodbury, Zhiwa. 2019. *Climate Trauma: Toward a New Taxonomy of Trauma.* Ecopsychology, 1–8. https://doi.org/10.1089/eco.2018.0021

17

TOXIC HERITAGE AND REPARATIONS: ACTIVATING MEMORY FOR ENVIRONMENTAL AND CLIMATE JUSTICE

Liz Ševčenko

Introduction

November 2022 marked a critical moment for heritage in the global effort to confront the climate crisis. The nations gathered at COP27 summit centered their discussions on "loss and damage," or how generations of emissions by the countries in the global north are now negatively impacting countries in the global south, and how that history should guide where money is invested. Vulnerable nations had been advocating for this historical approach for decades; but global north countries had consistently avoided facing their past, fearing the liability that would require them to assume (Pardikar 2021). At the 11th hour, the US – which has emitted more CO2 over time than any other country in the world – was the last COP member to concede to a "loss and damage" fund through which countries responsible for the most cumulative emissions would pay for the harms they caused.

If we understand the climate crisis as a problem of history and memory, then heritage work is central to environmental and climate justice. This chapter considers climate denial as a form of historical denial, largely driven by the US, rooted in this country's deep culture of refusing to acknowledge and assume accountability for its global leadership in structural racism. It explores the potential of participatory public memory to combat that denial, through the experience of teams of students, frontline environmental justice organizers, and community members, who came together across 20 cities to create public histories of environmental justice in their localities. Their experience suggests the potential for participatory public memory to support more than "loss and damage" claims, but to bring about a broader view of reparations, imagined, in the words of philosopher Olúfẹ́mi Táíwò as "a historically informed view of distributive justice, serving a larger and broader worldmaking project." (Táíwò 2022b, 74)

DOI: 10.4324/9781003365259-34

Climate as a Problem of History and Memory

Climate denial is often understood as a denial of science. But it is equally a denial of history: of generations of past human actions and the toll they have taken on the environment and people. As environmental justice scholars have long argued, the environmental destruction that produced the climate crisis is rooted in a larger history of global racial capitalism, one that extracted and expended both people and natural resources. Thus the histories of slavery and Indigenous genocide are inextricably linked to climate change (Principles of Environmental Justice 1991; Melosi 2005; Táíwò 2022a, 2022b).

As I have argued elsewhere, the US state's refusal to account for these historical harms has supported a culture of climate denial (Ševčenko 2022, Ch. 11). For nearly three centuries, despite consistent demands from generations of advocates, the US federal government refused to consider reparations to people who had been enslaved and their descendants or even officially apologize for this historical harm (Araujo 2017). It should come as no surprise, then, that while the US has historically contributed the most emissions that cause climate change, it leads the world in climate denial. In 2019, a poll across 23 nations found, only Saudi Arabia and Indonesia had a greater percentage of respondents denying climate change and human responsibility for it (Milman and Harvey 2019). This popular culture supports a US climate policy driven by historical denial, part of a pattern of refusing to officially reckon with past harms. Barack Obama came to power with a pledge to close Guantanamo, but rejected calls to officially account for the torture and abuse that took place there, arguing Americans should "look forward as opposed to looking backwards" (Johnston and Savage 2009). He also refused to assume responsibility for America's disproportionate historical role in the emissions that caused climate change, famously instructing his emissary to the 2011 Durban climate negotiations to reject any agreement that held the US to account for past actions (Klinsky and Brankovic 2018). As many climate policy analysts have noted, this refusal to reckon with the past has crippled global efforts to fight climate change, becoming the major block for countries like China and India, whose emissions are skyrocketing now but were tiny in the past, to agree to emissions restrictions (Klinsky and Brankovic 2018; Sengupta 2021; Táíwò 2022a, 2022b).

Historical denial has also shaped, and curtailed, the efforts of American environmentalists to stop climate change. Academic environmental historians initially focused on white scientists and activists as the main actors in the story of the environmental movement, erasing people of color from official history (Melosi 2005). The mainstream climate change movement was dominated for decades by affluent white activists who remained silent on the racist history of the environmental movement from which it emerged. Founding conservationists also supported eugenics, as part of an integrated vision of national purity (Purdy 2015). The post-war environmental movement cast overpopulation, especially of poor people of color, as its main enemy; strong factions within the Sierra Club's membership organized against immigration (Hopkins 2018). Labeling immigrants and people of color as threats to environmental purity denies the long traditions of environmental organizing and voting, sustainable practices, and resistance to environmental damage that many of these communities brought and developed, while wealthier white Americans radically expanded their consumption and waste. As a 2009 report on the state of the climate movement observed,

> most climate activists are primarily concerned with addressing the technological challenges of climate change. They frame climate change in terms of a scientific problem … . They tend to overlook the ways in which climate change is linked to historical exploitation and injustice.
>
> (Park 2009, 11)

By contrast, the environmental justice (EJ) movement, which emerged from civil rights struggles led by people of color, and formally coalesced in the First People of Color Environmental Leadership Summit in 1991, clearly framed both environmental and climate destruction as issues of historical justice. The EJ movement analyzed housing segregation, labor abuses, air and water contamination, and fossil fuel extraction as interrelated harms caused by larger systems of exploitation. Its approach to the climate crisis foregrounded history, articulating climate change as a historical process inextricably intertwined with centuries-long histories of racism, colonialism, and other structural inequalities (Murdock 2020).

By framing the climate crisis as a result of historical harms, the EJ movement made reckoning and repairing those harms central to confronting the crisis. The principles established at the 1991 Summit, recognized and referred to across the movement from then on, include that "Environmental Justice protects the right of victims of environmental injustice to receive full compensation and reparations for damages." (Principles of Environmental Justice 1991) But the EJ movement's calls for reparation are not limited to environmental reparations, but are grounded in a much more expansive vision of historical accountability, one that understands environmental racism as one prong of structural racism. The Indigenous Environmental Network, for instance, calls for "redressing past harms and creating new relationships of power for the future through reparations (Indigenous Environmental Network n.d.)." For the groups assembled in the Climate Justice Alliance, a Just Transition requires "reparations for land that has been stolen and/or destroyed by capitalism, colonialism, patriarchy, genocide and slavery." Here all historical harms from white supremacy undergird environmental harms; all reparations are environmental reparations. Building on this formulation, philosopher Olúfẹ́mi Táíwò argues that a just path through the climate crisis requires not only reparations for environmental harms but for the broader systems that caused them, recovering the anti-colonial vision of reparations as "central to the expansive project of building a more just world, not just as a material or symbolic mechanism of redress for past harms" (Táíwò 2022a, 3).

By 2020, mainstream environmental organizations with histories of white supremacy began to recognize broader racial reparations as necessary for their mission. That year the Sierra Club came out in support of HR 40, the federal bill to study reparations, explaining the organization now "believes it is impossible to create a healthy, safe, and sustainable future for all without acknowledging and materially addressing the past and present economic, cultural, psychological and spiritual impacts of racism" (Sierra Club 2020). The League of Conservation Voters signed on a year later, arguing, "an environmentally sustainable future can only stand on a foundation that dismantles the historical and contemporary legacies of racial oppression that the United States was built on" (Hinkson 2022).

So how should historical reparations that confront the climate crisis be achieved? Legal advocates have tried to adapt existing tools to force accountability for climate change: from charges of "loss and damage" to human rights abuse. But legal tools alone have proven insufficient for winning redress for environmental harms. Instead, legal advocates have begun arguing for a more interdisciplinary approach, where memory work plays a critical role. At the international level, where climate negotiations have been blocked by the US's refusal to assume responsibility for its historical emissions, some policy advocates turned to the idea of a climate truth commission, a space for people to share histories – both individual and structural – of climate impact. In return for public acknowledgment of responsibility and some form of symbolic reparations, polluters would be free from a legal or financial penalty (Klinsky and Brankovic 2018). At the local level, as legal scholar Catherine Millas Kaiman outlines, US legal tools are not equipped to address the needs of EJ communities, where the harms are most severe. Instead, she argues for a multi-pronged approach to environmental reparations that includes recognition and responsibility, driven through activities such as sharing memories of historical harm, as well as material reparations mandated through legislation (Kaiman 2016).

If we recognize that climate change is an accumulation of historical harms; that the world's inability to meaningfully confront the climate crisis is rooted in the refusal of the Global North – led by the US – to acknowledge and redress these harms; then saving the planet requires facing, and fixing, history. This places people who work with history and heritage at the center of climate solutions. But it requires them to use their tools in different ways than they traditionally do, and apply them to different contexts than they traditionally work in. Movements for material reparations require a combination of historical research and legal advocacy; oral history and organizing. They present a unique challenge to heritage and public memory workers to develop new practices and work with new partners.

Applying EJ Principles to Heritage Work: Participatory Public Memory for Climate Justice

In October 2017, nearly 50 historians, local environmental justice advocates, and public history faculty from 30 cities in the US, Latin America, and Europe came together to design a public history project – and process – to pursue climate justice. *Climates of Inequality: Stories of Environmental Justice* was the third project of the

Humanities Action Lab (HAL), a coalition of universities, issue organizations, and public spaces around the country that develops collective public memory projects around social justice issues. We gathered at the Rutgers University campus in Newark, NJ, HAL's home base, a city whose post-war history was forged from environmental racism and local resistance to it. Over the next three years, universities and community organizations in each city developed courses through which students and frontline community members dug into the history of environmental justice in their locality, and how it shaped local experiences of the climate crisis.

Local teams each co-created one "chapter" of a collective traveling exhibit and web platform integrating stories from all participating cities, which local teams then took turns hosting in public libraries, museums, or community organization offices, with public dialogues and actions connecting their local issues to shared experiences (Figure 17.1). Along the way, through virtual exchanges and convenings in person, participants grappled with how the project could promote a frame for discourse and

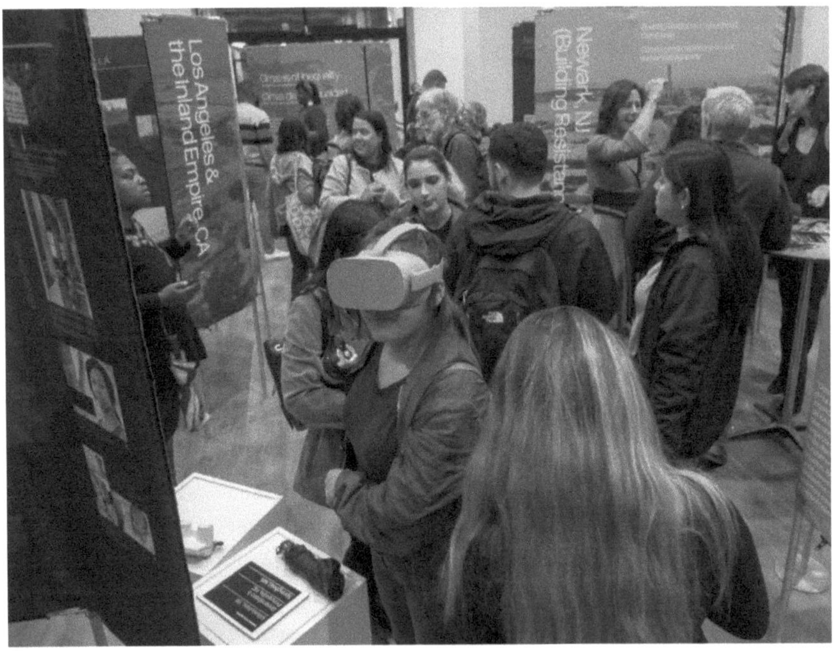

FIGURE 17.1 The Climates of Inequality traveling exhibit brought together all 21 local teams' environmental justice stories in a single installation, that included "nooks" about each locality, virtual reality "visits" to local sites, oral history interviews on Ipads, and interactive elements. As the exhibit traveled to each community that created it, community members could see their own stories and learn about others. Credit: Shelley Kusnetz.

action on the climate crisis grounded in the terms, concerns, spaces, and histories of communities disproportionately affected by it.

The project was created through participatory public memory: a process that applies the logic and modes of social movements to the ways we deal with the past. HAL's participatory public memory projects mobilize thousands of people from disparate locations and perspectives to research and exchange their local histories, then grapple with the collective implications – and identify collective demands – together. These projects invite all participants to serve simultaneously as authors and audience, researching and interpreting history through collaborations among students, community organizers, and faculty, continuously teaching and learning from each other while building a community around shared social change goals. These projects do not underestimate the power of historical denial. While they surface and circulate historical truths that have been suppressed, by doing research and sharing it in exhibits, they do not assume that stating historical truths more loudly or widely than those who would deny them, or legitimizing them in government truth commissions, can alone bring acceptance of those truths. Instead, they approach public history as an ongoing process of organizing, more analogous to movement building than to didactic education. In this way, they activate memory as a way both to establish claims and mobilize people to come together to make them.

The collaboration among EJ organizers and public historians suggested possibilities for a framework of public memory for climate justice that integrated movement organizing and memory work. EJ leaders' approaches were grounded in the Jemez Principles for Democratic Organizing, a foundational framework established in 1996. One is to "Let people speak for themselves" (Jemez Principles 1996). Applied to public memory, to ensure stories of frontline communities are more widely disseminated and understood, they should not be mediated by heritage professionals or students outside of frontline communities.

Another commitment EJ organizers introduced was that the relationships among people in the project should prefigure the ones in the world participants hoped to bring about. Moving the world from an extractive to a regenerative economy required moving from extractive to regenerative relationships. To apply this principle to a collaborative public memory project requires resisting exploitative structures and practices that remain entrenched in heritage and public history. For instance, despite public history programs' strong rhetorical support for "sharing authority" or "community collaboration," the vast majority of students are trained to "get stories" by conducting one-way interviews with directly impacted people and using them in a project, with no further involvement with the interviewees, their organizations, or their communities. A regenerative heritage practice would involve partners designing a project collaboratively for mutual benefit, which could include mutual interviewing that engaged both partners' relationship to the history being explored. The collaboration between IUPUI and Kheprw Institute in Indianapolis, for instance, taught students to read their city's social infrastructure; such that "as much as infrastructure and ecology are integral to the city's ecosystem, so too are the social relationships of its communities" (Kryder-Reid, Holzman, Nadaraj, and Humphrey 2022, 210).

The EJ commitment to relationships undergirded a theory of change that also challenges traditional heritage practice, even practice pursuing social justice. Most public history work designed to tell "untold" or "underrepresented" stories sought to raise awareness among people outside of directly impacted communities, with an unstated presumption that change would come when stories were told to those who had not listened for them before. But after decades of climate denial and environmental racism driven by white affluent voters, whose way of life depended on overconsumption and fossil fuels, EJ organizers did not put much stock in those constituencies as effective routes to change. Instead, participants in the *Climates of Inequality* project prioritized exchanging stories within environmental justice communities, to ground people in shared histories of both racism and resistance, and mobilize them to build a different future. For participants in the IUPUI/Kheprw Institute collaboration, "participants in the project came to new understandings of their own lived experience and positionality within the city's ecosystem" (Kryder-Reid, Holzman, Nadaraj, and Humphrey 2022, 210–211).

The EJ movement has long recognized that nation-states would not save the planet and would not advocate for the welfare of poor communities of color. Nor could local communities make a global change by working alone. Many groups have therefore organized translocally, bringing diverse local communities together into coalitions that highlighted their distinct experiences, while connecting them to common structural problems and mobilizing them into campaigns around shared goals for change (Baptista 2019, Mendez 2020). A participatory public memory project for climate justice must similarly work translocally, highlighting the specific histories of frontline communities while linking them together into the larger structural phenomena of which they are all a part. This translocal work can also link local claims for reparation into a national, or global, campaign, in at least two ways. First, it can provide solidarity and support across communities, and leverage national recognition to pressure local authorities. But perhaps more importantly, linking local experiences to shared national and global histories strengthens the basis for reparations claims, by demonstrating that current inequality is not the fault of individuals, nor a fluke of a single place or time, but rather rooted in structural racism of global capitalism.

Guided by EJ organizers, teams of students in frontline communities across the US dug into their communities' histories of environmental racism and how it was shaping their current and future experience of climate change. Connecting this history with their own memories, they created multimedia projects to share their stories with other communities who had distinct, but linked, historical experiences, and identify shared visions for the future.

Newark

In the fall 2018 Climates of Inequality class at Rutgers University-Newark, which focused on the city's Ironbound neighborhood, student after student spoke about asthma. One held up her pump; another showed a slide with all the paraphernalia she and her family used; another shared how in her middle school, "I remember before gym all the kids from the Ironbound would run to the nurse to go and get their inhalers." That semester, 1 in 4 residents of the Ironbound had asthma, three times the average rate in the rest of the state (USEPA 2015). For generations, the neighborhood's immigrant and Black residents were burdened with the toxic underbellies of systems the larger region enjoyed. From tanneries and breweries, to distribution of goods from the port through trucks spewing exhaust, to incinerating New York City's trash into plumes of pink poison gas, Newark's central role in production and waste brought myriad toxins to its air, soil, and water. Perhaps the most egregious is the Diamond Alkali company's manufacture of Agent Orange in the 1950s and 1960s – the chemical specially designed to decimate green landscapes in the Vietnam War – and its dumping of dioxin into the nearby Passaic river, causing cancer rates to skyrocket.

Students were in middle school when Superstorm Sandy tore through New Jersey in 2012. The storm churned up masses of toxic waste from generations of sediment that flooded streets, playgrounds, and living rooms. Climate change, by bringing higher water levels and increased chance and frequency of flooding, has brought the neighborhood's poisoned history into the present. Co-instructors Maria Lopez-Nuñez of the Ironbound Community Corporation (ICC) and filmmaker Julie Winokur of Talking Eyes Media recognized the need to connect students to the knowledge and experience they already possessed, or were in their own communities. One strategy was through a "Toxic Tour", a central practice of EJ organizers in Newark and other EJ communities, which ICC leader and environmental justice scholar Ana Baptista analyzes in Chapter 9 (Baptista, this volume) (Figure 17.2).

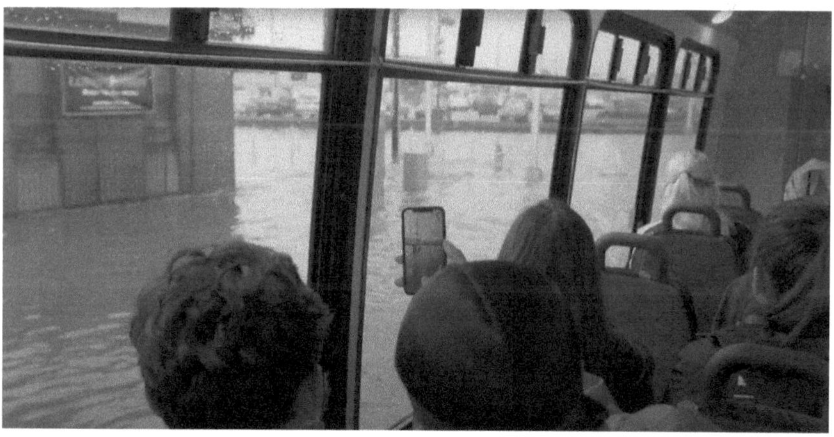

FIGURE 17.2 Rutgers-Newark students on a "Toxic Tour" given by the Ironbound Community Corporation, where a routine rainstorm caused severe flooding. Credit: Talking Eyes Media.

During this bus tour of sites like the incinerator and the superfund site, a heavy but not unusual rain, having drained fine elsewhere in the city, created miniature lakes on the route that rose higher than the bus door. As the bus plowed through, the brown water flooded in. "People think you need a hurricane to do this," explained Lopez-Nunez, who was giving the tour. "We don't. We just need a day full of rain. This is a problem. And not everyone lives this way, ok?" (HAL 2018) "This is the most craziest field trip I ever went on," one student later remembered. "That was my awakening moment" (HAL 2018).

The students in this class had almost all grown up in Greater Newark, many of them in the Ironbound, and literally bore its history in their bodies. Most students reported being unaware of the environmental hazards they had grown up with. And when asked at the outset of the class what environmental justice meant to them, students suggested recycling, changing light bulbs, and other acts of individual consumers. Despite growing up in an environmental justice community – characterized by extraordinary activism in response to egregious environmental burdens – the power of mainstream environmental discourse was such that it dominated students' framework, even as it was barely relevant to their own lives. To support students in building a new framework for understanding their own experiences, Lopez-Nuñez introduced students to generations of ICC activists in the Ironbound. Over the past 50 years, ICC had driven investigation and acknowledgment of the Diamond Alkali dioxin contamination; designation of a superfund site; and a lawsuit against the Covanta incinerator for spewing toxic iodine into the air, among many other fights. It also created an urban farm, a network of childcare, and other community resources (Baptista, this volume). After interviewing activists of different ages who fought and won these changes in the neighborhood, students then took water and soil samples from the places where they conducted the interviews, and sent them to a Rutgers-Newark lab to analyze lead levels. On this one metric alone, the levels far exceeded legal limits.

Connections with activists reframed students' own histories and potential futures. One student remembered that growing up in the Ironbound, "I was like I gotta get out of here. When I talked to [the activist she interviewed] it really opened my eyes because he was like I wanna go back in. That was the point when I stepped back and said we do need to be part of it to make a change." For another, conversations with activists "helped me realize that there needs to be change at a systematic level. We matter and if we continue to demand more we'll have better conclusions" (HAL 2018).

New Orleans

Shannon Rainey welcomed four University of New Orleans students lugging video equipment into her beautifully appointed home in Gordon Plaza, to explain how she came to lead a movement to abandon it. When she first learned of Gordon Plaza, a federally-subsidized development built in 1981 in New Orleans' upper

9th ward and marketed to Black families as an affordable path to homeownership, it seemed like a dream come true: "I was excited to find out that I was able to purchase one, so I did," she recounted to Lones Gagnard, Vickie Lacoste, Ella McIntire, and Daniel Lamplugh. What the outreach and advertising for Gordon Plaza didn't mention, and none of the majority Black residents knew, is that Gordon Plaza had been built on top of a toxic landfill. The ground beneath the tidy new houses consisted of over a half-century of the city's garbage, as well as its medical and industrial waste, covered with an additional layer of pesticide spray. "Things was going fine until we started digging in the yard," Rainey remembered. "We wound up digging up canisters with skeleton heads on it … we found out as time went on that we were living on a toxic landfill" (Gagnard et al. 2019).

After immediate advocacy by residents, a few years after the site opened the EPA began testing soil, and, Rainey explained, "they found like 150 chemicals that was back here that could cause us to die." (Gagnard et al. 2019) Indeed, decades after the first residents moved in, the census tract containing Gordon Plaza had the second-highest cancer rate in Louisiana (Maniscalco et al. 2019). Years after the soil samples revealed known carcinogens, in 1994 the area was declared a superfund site. The EPA's clean-up consisted of adding a layer of new soil on top of any exposed ground, separating the clean and contaminated soil with a porous mat.

Like in Newark's Ironbound, Gordon Plaza's layers of toxic history formed the foundation of its experience with climate change. When Hurricane Katrina hit in 2005 and the levees broke, massive floods devastated Gordon Plaza homes, and inundated the thin layer of topsoil the EPA had added. While the EPA came and tested for lead and found levels within legal limits, the residents' environmental scientist technical consultant found that groundwater leached through the topsoil and was exposing residents to a multitude of carcinogens.

Shannon Rainey was president of Residents of Gordon Plaza. For nearly four decades and over a half dozen mayoral administrations, multiple generations of residents have been fighting for the city to relocate them to homes that fulfill Gordon Plaza's original promise. Starting from the first class action lawsuit in 1994, residents launched multiple legal actions to claim restitution from the city. Nearly all have forced some acknowledgment of the horrific harm and secured minimal cash payments; but none yet has won the ultimate goal of a fully funded, just relocation (Figure 17.3).

The students interviewing Shannon Rainey didn't grow up in Gordon Plaza; but her story helped them understand their own. Lones Gagnard's father died of lung cancer after exposure to asbestos at the Avondale Shipyard, whose directors were later discovered to have been well aware of the toxic levels and their effects on human beings. Vickie Lacoste shared that she grew up two blocks from a paper mill in rural Louisiana, where "The sounds of toxic flares and transport trains mobilizing during the night were a common soundscape in my early growth years," but "I never gave much thought to the dangers that existed from living so near an industrial plant" (HAL 2019). If the students had resonant experiences with contamination, they had not grown up with exposure to community organizing around

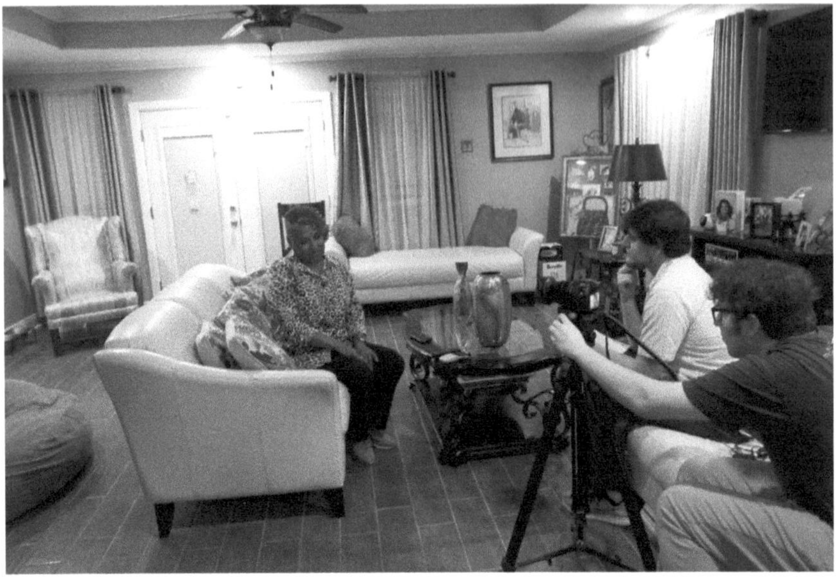

FIGURE 17.3 University of New Orleans students interviewing Shannon Rainey in her Gordon Plaza home. Credit: Ella McIntire.

it. When the students met Rainey and other Gordon Plaza activists, they were in the process of a major public awareness and political action campaign, which students wound up supporting, participating in actions like RISE for Cancer Alley, a rally connecting the histories and activist communities of Gordon Plaza residents with black homeowners living near the petrochemical plants constructed on former plantations along the Mississippi River.

Rainey was already a prominent spokesperson for the movement in the press; but students hoped to make the stories she told part of New Orleans' public history. Students integrated Rainey's video interview and contextual research on Gordon Plaza's history and current struggles into the New Orleans Historical app of digital city walking tours, and in the Humanities Action Lab's Climates of Inequality nationally traveling exhibit (New Orleans Historical and The Midlo Center for New Orleans Studies at the University of New Orleans n.d.; HAL 2019).

Puerto Rico

After Hurricane Maria tore through Puerto Rico in September of 2017, Ricia Ann Chansky, a professor at the University of Puerto Rico Mayagüez, spent the semester feeding her students and neighbors on a propane tank by candlelight, while opening a space for people to share stories, survival tips, and emotional support. The immediate physical damage was severe enough; but the deeper harm came from the total collapse of the island's infrastructure. The electrical grid was almost completely destroyed, leaving the majority of the island without power. Less than half had tap water; sewage treatment plants were out; cell service was gone; only a few radio stations could broadcast; fuel shortages grounded people in place; and there was a run on banks. The situation lasted for not days or even weeks but months, with deaths growing exponentially with each day the government failed to respond, and the psychological trauma growing deeper and deeper. But in this total vacuum of official support, communities constructed an extraordinary infrastructure of care, information, and resources.

Both the crumbling of the government grid and the emergence of the civic/communal infrastructure that replaced it were rooted in Puerto Rico's long history. Maria hit an island that was already ravaged by centuries of colonialism that left its government weak and under-resourced. The last decade had left it particularly vulnerable: a recent change in the tax system, compounded by the 2008 economic recession, created a debt crisis to which the Puerto Rican government responded with severe economic austerity measures. By 2017, the island's public infrastructure was little more than a house of cards. Maria exposed how the US had written off the entire island as a "sacrifice zone," a term coined to describe particular neighborhoods, almost always low-income communities of color, damaged by environmental harms or disinvestment. President Trump's denial of climate change, let alone climate inequality, included supporting the government of Puerto Rico in an erasure of the thousands of people who lost their lives in this disaster: while the island government insisted the death toll was no more than 64, after massive pressure from activists and research from investigative journalists, a year later the government officially revised the number to nearly 3000 (Klein 2018).

The government's total abandonment of its citizens was both physically and physiologically devastating. As predictable and consistent with history as this abandonment may have been, it was still deeply traumatizing. But citizens had another tradition to build on: one of organizing protest and communal care in crisis, by what University of Puerto Rico at Mayagüez students called "citizen responders". One initiative that emerged from the wreckage was "Mi María", a disaster response based on sharing stories. Chansky brought together over 150 UPR students to interview people in their home communities. Where Newark students grew up with slow violence all around them, students at UPR were struggling for immediate survival after the sudden destruction of their homes and neighborhoods. Both struggled with the revelation or the confirmation that the state was not there for them, even in the direst circumstances. The Mi María project gave students agency and community as "citizen responders" using the unique tool of storytelling (Figure 17.4).

By inviting people to exchange and record what they had seen and experienced – including the students themselves – the project provided a multi-layered resource including healing from trauma; circulating vital information for survival; citizen science documentation of impacts and needs; and organizing to demand justice. Students exposed an abandoned warehouse of donations that had never been distributed; created and circulated information booklets; and began building a multimedia archive of stories of survival (Chansky and Denesiuk 2021).

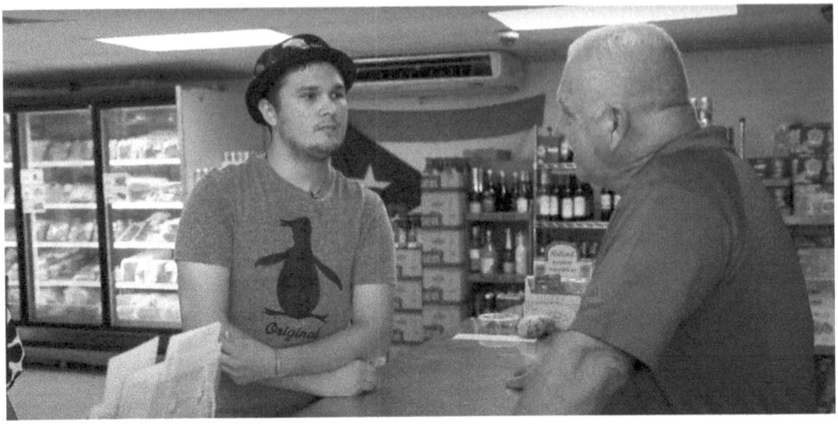

FIGURE 17.4 University of Puerto Rico at Mayagüez student interviewing a shop owner about his experience trying to support his community in the aftermath of Hurricane Maria. "The merchandise started to run out, since we were the only open business," he remembered. Credit: Talking Eyes Media.

Conclusion

If the climate crisis is in large part driven by historical harms, and their denial, then those who work historically – who understand confronting history as a precondition for building a just future – stand at the center of the fight against climate change. This is not to say that those named as history or heritage professionals have an exclusive role or responsibility. Rather, activating heritage to combat climate change requires people working in every sector to grapple with their pasts. History and heritage professionals can seek collaborations with advocates, scientists, policymakers, and others to connect people with their historical experiences and mobilize to redress their legacies. This vision of reparations includes material and legal restitution, as well as the acknowledgment, healing, and education at the core of public memory work. Connecting communities and their granular local histories in participatory public memory can build a dynamic, national understanding of the historical harms that need to be righted for a Just Transition.

References

Araujo, A. L. 2017. *Reparations for Slavery and the Slave Trade: A Transnational and Comparative History*. Bloomsbury Academic; HPOD edition.

Baptista, A. L. "Environmental Justice Tours: Transformative Narratives of Struggle, Solidarity, and Activism," Chapter 9, this volume.

Baptista, A. L. 2019. "The Evolution of the Environmental Justice Movement: Translocal Voices for Systemic Transformations," In Canty, J. (Ed.), *Globalism and Localization: Emergent Solutions to Ecological and Social Crises*. Taylor & Francis Group.

Chansky, R. A., and Denesiuk, M. 2021. *Mi María: Surviving the Storm: Voices from Puerto Rico*. Haymarket Books.
Gagnard, L. III, Lacoste, V., Lamplugh, D., and McIntire, E. 2019. "Oral History with Shannon Rainey, Gordon Plaza," https://climatesofinequality.org/project/oral-history-with-shannon-rainey-gordon-plaza/
Hinkson, L. 2022. "The Time for a Federal Reparations Commission is Now," League of Conservation Voters Blog, accessed at https://www.lcv.org/article/the-time-for-a-federal-reparations-commission-is-now/
Hopkins, H. 2018. "How the Sierra Club's History With Immigrant Rights Is Shaping Our Future," accessed at https://www.sierraclub.org/articles/2018/11/how-sierra-club-s-history-immigrant-rights-shaping-our-future
Humanities Action Lab. 2018. "Documentary," accessed at https://youtu.be/TCkodLRCG7M
Humanities Action Lab. 2019. "International Exchange Post Spring 2019, student reflection on Notion platform, maintained by Humanities Action Lab."
Humanities Action Lab. 2018. International Exchange Post Fall 2018, student reflection on Notion platform, maintained by Humanities Action Lab.
Indigenous Environmental Network, n.d. "Indigenous Principles of Just Transition," http://www.ienearth.org/wp-content/uploads/2017/10/IENJustTransitionPrinciples.pdf
"Jemez Principles for Democratic Organizing" 1996. accessed at https://www.ejnet.org/ej/
Johnston, D., and Savage, C. 2009. "Obama Reluctant to Look into Bush Programs," *New York Times*, January 11, accessed at https://www.nytimes.com/2009/01/12/us/politics/12inquire.html
Kaiman, C. M. 2016. "Environmental Justice and Community-Based Reparations," *Seattle University Law Review*, Vol. 39, 1327–1374, accessed at https://digitalcommons.law.seattleu.edu/cgi/viewcontent.cgi?article=2351&context=sulr
Klein, N. 2018. *The Battle for Paradise: Puerto Rico Takes on the Disaster Capitalists*. Haymarket Books.
Klinsky, S., and Brankovic, J. 2018. *The Global Climate Regime and Transitional Justice*. Routledge.
Kryder-Reid, E., Holzman, L. M., Nadaraj, A., and Humphrey, L. 2022. "An Environmental Justice Lens on Indianapolis's Urban Ecosystem: Collaborative Community Curation." In Mutnick, D., Cuonzo, M., Griffiths, C., Leslie, T., and Shuttleworth, J. M. (Eds.), *The City is an Ecosystem: Sustainable Education, Policy, and Practice* (1st ed.). Routledge, pp. 205–219.
Maniscalco L., Zhang L., Yi Y., Lefante C., Rosales C., Hsieh M. C., and Wu X. C. (eds). 2019. *Cancer Incidence in Louisiana by Census Tract, 2005-2015*. Louisiana Tumor Registry.
Melosi, M. 2005. "Environmental Justice, Ecoracism, and Environmental History," In Glave, D. D., and Stoll, M. (Eds.), *To Love the Wind and the Rain: African Americans and Environmental History*. University of Pittsburgh Press, 121–122.
Mendez, M. 2020. *Climate Change from the Streets: How Conflict and Collaboration Strengthen the Environmental Justice Movement*. Yale University Press.
Milman, O., and Harvey, F. 2019. "US Is Hotbed of Climate Change Denial, Major Global Survey Finds." *The Guardian* May 7. https://www.theguardian.com/environment/2019/may/07/us-hotbed-climate-change-denial-international-poll. For methodology, see https://www.theguardian.com/world/2019/may/01/populism-what-is-yougov-cambridge-globalism-project-methodology
Murdock, E. 2020. "A history of environmental justice: Foundations, narratives, and perspectives," In Coolsaet, B. (Ed.), *Environmental Justice: Key Issues*. Taylor & Francis Group.
New Orleans Historical and The Midlo Center for New Orleans Studies at the University of New Orleans, n.d. "Gordon Plaza: Residents Fight for Fair and Just Relocation," accessed at https://neworleanshistorical.org/tours/show/100

Park, A. 2009. "Everybody's Movement: Environmental Justice and Climate Change." Environmental Support Center.
Pardikar, R. 2021. "The US Is Blocking Climate Reparations." *Jacobin* November 17.
"Principles of Environmental Justice." 1991. accessed at https://www.ejnet.org/ej/
Purdy, J. 2015. "Environmentalism's Racist History." *The New Yorker*, August 13, accessed at https://www.newyorker.com/news/news-desk/environmentalisms-racist-history
Sengupta, S. 2021. "Calls for Climate Reparations Reach Boiling Point in Glasgow Talks," *The New York Times* November 11.
Ševčenko, L. 2022. *Public History for a Post-Truth Era: Fighting Denial through Memory Movements*. Routledge.
Sierra Club. 2020. "Sierra Club's Statement on Reparations," accessed at https://www.sierraclub.org/sierra-clubs-statement-reparations
Táíwò, O. O. 2022a. "The Fight for Reparations Cannot Ignore Climate Change," *Boston Review*, January 10, 2022.
Táíwò, O. O., 2022b. *Reconsidering Reparations*. Oxford: Oxford University Press.
United States Environmental Protection Agency. 2015. "EPA Agreement with Port Authority and Port Terminal Operators Will Cut Harmful Pollution from Idling Trucks in Newark and Beyond," July 23 accessed at https://archive.epa.gov/epa/newsreleases/epa-agreement-port-authority-and-port-terminal-operators-will-cut-harmful-pollution.html

CASE STUDY 9

FROM LEFTOVERS TO TAKEOVER: LATENT INSURGENCY AMIDST THE SYSTEM'S REMNANTS

Ana Valderrama

Saladillo stream is one of the two urban streams that gave the city the original name of Rosario city: *Pago de los Arroyos* – in English, *site of the streams*. During the first 50 years after Rosario's virtual foundation in 1835, the stream gave place to multiple productive and recreational activities located on the peripheries of the city. Meat production, fishing, and the port attracted local people and foreign immigrants, affirming the identity of what is known as "The South of Rosario." However, in the past 20 years, the stream has been increasingly polluted and abandoned by official policies, turning a blind eye to the spill of glyphosate and other agricultural chemicals, the leak of industrial waste, the location of dumps and landfills, and the illegal production of bricks on its banks. This process of intoxication has been accompanied by rapid growth and overcrowding of informal settlements. Nowadays, around 1,000 families are settled in flood-prone areas, without any physical, social or environmental infrastructure. Besides, 30% of the people who live on the basins collect, classify and recycle waste inside their homes or in the immediate surroundings. This generation and everyday contact with toxicities in the stream basins could spread and have irreversible negative effects on the health of human and non-human entities (Figure 1).

The *Matéricos Periféricos*' production on Saladillo Stream began with the research project called "Architecture of the Periphery" that emerged after the fall of the Berlin Wall, with the consequent reappearance of neoliberal policies in Latin America and the dissolution of the Welfare State in Argentina. At the end of 1999, the housing deficit in Rosario had reached 20.000 units, and peripheries began to transform into a fabric of polluted green infrastructures, dumps, landfills, encroached informal settlements, and new hybrid uses. The group decided to abandon the formal city and propose possible futures for Rosario's peripheries through architectural and urban projects.

DOI: 10.4324/9781003365259-35

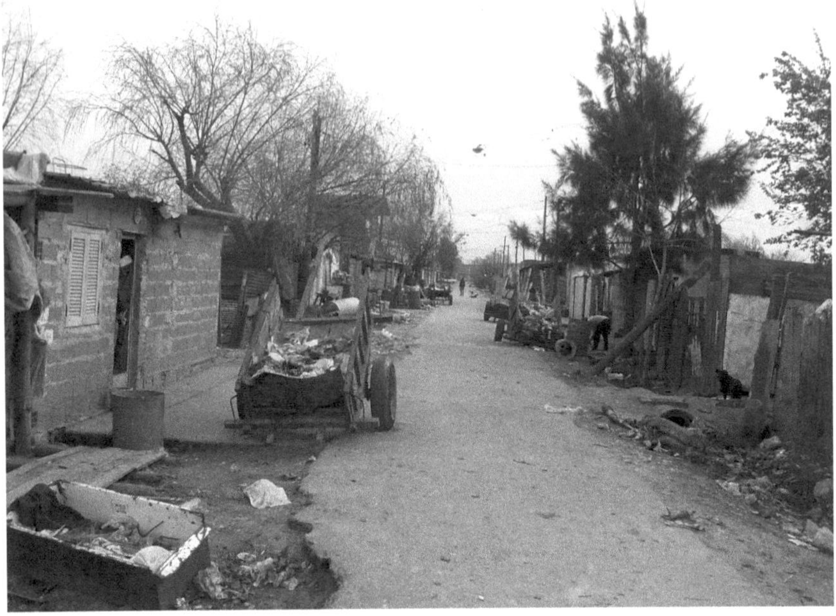

FIGURE 1 Ana Valderrama, *Las Flores (flowers) Neighborhood*, 2005. Digital photograph.

The *Great Argentinian Crisis*[1] – which reached its peak in 2001 with the resignation of the president – anticipated a future of accumulative poverty and toxicity and of definitive abandonment of the periphery by the Government. After 2001, the research turned into a platform to monitor the evolution of the social production of habitat and the intersections with material mobilizations. The periphery's materials, that is, the waste, and the leftovers of the system were the main sources for research and design. *Matéricos Periféricos* developed interpretative tools such as maps and diagrams to record those processes, as well as participatory methodologies to work with communities. From then until now, those recording and participatory tools contributed to producing land-based and situated proposals related to social and environmental justice.

Between 2005 and 2008, the "soy boom" provoked new displacements of the poor towards the peripheries. The group signed an agreement with the Rosario Government to explore alternatives on the banks of Saladillo Stream where informal settlements, dumps, landfills and illegal production of bricks increased and interwoven dramatically while glyphosate and industrial pollutants circulated through them. The group led a series of workshops that proposed a chain of green spaces that would recover 131 hectares of land for public use along with social housing, garbage separation plants, health, educational, religious, and sportive facilities, clean brick production plants, and community gardens. Those projects constituted the official Master

Plan of Saladillo Stream, although none of them ever materialised. The Plan became the very fetish of local urban planning.

Around the year 2008, with a certain pessimism about the capacity of the Government and urban planning to resolve the issues, *Matéricos Periféricos* appealed to the culture of "doing with our hands, and doing with what it is at hand." The group decided to actively participate in the transformation of territories in social or environmental emergencies and created design-build workshops. During the workshops, professors, students, social organisations and local communities co-produced community infrastructures made of recycled, donated and discarded materials collected from the place. The workshops were a highly effective instrument with which to reinforce awareness of environmental pollution and risks, underpin local institutions in their community activities, challenge state agency, and activate cooperative processes. Over time, some of the artefacts have forced public interventions otherwise unimagined. In addition, the processes involved in the workshops helped to build community bonds and to unveil insurgent features latent in the ground.

The extraordinary foreign debt acquired by the Argentinian government in 2018 and the Global Pandemic increased the number of people below the poverty line to 45% of the Argentinian population and the housing deficit in Rosario had reached 50.000 units. In 2019, the Rosario government (Municipalidad de Rosario 2002) officially recognized that they were burying 800 tons of Solid Urban Waste (SUW) per day in landfills. Of the 800 tons of that SUW, 58% were organic material, 32% were reusable material (plastic, paper, glass, metals, cartons and wood), and 10% were non-recyclable metals. This meant that 250 tons per day of material were "missed" in the cycle of circulation. Nevertheless, as Lecitra (2010) pointed out, there was a parallel informal circuit of SUW that mobilised 100 tons of SUW per day. This parallel informal circuit is carried out by *Cartoneros, Carreros y Recicladores*—in English, *cardboard collectors, cart collectors and recyclers*. In 2019, there were approximately 3000 in Rosario, and a third of them lived on the banks of Saladillo Stream.

At the end of 2019, *Matéricos Periféricos* started a new project called "Map of industrial solid waste. Towards a circular model of re-insertion of urban waste into the productive system." The project (in progress) consists of a georeferenced survey, a catalogue, a digital application for the management of solid waste located in the peripheries, and a series of technological developments to contribute to shifting the traditional *cradle to grave* as defined by McDonough and Braungart (2002). It is expected to contribute to a continuous industrial cycle in Rosario by providing precise information about the materials available in the periphery and innovative technologies for the construction and design industry. Equally important, the project plans participatory engagement and collaboration with *Cartoneros, Carreros y Recicladores* to ensure articulation of knowledge, mutual access to scientific results and the generation of safe conditions for their work.

But *Cartoneros, Carreros y Recicladores* are not only individual collectors and recyclers. During the Argentinian crisis in 2001, they had an important political role:

through demonstrations in the streets and solidarity activities, they were one of the main organised groups who contributed to the end of Neoliberal project of the nineties in Argentina, along with *Piqueteros* and other social and political organisations.[2] Since 2011, they have been part of the Confederation of Workers of the Popular Economy (CTEP).[3] CTEP is nowadays one of the main organised forces of resistance to the violence that the colonial-imperial system exerts through the practices of accumulation by dispossession combined with the environmental pollution in our country.

Notes

1 For more information about the *Great Argentinian Crisis* see Brenta, Noemí and Mario Rapoport. (2003). "El FMI y la Argentina en los años 90: de la hiperinflación a la hiperdesocupación." In *Los guardianes del dinero. Las políticas del FMI en la Argentina,* 67-100. Ciudad Autónoma de Buenos Aires: Editorial Norma.

2 For more information about the role of *Piqueteros* see Retamozo, Martín. (2006). "El movimiento de los trabajadores desocupados en Argentina: cambios estructurales, subjetividad y acción colectiva en el orden social neoliberal." *Argumentos* 19 (50): 145–166, https://www.scielo.org.mx/scielo.php?script=sci_arttext&pid=S0187-57952006000100007

3 For more information about CTEP see https://ctepargentina.org/nacio-la-utep/

Bibliography

Brenta, Noemí and Mario Rapoport. 2003. "El FMI y la Argentina en los años 90: de la hiperinflación a la hiperdesocupación." In *Los guardianes del dinero. Las políticas del FMI en la Argentina,* 67–100. Ciudad Autónoma de Buenos Aires: Editorial Norma.

CTEP, "Nació la UTEP," https://ctepargentina.org/nacio-la-utep/

Harvey, David. 2003. *The 'new' Imperialism: Accumulation by Dispossession.* Oxford: Oxford University.

Lecitra, Micaela. 2010. "Reducir, Reutilizar y Reciclar: El problema de los residuos sólidos urbanos." *Derechos Humanos y Ambiente AI 041.* https://geic.files.wordpress.com/2010/10/reducir-reutilizar-y-reciclar.pdf

Liboiron, Max. 2021. *Pollution is Colonialism.* Durham: Duke University Press.

McDonough, William and Michael Braungart. 2002. *Cradle to Cradle: Remaking the Way We Make Things.* New York: North point press.

Municipalidad de Rosario. 2002. "Plan General Integral de Residuos Sólidos Urbanos (PGIRSU) *Municipalidad de Rosario.* Accessed March 28 2022. https://www.rosario.gob.ar/ArchivosWeb/libro%20PAR_10%2002.pdf

Retamozo, Martín. 2006. "El movimiento de los trabajadores desocupados en Argentina: cambios estructurales, subjetividad y acción colectiva en el orden social neoliberal." *Argumentos* 19 (50): 145–166. https://www.scielo.org.mx/scielo.php?script=sci_arttext&pid=S0187-57952006000100007

VISUAL ESSAY 4

TAKING CARE OF NUCLEAR WASTE

Cornelius Holtorf

Introduction

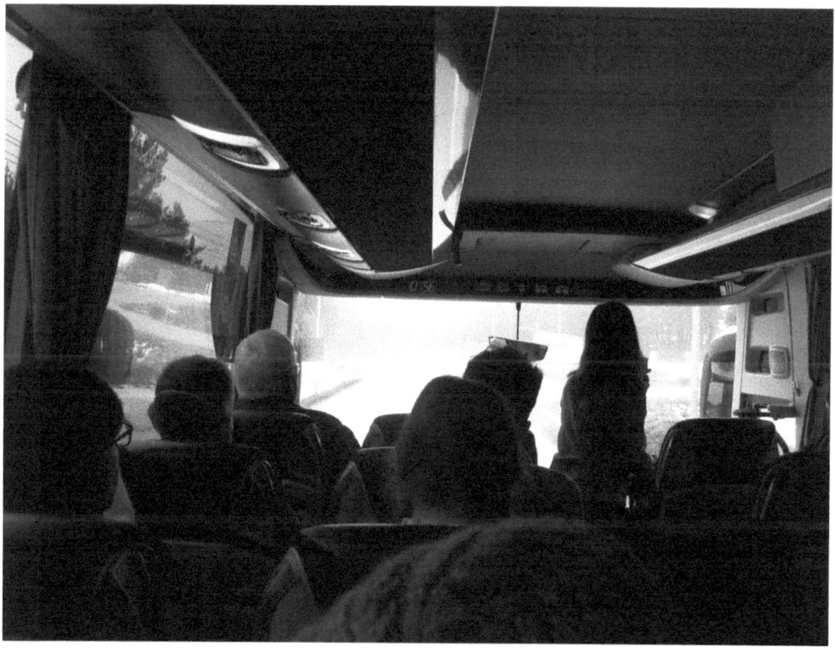

FIGURE 1 After passing the gates to the nuclear facilities at Olkiluoto, Finland, photographs are permitted only in designated places but mental images can be taken anywhere you go. Photograph: Cornelius Holtorf.

DOI: 10.4324/9781003365259-36

This visual essay is based on a trip to Helsinki and Olkiluoto in Finland in April 2022. The occasion was the Sixth International Conference on Geological Repositories (ICGR), organised by the Nuclear Energy Agency (NEA) at the OECD and this time hosted by the Ministry of Economic Affairs and Employment of Finland in Helsinki. After the conference, my colleague Leila Papoli-Yazdi and I took part in a site visit to Olkiluoto on the west coast of Finland. This is the site of three nuclear power stations, an operational underground repository for low and intermediate-level radioactive waste and the building site of ONKALO, the Finnish geological repository for spent nuclear fuel. Onkalo is widely known as the first repository of its kind being built in the world, and as the subject of Michael Madsen's powerful documentary *Into Eternity* (2010). The site is operated by the company Posiva.

From existential risk to global climate action

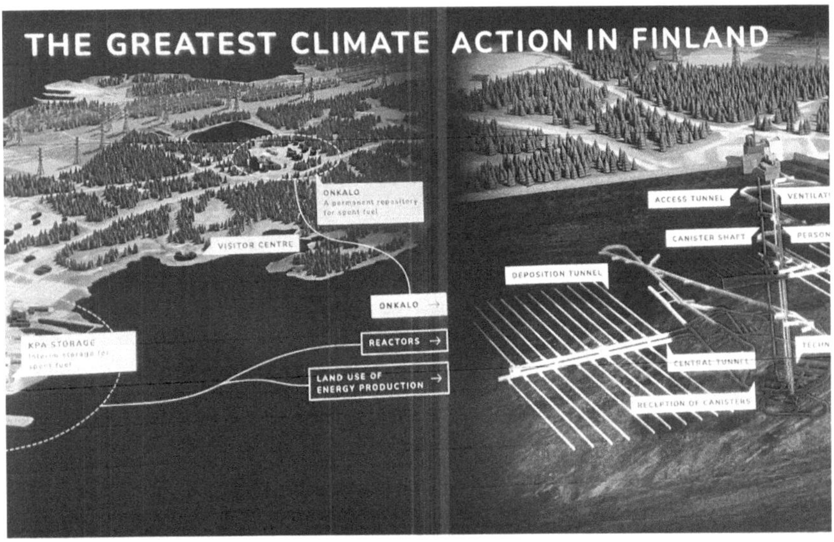

FIGURE 2 "The Greatest Climate Action in Finland" – display in the Visitor Centre at Olkiluoto. Photograph: Pasi Tuohimaa.

100,000 years is a very long time and some spent nuclear fuel and other radioactive waste decay very slowly. But understandings in society of the significance of nuclear energy, the roles and responsibilities of the nuclear energy sector, and thus the appreciation of its legacy change fast. Radioactive waste had not been given much attention when nuclear energy was first introduced, but in the 1970s the anti-nuclear movement came to see the long-lived legacy of burnt nuclear fuel to be the Achilles heel of nuclear energy, sensitising citizens and politicians for environmental concerns. Today, some people in the environmental movement have started to reconsider this position, as nuclear energy (unlike coal, oil, and gas) does not emit carbon into the atmosphere and can thus contribute to minimising climate change threatening human futures. Radioactive waste increasingly appears as a smaller problem than climate change. Unsurprisingly, the nuclear energy sector and Posiva now pride themselves in being able to contribute to sustainable energy production, protecting the biosphere.

The art of forgetting

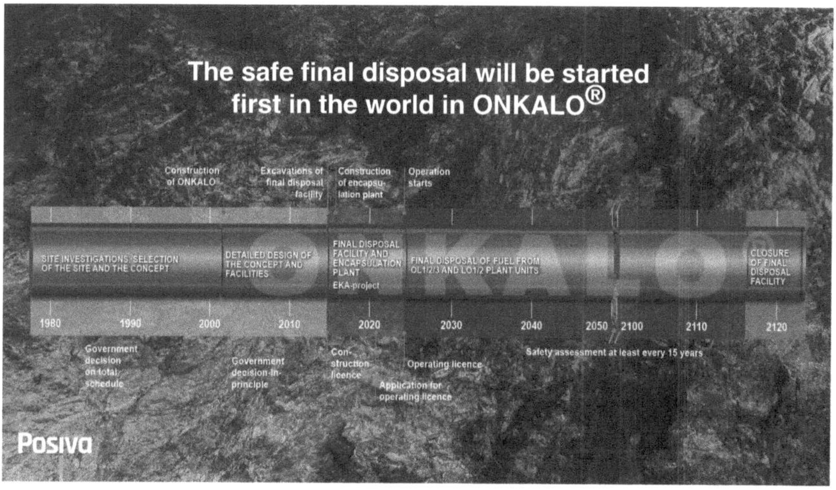

FIGURE 3 ONKALO Timeline. Graphic provided by Pasi Tuohimaa.

On the long coach journey to Olkiluoto, I am reading Ian Hodder's book *Entanglement* (2012). Things are usually taken for granted by people, their own lives and histories being forgotten. "But suddenly the things return or transform and have to be dealt with", argues Hodder (2012, 103), reminding us that people and things tend to get entangled with each other, sometimes unwittingly and unexpectedly. According to the ONKALO timeline, the geological repository at Olkiluoto is expected to be closed approximately 100 years from now. From then onwards, the design of the repository is intended to ensure passively the safety of people and other living beings. During the site visit, we listened to a presentation by Mika Pohjonen, Managing Director of Posiva Solutions, Posiva's offshoot company selling know-how in nuclear waste management and final disposal. When I asked him publicly when the site will be completely forgotten, he first apologised about his poor memory but then responded that it will definitely be forgotten at the start of the next Ice Age, in about 10,000 years. According to Posiva, no active oversight or other actions will be required in the long term. But what does it mean that in the ONKALO Research Gallery exhibition, the visualisations of the site 4,000 and 100,000 years ahead were entirely devoid of any trace of living beings other than trees? Does this vision represent a worrying post-human dystopia or a welcome Earth-centric utopia?

Taking a people-centred approach

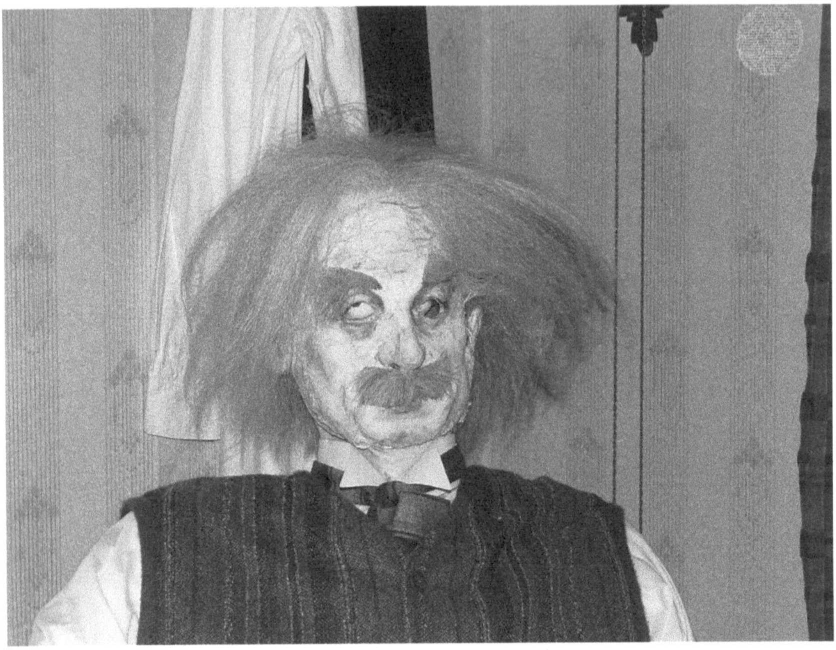

FIGURE 4 Would you buy a used car from him? Scientific genius Albert Einstein in a didactic display in the Olkiluoto Visitor Centre, presumably intended to reinforce the credibility of science. Photograph: Cornelius Holtorf.

During the ICGR conference in Helsinki, there was a session on building and maintaining trust in society, run by the NEA Forum on Stakeholder Confidence. Speakers reflected on the importance of communication and dialogue, especially with local communities near (possible) sites of geological repositories. There is something of a consensus in the global radioactive waste sector that the biggest challenges to be solved are not technical and scientific but about public acceptance, i.e. social and cultural. Accordingly, one slide of Mika Pohjonen's presentation emphasised the importance of trust and transparency for public acceptance: "it takes years to earn the trust, and only minutes to lose it – we do not risk this under any circumstances." Yet at Olkiluoto, such people-centred views were surprisingly absent. Although Posiva catered well for the visitors' needs, the displays and messages were thoroughly object- and techno-centred, largely devoid of prominent consideration for people. The Visitor Centre features a much-neglected puppet of Albert Einstein, apparently embodying the science people were asked to trust. Elsewhere, the silhouette of a mother with a daughter was drawn to indicate the scale of nuclear fuel rods inside the nuclear reactor, as if it was perfectly logical and safe for them to be there. In the ONKALO Research Gallery exhibition, we met a dummy of a dehumanised worker, reduced to his high-viz clothes and safety accessories, placed next to a metal fence with a red "keep out" sign.

Uncertainty as an opportunity for care

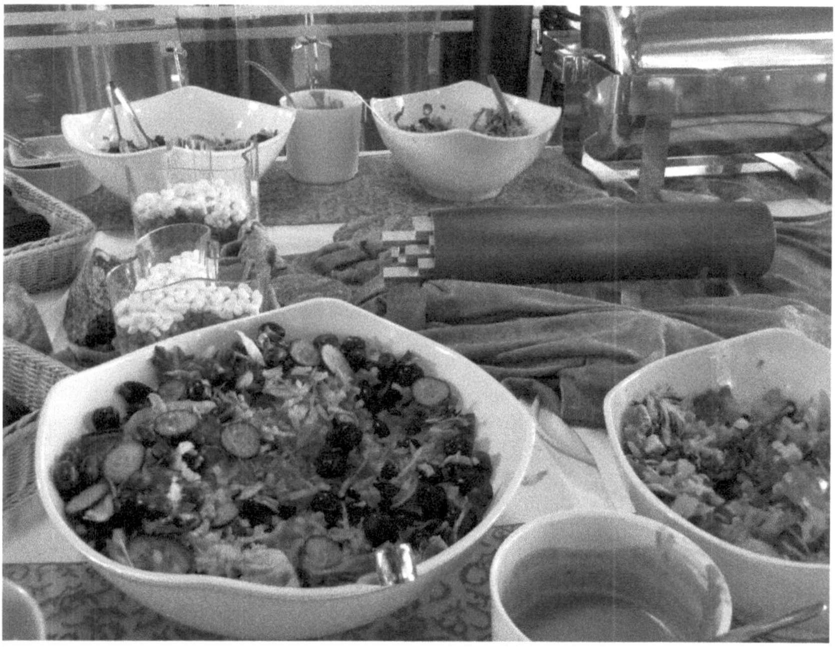

FIGURE 5 In the Visitor Centre at Olkiluoto, Posiva Solutions is taking good care of its visitors. Instead of flowers, the buffet table is embellished with a reference to technical beauty. Photograph: Cornelius Holtorf.

Another session at the ICGR conference focused on communicating scientific uncertainty from the scientists to the community. One senior scientist representing the European Joint Programme on Radioactive Waste Management (EURAD) asked with consternation what had gone wrong that people were so fearful of geological repositories when in fact there are far greater uncertainties (and risks) in daily life that do not cause corresponding reactions. Several speakers argued that better ways of communicating scientific reasoning and an improved understanding of science will lead to wider acceptance by the public. But ironically, trust and acceptance may also result from naivety and ignorance, whereas comprehensive knowledge of science may foster scepticism and opposition to nuclear engineering. The so-called deficit model of the public understanding of science dismisses people's genuine concerns. In some languages, including German and Swedish, certainty and safety are the same word, but perceived uncertainty can advance the safety of geological repositories too. Uncertainty is an invitation to let your action be informed by sympathy and care for local communities and other people who are deeply worried. It provides an opportunity to take responsibility for the future, prepare for change ahead, and design creative responses (see also Holtorf and May 2020).

Toxicity of cultural heritage

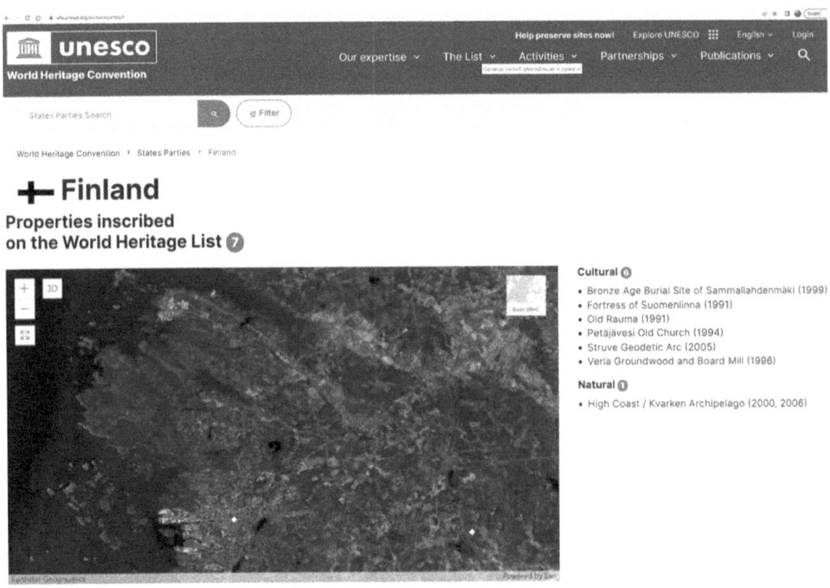

FIGURE 6 Seven. Natural and cultural legacies in Finland, inscribed in UNESCO's World Heritage List. Light points at the bottom of the picture: World Heritage sites Old Rauma and Sammallahdenmäki. Buildings at the top: Olkiluoto. Source: UNESCO, https://whc.unesco.org/en/statesparties/fi.

Like radioactive waste, UNESCO World Heritage is of global significance but managed nationally. Not far from Olkiluoto, the city of Rauma has two World Heritage properties: the historic city of Old Rauma and the Bronze Age Burial Site of Sammallahdenmäki. Cultural heritage like these sites is being conserved for the benefit of future generations, just like radioactive waste is being safely locked away for the benefit of future generations. Arguably, radioactive waste may be seen as a form of cultural heritage (Holtorf and Högberg 2021), and cultural heritage, in a way, can be considered toxic, too (Wollentz et al. 2020). There are currently 1154 inscribed World Heritage sites deemed to be of "outstanding universal value". But too many of the States Parties to the World Heritage Convention have turned the act of inscription into a matter of competition and prestige. UNESCO World Heritage has become a tool for nation-building and is frequently misappropriated as "World-level National Heritage" (Yan 2016, 239). In many contexts, counting World Heritage sites (Finland has seven) is most important, contaminating the Convention's original ambition to increase and diffuse knowledge about the peoples of the world and to advance global peace and collaboration. Fortunately, the global participants at the ICGR and passing Rauma and Sammallahdenmäki on the trip to Olkiluoto exchanged knowledge, met friends, and instigated collaborations, benefitting everybody.

Acknowledgements

For company, during the visit to Olkiluoto, I am grateful to Leila Papoli-Yazdi and for comments on an earlier draft, I would like to thank Petri Paju and the editors. For help with Figures 2 and 3 and permission to use them, I am very grateful to Pasi Tuohimaa at Posiva Oy; for permission to use Figure 6, I would like to thank Diana Sharafieva at UNESCO.

References

Hodder, Ian. 2012. *Entanglement*. Chichester: Wiley-Blackwell. 10.1002/9781118241912

Holtorf, C. and A. Högberg. 2021. "What lies ahead? Nuclear waste as cultural heritage of the future." In *Cultural Heritage and the Future*, edited by C. Holtorf and A. Högberg, 144–158. London and New York: Routledge. 10.4324/9781315644615-10

Holtorf, C. and S. May. 2020. "Uncertainty, collaboration and emerging issues." In *Heritage Futures. Comparative Approaches to Natural and Cultural Heritage Practices*, by R. Harrison, C. DeSilvey, C. Holtorf, S. Macdonald, N. Bartolini, E. Breithoff, H. Fredheim, A. Lyons, S. May, J. Morgan, and S. Penrose, 336–343. London: UCL Press. 10.1080/00293652.2022.2083980

Madsen, Michael. 2010. *Into Eternity: A Film for the Future*. Documentary, 75 min. https://studiomichaelmadsen.com/film/into-eternity/

Wollentz, G., S. May, C. Holtorf and A. Högberg. 2020. "Toxic heritage: Uncertain and unsafe." In *Heritage Futures. Comparative Approaches to Natural and Cultural Heritage Practices*, by R. Harrison, C. DeSilvey, C. Holtorf, S. Macdonald, N. Bartolini, E. Breithoff, H. Fredheim, A. Lyons, S. May, J. Morgan, and S. Penrose, 294–312. London: UCL Press. 10.2307/j.ctv13xps9m.25

Yan, Haiming. 2016. "World Heritage and National Hegemony: The Discursive Formation of Chinese Political Authority." In *A Companion to Heritage Studies*, edited by W. Logan, M. Nic Craith, and U. Kockel, 229–242. Chichester: Wiley Blackwell. 10.1002/9781118486634.ch16

18

TOXIC AND WASTED: ARTISTS THINKING ABOUT HOW TO ENGAGE WITH MATERIAL FUTURES

Rosemary A. Joyce

The artist Michael Heizer is quoted as saying "the history of sculpture … consists mostly of remains and fragments, damaged either by men or by natural phenomena" (Kastner 2009, 40). Productively blurring the line between art and waste, this assertion points us toward considering the essential unity of things that humans make that outlast their makers. Every material humans enlist, whether aesthetically pleasing or seen as exhausted, understood as neutral, or feared to be destructively active, has the potential to end up discarded. And everything can return from discard to raw material for new purposes.

These points can be exemplified even by what are viewed as some of the most toxic of contemporary discards, radioactive wastes. Some day, even these may circle back as useful. As Cornelius Holtorf and Anders Högberg (2014:5) wrote, "new technologies, such as transmutation, may allow using nuclear waste to generate further energy or for other purposes, so that this waste becomes a precious resource". Artists have a long history of undertaking the aesthetic conversion of materials viewed as waste or toxic. Holtorf and Högberg mention the case of James Acord (2009), who fought to obtain a radioactive handling license to use radioactive uranium in his sculptures.

Gabrielle Decamous (2011) has traced artists' responses to nuclear technology from the very beginning of the nuclear age, identifying works produced as early as 1946. These extended from the use of radioactive materials to works designed to give a sense of otherwise invisible nuclear facilities. Decamous (2011) draws particular attention to the ways art has been used to represent the invisible, showing radioactivity as an active force. She cites the work of Piotr Kowalski, a Polish artist resident in France, who in 1968 "used luminescent radioactive gas encapsulated in nine transparent marbles for *9 ans* (9 Years)" (Decamous 2011, 129). In a related experiment, artist Andy Weir produced sound files made in deep nuclear waste

DOI: 10.4324/9781003365259-37

repositories (Allen 2012). These artists give contemporary people a sense of buried but still present, invisible but still active, radioactive materials.

Other artists, like Acord, demonstrate the way that these notionally, sometimes actually, toxic materials have already penetrated everyday life. In a provocative work, Jonathon Keats created a musical instrument that would sound when objects he bought on eBay emitted gamma rays (Strickland 2018). The same kind of objects activated a work called "Universes Unlimited" that Keats installed in 2008 in a San Francisco gallery. When a radiation source was detected, a visible spark was set off (Keim 2008).

The raw materials Keats used came from the same sources used by Acord: kitchenware made using radioactive materials to produce distinctive colors. Keats described his acquisitions, demonstrating the way that toxicity cannot be separated from everyday life:

> I went onto eBay, and bought some uranium-doped glass. Uranium was used as a colorant in classic red glass Fiesta Ware in the 1950s, and in light green glass made in the 1920s and 1930s. That provide[d] my uranium. Then I found a guy who sells scintillating crystals, which are used by Homeland Security as a simple way to detect whether someone has a nuclear bomb under their coat. When a gamma ray hits the crystal, a photon is produced, so it glows a beautiful blue.
>
> (Keim 2008)

Fiesta Ware is a staple of 20th-century interior design that is simultaneously a carrier of toxicity (Sheets and Thompson 1995; Sheets and Turpen 1998). Other ceramics made with non-radioactive elements, including lead, cadmium, and cobalt, that provide colorants or stabilize glazes can also be toxic (Sheets 1998). None of these ceramics are considered appropriate for use in meals today. Yet they are still prized as collectibles, so are not trash either; they are domestic heritage. In work by Acord and Keats, Fiesta Ware dishes moved from collectible heritage property to art.

These artists directly engaged with efforts by the US government to grapple with the much larger legacies of nuclear industries left by weapons development and power plants. Commenting on the work of Acord, Decamous (2011, 129) underlines both his direct engagement with the Hanford site in Washington, a now-decommissioned site of production of plutonium for nuclear weapons and his motivation to create a work that would be a warning "of the risk of contamination 10,000 years from now, when even language will have changed totally and image would be the sole means of communication".

Keats linked his project "Universes Unlimited" to specific proposals under consideration by the US government to warn future human beings against intruding on buried nuclear waste sites. Writing to the US Department of Energy, he proposed installing the same kind of interactive device he demonstrated in the San Francisco gallery at the site of Yucca Mountain, at the time the US government's proposed high-level nuclear waste repository, writing "with this, you'd see Yucca Mountain at work, making new possibilities for us":

> It would be quite beautiful: the idea is to sink two-mile-deep scintillating crystal stacks into the mountain, sticking out like chimneys, looking like a factory. But instead of sending out smoke, they'd glow in the night. I don't know if the government would go for it, but it'd be less expensive than other things that they've done in the past.
>
> (Keim 2008)

In the pages that follow, I want to pursue some of the connections between waste and art that engagements like this reveal, considering what they can tell us about thinking about toxicity as an expected byproduct of human activities. I begin by outlining proposals for marking nuclear waste repositories developed in parallel, one by experts advising the US Department of Energy, others by artists, participants in a long tradition of artists' efforts to reclaim or commemorate landscapes affected by industrial development.

Planning for Nuclear Waste Repositories: Intended and Unintended Monuments

In 2002, the Desert Space Foundation sponsored a competition inviting artists to design alternative ways to mark the proposed repository at Yucca Mountain, Nevada, intended to house waste from civilian nuclear power generation (Auer 2002). The competition was inspired by the existence of a mandate for the US Department of Energy (DOE) to design a "marker system", required for the only already-approved nuclear waste repository in the US, the Waste Isolation Pilot Plant (WIPP) in New Mexico (Trauth, Hora, and Guzowski 1993).

In the planning process for WIPP, distinctions between markers and monuments, between functional signs and works of art that might become cultural heritage, were erased (Joyce 2020). Two approaches were taken by different teams of experts advising the DOE. In one, archaeological sites recognized today as cultural heritage monuments were cited as prototypes for a surface marker system that would be a modern analog to Stonehenge. A second team proposed instead that universal human responses could be elicited by building large-scale assemblages of stone, earth, and other materials forming spikes, blocks, and other abstract forms. Advocated by an environmental design scholar, this plan forcefully invokes the aesthetics of contemporary non-representational sculpture but justifies it with appeals to supposed universal responses to ancient monuments.

The second proposal put monuments and waste into relation with a third term: art. It invited the kind of responses seen in the Desert Space Foundation competition and other projects in which artists respond to the challenge of revealing the invisible presence of material potentially deadly for more than 10,000 years.

In these projects, artists propose productive approaches to sites considered to be damaged by human activity through the development of modern industry and urbanism.

Both the experts advising DOE and many artists adopted a landscape scale for their proposals. In this, the experts and artists shared a view of the post-industrial landscape as an empty canvas. They echo perspectives that emerged in the 1970s as a movement called Land Art. The experts advising the DOE actually included a recommendation that a contemporary artist be commissioned to produce a piece of Land Art as part of a marker system for the WIPP site, demonstrating that for them, their proposed markers were like works of art.

The initial wave of Land artists wrote about what they thought the projects they were creating would become in the future. This writing provides a source for thinking about the intersections of art and waste. These provide a context for the understandings shared by experts involved in the planning effort for nuclear waste markers (Joyce 2020). They cast light as well on the continuing engagements of artists with such sites and materials.

Art, Waste, and Heritage: Land Art and Industrial Landscapes

Land Art and Earth Art are labels applied to works that began to be created in the late 1960s. A 1968 exhibition in New York City featured photographic documentation of some of what are now recognized as the first Land Art projects (Rigaud 2012). Many blurred lines between natural features and forces, as in the use of metal poles as attractors for lightning strikes in Walter de Maria's "Lightning Field", orientation toward the sun in Nancy Holt's "Sun Tunnels", or the construction of sighting lines in an extinct volcano in James Turrell's "Roden Crater" (Hobbs 1982; Matts and Tynan 2012; Nisbet 2013; Saad-Cook 1988).

Many Land Art installations were sited in the US southwest as a kind of untouched canvas. Yet there is more to Land Art than simply using a purportedly empty landscape as a site. Robert Smithson, who became perhaps the iconic practitioner of Land Art, made clear connections between Land Art and the questions raised when we consider how human activity concentrates toxic materials in certain locations. Smithson insists his work is about entropy, universal processes of decay and change (Smithson 1996a, 1996b). Perhaps his best-known work is "Spiral Jetty" in the Great Salt Lake (Cooke et al. 2005). Entropy is built into Spiral Jetty by the activity of the lake itself. Rising and falling, covering and uncovering the spiral, and leaving deposits on its surface, the lake works beyond the artist's original control and intention, exemplifying the unpredictable future that lies ahead once humans have relocated matter in space. It creates a geological concentration of materials with specific properties, crystals of selenite gypsum or hydrous calcium sulfite ($CaSO4{\cdot}2H2O$) that have made Spiral Jetty a point of reference for contemporary mineral collectors (Sikorski 2019).

Smithson began to imagine his large-scale projects in places on the landscape already affected by human industry, proposing works for a Kennecott Copper mining site in Utah, a Hanna Coal extraction site in Ohio, and a Minerals Engineering site in Colorado (Hobbs 1982; Ryan 2007). Andrew Menard (2014) argues that Smithson saw any imagination of the US west as untouched by human projects as historically unrealistic. It was a romanticization to be countered by engaging with industrial development sites. Writing about a provocative work entitled "Tour of the Monuments of Passaic, New Jersey" in 1967, Smithson characterized contemporary buildings as "ruins in reverse", saying they were "the opposite of the 'romantic ruin' because the buildings don't *fall* into ruin *after* they are built but they rather *rise* into ruin before they are built" (Smithson 1996c, 72).

A connection with ruination is evident in some of the early work of a second noted Land Art movement participant, Michael Heizer. "Effigy Tumuli", created in 1985 in Illinois, consists of five earthen mounds in the shape of animals (Doss 2011; McGill 1990). The work is notable for its location, on land reclaimed after coal mining. In this case, Heizer attempts to commemorate the indigenous relationship to the landscape after it was interrupted and distorted by an extractive industry. In this work Heizer, as did Smithson, imagined the reclamation of a waste site, something contaminated brought back into active and positive engagement with living people.

Heizer's use of the word "tumulus" identifies the effigy mounds that dot the midwest US with tombs, relegating them to the past. This explicit use of an indigenous vocabulary of form associated with the end of tradition forcefully raises questions of the artistic imagination of historical continuity and discontinuity. Indigenous artists respond to the same sites quite differently, in ways that we shall see also characterize Indigenous responses to the wicked problem of curating radioactive waste.

Reclaiming Landscape: Indigenous Artists

Native American scholars and activists often treat landscape features as having continuing life. In their art practice, these artists shift from treating landscape scale art as a monument to enabling their activity in art as a performance. In a particularly rich project, an indigenous poet, Allison Hedge Coke, wrote a cycle of poems inspired by earthen mounds at a place originally called Blood Run, now part of a South Dakota State Park, Good Earth (Allen 2015; Hedge Coke 2006). Hedge Coke's poems explore the relations of a serpent-shaped effigy mound recorded there in 1889, destroyed by later extractive land use, to other beings, invoking the voices of the nonhuman participants making up the place. The poems are meant to be performed, to enact the recognition of the place and its history through sound. Indigenous artists LeAnne Howe (Choctaw) and Monique Mojica (Guna/Rappahanock) also produced performative work in relation to indigenous earthen mounds (Allen 2015).

These indigenous artworks reflect an ontology in which nonhumans, effigy mounds, and even features that government experts and some Land Artists treat as inert natural objects, are not dead, but lively. For example, Yucca Mountain, the proposed container for civilian nuclear waste, while not a human construct, is understood by Shoshone and Paiute people who historically have lived in relation to it as alive, some naming it with terms that can be glossed in English as "Serpent Swimming West" (Endres 2013). The placement of radioactive waste within the depths of Yucca Mountain, selected as a geologically inert feature from the perspective of US agencies, will actually poison a living being, indigenous critics argue (Endres 2012; Zabarte 2002).

Remaining Connected: From Marking to Living with Waste

One brief suggestion in the proposals for markers for WIPP called for something more in line with the performativity of engagements with heritage places through which indigenous poets and artists maintain their liveliness. This was a call for what the planning team described as "aeolian structures" that would "resonate in the wind" making "dissonant and mournful" sounds, conveying that it was "a place of great foreboding" (Ast et al. 1993, F-136). The invitation to imagine installing such things was taken up by one of the artists who participated in yet another art project reacting to the planning for marking waste repositories, commissioned by the journal *Columbia* (Wong et al. 2011). The eight artists involved each proposed their own ideas for markers for the WIPP site. Among these, Charles Alwakeel designed "razor-sharp … beaks" that would "create piercing sounds when wind passes through them".

Here, there is an echo of two kinds of engagements with a place that were systematically de-emphasized both in the approved design schematics for a faux monument for the WIPP site, and in the alternative installations that echo the work of Land artists. Both of these rely on visuality to claim the attention of future humans. The visual was so critical for WIPP that the proposal included a call to cover the surfaces of the stone elements with explicit messages. Some were in the form of serial drawings, like comic strips showing death as the consequence of digging up buried waste, relying on claims that a universal language of forms has been recognized by humans since at least the Palaeolithic. Other visual messages that would be inscribed on the marker elements would be texts, repeated in multiple languages, ultimately considered the most reliable means to convey the warnings required.

In contrast to this emphasis on visuality, indigenous scholars and activists have called for the transmission of stories of buried waste over generations (TallBear 2001; Weatherdon 2017). Kim TallBear (2001, 4) argued that including "stories, songs … culturally significant place names, and theater" would be more effective in ensuring the survival of memories of buried toxic wastes than monuments alone. Indigenous critics argue that the formation of lasting memory requires attention to, and acknowledgment of, the failures of the present generation to think ahead before creating a burden for future generations. The tenor of such stories might, then, be elegiac.

In that way they converge with one of the scripts one expert team felt the WIPP markers should convey: "This is not a place of honor … no highly esteemed deed is commemorated here … nothing valued is here" (Trauth, Hora, and Guzowski 1993, F-49). Other contributors to the planning of markers also called for orality as a tool to preserve knowledge of the repository. In an early contribution to thinking about commemorating nuclear waste, semiotician Thomas Sebeok actually proposed the deliberate creation of a cult guarded by an "atomic priesthood", to be passed down within that group as a sacred secret, with only warnings shared with the people at large, memory to be conserved via "folkloristic devices, in particular a combination of an artificially created and nurtured ritual- and- legend" (Sebeok 1984, 24). A distinct expert group employed by the US Department of Energy to assess the probabilities of intrusion in the waste repository in the future chose to embody some of their thinking in narrative form, writing scenarios for imagined futures. Most predicted the breach of the buried site. The sole vision they provided that suggested success imagined a museum built on the site, complete with an invented children's culture around the character of Nicky Nuke – a Smokey the Bear for nuclear waste circulating in "children's books, stories, narrative poems, puzzles, animated films, live theatre productions, and other media" (Joyce 2020, 222).

These interventions, from inside and outside the government planning effort, open up the question of how to preserve memories of toxic materials to wider participation, an invitation taken up by artists. In an award-winning response to a competition to imagine a future marker system for WIPP, Germain Canon produced a series of drawings called "The Time Between the Stones – A Ritual" (Canon 2017). His drawings and captions describe future residents systematically disassembling a construction composed of 10,000 stone slabs, one every year, accompanied by story-telling. Canon wrote "Memory is not a passive action, to be understood and trusted it must be passed on rather than discovered ... This marker does not directly warn about the danger underground, but rather about the interruption of the ritual: the message is to pass on the message" (Canon 2017).

Art and Toxic Memories

Multiple competitions for alternative markers for nuclear waste have inspired artists to engage with these issues. The Desert Space Foundation selected "Blue Yucca Ridge", a work by Ashok Sukumaran that proposed planting a genetically altered field of yucca plants that would glow blue as a kind of unnatural sign, as the winner of its competition (Auer 2002). Artistic competitions call attention to the way that waste is assigned to spaces out of the sight of the powerful, as in the winning design for a "Plutonium Memorial" to be built near the White House, countering the concealment of projects like these from the powerful by their common location in what the sponsors called "the backyard of the poor" (Rogers 2002). Other designs for this memorial proposed to incorporate human skeletons and bones. This acknowledgment of the potential human cost of waste echoed a comment by one of the experts advising the DOE, who argued that "the most effective 'marker' for any intruders will be a relatively limited amount of sickness or death caused by the radioactive waste" (Ast et al. 1993, F-143).

Artists routinely insist that rather than tucking radioactive waste out of sight and out of mind, humanity needs to maintain an ongoing dialogue about and with these materials. So, as part of the *Columbia* journal response to the WIPP site, an artist named Emcee C. M. proposed a conversation to be carried on for millennia by a "Committee for Radical Inquiry into the Earth's Regretful Situation" (Wong et al. 2011). Claire Jamieson proposed a 10,000-year-long project of "sequenced ruination" through which a Stonehenge-like monument would gradually be uncovered by erosion, followed by the emergence of an "observatory-like chapel" (Wong et al. 2011). This, she proposed, might promote the circulation of oral traditions keeping alive the memory of the danger of buried nuclear waste.

These are all interventions by artists who are skeptical of the project of containing nuclear waste, and critical of the society that produced it without a plan to manage it safely. Yet even when artists are committed unironically to marking nuclear waste sites, a rejection of universal assumptions and an insistence on the local, the site-specific, is notable. Belgian artist Cecile Massart is one of four artists who participated in a 2014 conference on constructing the memory of such sites. She proposed that after closure, a living space, a laboratory for the production of art, music, and dance, should be established on the site (Massart 2015). Rather than concealing the buried waste, her proposal would mark the locations, including by installing surface marks. But these are not imagined as remaining static or preventing engagement. While not proposing to recycle the materials that are byproducts of the nuclear industry, her proposal, and others by artists imagining large-scale installations, recycles the places contaminated by these activities, in precisely the way Smithson called for in his proposals for Land art on former industrial sites.

Nothing is Wasted; Everything is Toxic; Heritage is Everything

Proposals to mark buried nuclear waste aim to cut it off from further interaction with humans. Yet scenarios of the very near future proposed by experts suggested that geological materials today considered to be uninteresting might rapidly achieve economic importance, belying attempts to find sites where radioactive materials would remain undisturbed. We might consider this as an affirmation of the idea that ultimately, nothing is waste in the long history of humankind. If nothing is truly wasted, what might this imply for trying to think about a category of "toxic heritage"?

The experts who participated in proposing designs for markers for the US Waste Isolation Pilot Plant and any successor to be built actively debated the relationship of the site to concepts of heritage as they understood them. They repeatedly suggested that the monumental markers they proposed would work because they would be perceived as important heritage sites in the future. They wrote that "with age" these "could become recognized as a preservable, historical resource" (Baker et al. 1993, G45). A linguist involved in the project, Frederick Newmeyer, suggested "the WIPP site will quickly become known as one of the major architectural and artistic marvels of the modern world", attracting mass tourism (Ast et al. 1993, F-149). It was from this perspective that it made sense to propose commissioning a major piece of Land Art as part of the proposed marker. These suggestions, though, did not extend to viewing the buried radioactive materials themselves as heritage.

Heritage in its present use combines the notion of inheritance from the past and value in the present sufficient to require or justify attention and investment for management for the future. It is no accident that concepts of heritage were developed alongside nationalist projects, and today are fostered by international agencies that assign the rights to determine heritage to nations, rather than to people. Artists who work directly with radioactive materials, who imagine or implement artworks on industrial landscapes, show us that everything inherited from the past can be an occasion for some kind of response, escaping the intentions of control embraced by governments. Whether future human responses to inherited toxic materials are aesthetic, narrative, visual, or aural may truly be unpredictable. What they inevitably will be, however, is active and emergent, defying the logic of nationalism and the cultural heritage enterprise in which it is immersed that demands fixity of form or meaning.

References

Acord, James. 2009. "Q&A: The Art of Transmutation." *Nature* 458: 577.

Allen, Chadwick. 2015. "Performing Serpent Mound: A Trans-Indigenous Meditation." *Theatre Journal* 67: 391–411.

Allen, Jamie. 2012. "Deep Time Contagion: A Project by Andy Weir." *Continent* 2, no. 3: 167–169.

Ast, Dieter G., Michael Brill, Ward Goodenough, Maureen Kaplan, Frederick Newmeyer and Woodruff Sullivan. 1993. "Appendix F: Team A Report: Marking the Waste Isolation Plant for 10,000 Years." In *Expert Judgment on Markers to Deter Inadvertent Intrusion into the Waste Isolation Pilot Plant.* Sandia Report SAND92-1382. Albuquerque, NM: Sandia National Laboratories.

Auer, Catherine. 2002. "Just a Matter of Time," *Bulletin of the Atomic Scientists*, September/October.

Baker, Victor, Frank Drake, Ben Finney, David Givens, Jon Lomberg, Louis Narens, and Wendell S. Williams. 1993. "Appendix G: Team B. Report: The Development of Markers to Deter Inadvertent Human Intrusion Into the Waste Isolation Pilot Plant (WIPP)." In *Expert Judgment on Markers to Deter Inadvertent Intrusion into the Waste Isolation Pilot Plant.* Sandia Report SAND92-1382. Albuquerque, NM: Sandia National Laboratories.

Canon, Germain. 2017. "The Time Between the Stones: A Ritual". Accessed October 30, 2022. https://www.germaincanon.com/the-time-between-the-stones

Cooke, Lynne, and Karen Kelly, withBettina Funcke and Barbara Schrôder, eds. 2005. *Robert Smithson: Spiral Jetty*. Berkeley: University of California Press.

Decamous, Gabrielle. 2011. "Nuclear Activities and Modern Catastrophes: Art Faces the Radioactive Waves." *Leonardo* 44 (2): 124–132.

Doss, Erica. 2011. "Public Art Chronicles: Michael Heizer's *Effigy Tumuli*." *Public Art Dialogue* 1 (2): 241–246.

Endres, Danielle. 2012. "Sacred Land or National Sacrifice Zone: The Role of Values in the Yucca Mountain Participation Process." *Environmental Communication: A Journal of Nature and Culture* 6 (3): 328–345.

Endres, Danielle. 2013. "Animist Intersubjectivity as Argumentation: Western Shoshone and Southern Paiute Arguments Against a Nuclear Waste Site at Yucca Mountain." *Argumentation* 27 (2): 183–200.

Hedge Coke, Allison Adelle. 2006. *Blood Run: Free Verse Play*. Cambridge, MA: Salt.

Hobbs, Robert. 1982. "Editor's Statement: Earthworks: Past and Present." *Art Journal* (Fall): 191–194.

Holtorf, Cornelius, and Anders Högberg. 2014. "Nuclear Waste as Cultural Heritage of the Future." In *Proceedings of the WM2014 Conference "40 Years of Meeting Global Radioactive Waste Management Challenges"*, Phoenix, AZ, March 2-6, 2014. Phoenix, AZ: WM Symposia.

Joyce, Rosemary A. 2020. *The Future of Nuclear Waste*. Oxford: Oxford University Press.

Kastner, Jeffrey. 2009. "What Remains." *Public Art Review* 20 (2): 40.

Keim, Brandon. 2008. "Artist Wants Nuke Waste Dump to Make New Universes", *WIRED* November 20. Accessed October 15, 2015, http://www.wired.com/2008/11/create-your-own-2/

Massart, Cécile. 2015. "Constructing Memory Through Artistic Practices." In *Radioactive Waste Management and Constructing Memory for Future Generations: Proceedings of the International Conference and Debate, 15-17 September 2014, Verdun, France*, 125–131. Nuclear Energy Agency No. 7259. Paris: OECD Publishing. 10.1787/9789264249868-en

Matts, Tim, and Aidan Tynan. 2012. "Earthwork and Eco-clinic: Notes on James Turrell's 'Roden Crater Project'." *Design Ecologies* 2: 189–211.

Menard, Andrew. 2014. "Robert Smithson's Environmental History." *Oxford Art Journal* 37: 285–304.

McGill, Douglas C. 1990. "Michael Heizer's *Effigy Tumuli*." In *Michael Heizer's Effigy Tumuli: The Reemergence of Ancient Moundbuilding*. New York, NY: Harry N. Abrams.

Nisbet, James. 2013. "A Brief Moment in the History of Photo-energy: Walter de Maria's *Lightning Field*." *Grey Room* 50: 67–89.

Rigaud, Antonia. 2012. "Disorienting Geographies: Land Art and the American Myth of Discovery." *Miranda* 6, http://miranda.revues.org/2955

Rogers, Paul. 2002. "150 Solutions, One Fun Contest." *Bulletin of the Atomic Scientists* (May/June): 8–11.

Ryan, Leslie A. 2007. "Art + Ecology: Land Reclamation Works of Artists Robert Smithson, Robert Morris, and Helen Mayer Harrison and Newton Harrison." *Environmental Philosophy* 4: 95–116.

Saad-Cook, Janet. 1988. "Touching the Sky: Artworks using Natural Phenomena, Earth, Sky, and Connections to Astronomy (with Charles Ross, Nancy Holt and James Turrell)." *Leonardo* 21: 123–134.

Sebeok, Thomas A. 1984. *Communication Measures to Bridge Ten Millennia*. Technical Report, Human Interference Task Force. April 1984. BMI/ONWI-532. Columbus, OH: Batelle Memorial Institute, Office of Nuclear Waste Isolation.

Sheets, Ralph W. 1998. "Release of Heavy Metals from European and Asian Porcelain Dinnerware." *The Science of the Total Environment* 212: 107–113.

Sheets, Ralph W., and Clifton C. Thompson. 1995. "Accidental Contampination from Uranium Compounds through Contact with Ceramic Dinnerware." *The Science of the Total Environment* 175: 81–84.

Sheets, Ralph W., and Sandra L. Turpen. 1998. "Release of Uraniium and Emission of Radiation from Uranium-glazed Dinnerware." *Journal of Radioanalytic and Nuclear Chemistry* 235 (1-2): 167–171.

Sikorski, Alice. 2019. "Utah's 'Dirty Diamonds': Gypsum and Other Minerals From the Great Salt Lake." *Rock & Gem* https://www.rockngem.com/utahs-dirty-diamonds/
Smithson, Robert. 1996a. "Conversation in Salt Lake City (1972)." In *Robert Smithson: Collected Writings*, edited by Jack Flam, 297–300. Berkeley, CA: University of California Press.
Smithson, Robert. 1996b. "Entropy and the New Monuments (1966)." In *Robert Smithson: Collected Writings*, edited by Jack Flam, 10–23. Berkeley, CA: University of California Press.
Smithson, Robert. 1996c. "Tour of the Monuments of Passaic, New Jersey (1967)." In *Robert Smithson: Collected Writings*, edited by Jack Flam, 68–74. Berkeley, CA: University of California Press.
Strickland, Eliza. 2018. "Music for Aliens: Jonathon Keats Creates Instruments for Nonhuman Senses." *IEEE Spectrum* 55 (9): 20.
TallBear, Kimberly. 2001. *Tribal Social & Cultural Institutions for Long-Term Stewardship of Hazardous Sites*. Report presented at the National Academy of Sciences National Research Council Board on Radioactive Waste Management, Washington, D.C., 3 April 2001. Denver, CO: International Institute for Indigenous Resource Management. Accessed 15 February, 2018, http://www.iiirm.org/publications/Articles%20Reports%20Papers/Environmental%20Protection/socialcul.pdf
Trauth, Kathleen M., Stephen C. Hora, and Robert V. Guzowski. 1993. *Expert Judgment on Markers to Deter Inadvertent Intrusion into the Waste Isolation Pilot Plant*. Sandia Report SAND92-1382. Albuquerque, NM: Sandia National Laboratories.
Weatherdon, Meaghan Sarah. 2017. "Indigenous Knowledge and Contested Spirituality in Canadian Nuclear Waste Management." *Journal for the Study of Religion, Nature and Culture* 11 (1): 86–108.
Wong, Willy, Andrew Liebchen, Bryony Roberts, Charles Alwakeel, Christopher Garcia, Stuart Henley, C. M. Emcee, and Claire Jamieson. 2011. "Passive Institutional Controls Revisited." *Columbia: A Journal of Literature and Art* 48: i–xv, 86.
Zabarte, Ian. 2002. *Tribal Concerns about the Yucca Mountain Repository: An Ethnographic Investigation of the Moapa Band of Paiutes and the Las Vegas Paiute Colony*. October 20, 2002. Prepared for the Clark County Department of Comprehensive Planning, Nuclear Waste Division. Scottsdale, AZ: Urban Environmental Research, LLC.

CONCLUSION: WHY TOXIC HERITAGE MATTERS

Elizabeth Kryder-Reid and Sarah May

The central premise of this collection is that exploring toxic heritage as both a material phenomenon and a concept is a critical part of grappling with the damage humans have caused to the planet. In order to address this damage, we need a shared understanding of the material realities of the world, and that requires embracing environmental harm as a central heritage of modern human history (Gan et al. 2017; Garcia & Bauzà 2021; Fiske 2020; McNeill and Engelke 2014). It means naming historical conditions that brought us to this present – extractivism, carbon-based economies, environmental racism, settler-colonialism, and ongoing geo-political exploitation (Liboiron 2021; Moore 2017; Pulido 2017). It also requires paying attention to the politics of toxic heritage and the ways it traces the faultlines of social inequalities and other exercises of power. These are central issues for critical heritage studies, as other volumes in this series demonstrate, and this collection locates toxicity and environmental harm squarely in that critical heritage conversation. While heritage is historically founded with imperialism and settler colonialism (Bennett et al. 2017), critical perspectives have developed to reconfigure heritage as a tool for constructing a just future (Turunen 2020). This collection addresses a key component of any just future, namely making visible the harms of unjust, exploitative practices for people and the planet and framing them as a central part of modern human heritage.

This idea that the toxic pollution of the planet is largely invisible and that we are habituated to it experientially and ideologically creates a framework for considering the idea of toxic heritage, particularly through a critical heritage lens which is at its heart about memory and politics. Environmental justice scholars (e.g. Bullard 1993, 1994, 2018; Davies 2022; Liboiron 2021; Pulido 2015; Sze 2018) have long argued that the suppression of knowledge of toxic contamination is a political exercise. Imperceptibility is a form of privilege, and habituation is a strategy of survivance.

DOI: 10.4324/9781003365259-38

Consequently, the work of remembering, commemorating, and publicly surfacing these hidden toxic histories is a potent form not merely of environmental advocacy but of political resistance. Toxic heritage stands, therefore, as a counternarrative to the denial and amnesia that often serve corporate and state interests, just as it has the potential to activate citizen awareness and advocacy. The stories of pollution, contamination, and their effects on people's health and livelihoods are particularly compelling when they engage those affected populations in participatory heritage strategies. For example, authors in this volume document the efficacy of toxic tours (Baptista), community-based design (Valderrama), citizen science (Filippelli), virtual storytelling through Climate Museum UK (McKenzie), and public humanities projects (Sevcenko). These interventions are collectively a response to what Amelia Fiske has called our "chemically saturated present" and part of "a reconfiguration of toxicity – as a socio-material process, epistemic concept, and embodied experience – in order to work towards political and environmental, as well as epistemological, justice" (Fiske 2020, 1).

The authors in this collection approach the topic from a variety of disciplinary perspectives including history, archaeology, geography, and artistic practice. Across the chapters, case studies, and visual essays, they investigate a wide array of sites from six continents and toxic materials including chemicals, ocean plastics, e-waste, building materials, sewage, radioactive materials, mortuary waste, heavy metals, and mine tailings. The contexts are widely divergent as well – agricultural plantations, mining sites, factories, dry cleaners, nuclear plants and storage facilities, petrochemical plants, public housing, museum collections, household trash, dumps, military sites, underwater shipwrecks, and battlefields. The authors reveal the complex and creative engagements with these toxic legacies, as well as the structures and processes complicit in how they are managed and remembered.

As diverse as the particular materials and circumstances are, the collection is also unified in its centering of the people impacted by the environmental harms. It highlights the urgency and ubiquity of toxicity that is threatening the planet's livability. The weight of the findings also exposes the complex intersections of toxicity with memory practices in formal and informal heritage settings. The contributions comprise both an examination of the politics of toxicity and an exposé of its extractive roots and often exploitative consequences.

The hope is that this collection does not merely advance contemporary scholarship on the emerging topic of toxic heritage, but that it spurs conversation among heritage professionals, affected communities, environmental scientists and advocates, and anyone else who is interested in creating change. Our premise is that thinking about places of environmental harm as a heritage will help engage public audiences in the complex history of environmental harm, its consequences today, and possibilities for a more just, sustainable future. To the extent that heritage is the stories we tell ourselves about our past, naming these environmental harms is a potent way to register and even claim damage to the earth as our inheritance. David Lowenthal provocatively asserted that "History is for all, heritage for us alone … .

Heritage passes on exclusive myths of origin and endurance, endowing us alone with prestige and purpose." (Lowenthal 1998: 8). These stories from around the globe attest to a collective origin story for our modern world. Lowenthal's ostensible inclusionary premise of heritage "for us" demands that we reckon not only with a planetary human heritage, but that we understand the very biome is an integral part of the "we" in the stories we tell about ourselves. This notion of toxic heritage compels us to recognise the cumulative historical burden and ongoing everyday ecocide around us.

The collected work in this volume demonstrates that heritage need not be an authorising practice, reinforcing dominant narratives, and legitimising status quo power dynamics. Several of the contributors here explore ways that heritage allows critical engagement with pasts that leave toxic legacies. Schofield and Pocock explore how engaging in archaeological research can support communities to see the toxic harms of plastic as a legacy which can be challenged. Rankin et al. discuss how the authorising framework of heritage management can surface toxic harms to indigenous communities which have been hidden through centre/periphery dynamics of isolation. Fiske uses both tours and graphic narrative techniques more commonly associated with valourising heritage to reveal harmful pasts in the Ecuadorian Amazon, and Baptista's toxic tours similarly expose the intersections of unjust practice that have created Newark's sacrifice zone. Engaging with toxic materials as heritage challenges the relationship between history and heritage when, as many of this volume's contributions attest, the past is intentionally deployed in the present to highlight environmental injustice and to advocate for more equitable futures.

Just as the heritage framing for environmental harm allows us to understand the human stories of environmental harm, the toxic framing of the heritage described here also brings strength to the discipline. Many of the contributors describe circumstances where heritage practice is assumed to be valourising. To name something as heritage is to select it for continuity. If heritage is a gift from the past to the future, toxic heritage is the unwanted gift (May 2020). We understand practices of decontamination, reclamation, and the forgettings explored in this volume to be as central to heritage as the more commonly discussed designation, conservation, valourisation, and remembrance. They are linked. Whether toxic materials are treated as heritage, or that status is resisted, the practices of heritage often contribute to narratives of progress, purity, and productivity that exacerbate the material harms of these sites and perpetuate the impact on communities.

References

Bennett, Tony, Fiona Cameron, Nélia Dias, Ben Dibley, Rodney Harrison, Ira Jacknis, and Conal McCarthy. 2017. *Collecting, Ordering, Governing: Anthropology, Museums, and Liberal Government*. Durham: Duke University Press.

Bullard, Robert D. 1993. "Anatomy of Environmental Racism." In *Toxic Struggles: The Theory and Practice of Environmental Justice*. Edited by Richard Hofrichter. New Society Publishers.

Bullard, Robert D. 1994. *Unequal Protection: Environmental Justice and Communities of Color.* San Francisco: Sierra Club Books.
Bullard, Robert D. 2018. *Dumping in Dixie: Race, Class, and Environmental Quality.* Routledge.
Davies, Thom. 2022. "Slow Violence and Toxic Geographies: 'Out of Sight' to Whom?" *Environment and Planning C: Politics and Space* 40 (2): 409–427. 10.1177/2399654419841063
Fiske, Amelia. 2020. "Naked in the Face of Contamination: Thinking Models and Metaphors of Toxicity Together." *Catalyst: Feminism, Theory, Technoscience* 6(1): 1–30.
Gan, Elaine, Anna Lowenhaupt Tsing, Heather Swanson, and Nils Bubandt. 2017. "Introduction: Haunted Landscapes of the Anthropocene." In *Arts of Living on a Damaged Planet: Ghosts and Monsters of the Anthropocene*, edited by Elaine Gan, Anna Lowenhaupt Tsing, Heather Swanson, and Nils Bubandt (pp. G1–G14). Minneapolis: University of Minnesota Press.
Garcia, B. S., and G. B. Bauzà 2021. "The Materiality of Education: Heritage Conservation and New Approaches to the History of Education in Spain (1990-2020)." *Educació i Història: revista d'història de l'educació*, 105–133.
Liboiron, Max. 2021. *Pollution Is Colonialism.* Durham, NC: Duke University Press.
Lowenthal, David. 1998. *The Heritage Crusade and the Spoils of History.* Cambridge University Press.
May, S., 2020. "Heritage, Thrift, and Our Children's Children." In *Cultural Heritage and the Future*, edited by C. Holtorf and A. Högberg (pp. 46–58). Taylor and Francis.
McNeill, John Robert, and Peter Engelke. 2014. *The Great Acceleration: An Environmental History of the Anthropocene Since 1945.* Cambridge, MA: The Belknap Press of Harvard University Press.
Moore, Jason W. 2017. "The Capitalocene, Part I: On the Nature and Origins of Our Ecological Crisis." *The Journal of Peasant Studies* 44(3): 594–630. 10.1080/03066150.2016.1235036
Pulido, Laura. 2015. "Geographies of Race and Ethnicity 1: White Supremacy vs White Privilege in Environmental Racism Research." *Progress in Human Geography* 39(6): 809–817. 10.1177/0309132514563008.
Pulido, Laura. 2017. "Racism and the Anthropocene." In *Future Remains A Cabinet of Curiosities for the Anthropocene*, edited by Gregg Mitman, Marco Armiero, and Robert S. Emmett (pp. 116–128). Chicago: University of Chicago Press.
Sze, Julie, ed. 2018. *Sustainability: Approaches to Environmental Justice and Social Power.* New York: New York University Press.
Turunen, Johanna. 2020. "Decolonising European minds through Heritage." *International Journal of Heritage Studies* 26(10): 1013–1028, 10.1080/13527258.2019.1678051

INDEX

Kolibri 360 GmbH
Ferdinand-Jühlke-Straße 7
99095 Erfurt, Deutschland
produktsicherheit@kolibri360.de